ANIMAL SCIENCE, ISSUES AND RESEARCH

CHORDATE

ORIGIN, EVOLUTION AND AFFINITY

ANIMAL SCIENCE, ISSUES AND RESEARCH

Additional books in this series can be found on Nova's website
under the Series tab.

Additional e-books in this series can be found on Nova's website
under the e-book tab.

ANIMAL SCIENCE, ISSUES AND RESEARCH

CHORDATE

ORIGIN, EVOLUTION AND AFFINITY

PRANVEER SINGH

nova publishers
New York

Additional color graphics may be available in the e-book version of this book.

Library of Congress Cataloging-in-Publication Data

Chordate : origin, evolution, and affinity / editor, Pranveer Singh (Department of Zoology, Indira Gandhi National Tribal University, Amarkantak).
pages cm. -- (Animal science : issues and research)
Includes index.
ISBN 978-1-63463-907-1 (hardcover)
1. Chordata--Evolution. I. Singh, Pranveer.
QL605.C455 2015
596--dc23
2015000083

Published by Nova Science Publishers, Inc. † New York

To…..
Mumy & Aazi

CONTENTS

PREFACE

Evolution is the GUT (GRAND UNIFYING THEORY) of biology. It is the bedrock principle of our scientific understanding of the natural laws that govern life. Furthermore, it is logically necessary for life's survival in a changing world environment. To deny this scientific principle is analogous to believing that the earth is the center of the universe!

Dr. J. William Schopf, paleobiologist and director of UCLA's Center for the Study of Evolution and the Origin of Lifes (Mader 1998; page 5)

The current book entitled "**Chordate: Origin, Evolution and Affinity**" is structured into 8 chapters starting from introductory chapters throwing light on basic evolutionary principles to sensitize readers about different aspects of chordate evolution. Book is written keeping in mind to accommodate the needs of diverse readers that range from undergraduate, postgraduate to research scholars.This book spans diverse but related disciplines that range from Morphology, Anatomy, Taxonomy, Physiology, Evolution and Diversity and has updates from latest of the chordate's research.

First time Evolution of Chordates is discussed on comparative basis to better the understanding related to chordate origin, affinity and evolution. Emphasis is given on the so far neglected but important topics like Archaeopteryx as a connecting link between reptiles and birds and another topic like bird migration. A separate chapter on Archaeopteryx is added in exhaustive detail (earlier this was used to embed with bird or reptile chapter). Similarly, separate chapter on Bird's migration covering all relevant details is also added instead of clubbing it as a small sub-section in chapter on Class Aves. Mammals, being the climax class is given special emphasis with a separate chapter devoted to their origin and evolution leading to primates and finally to humans and another one on their diversity.

It was an endeavor to present the book in simple, lucid and comprehensible language. Concepts related to chordate's origin and evolution is explained in easy to understand manner for readers from non-science background. Present work is fruition of sincere and honest efforts invested to present the literature in holistic and comprehensible manner but science is a collective activity and readers are welcome with their constructive and creative inputs to synergize this endeavor into a grand reality.

Lastly, a privileged opportunity to thanks all who directly or indirectly were the part and parcel of the long gestation period before the concept 'Chordates: Origin, Affinity and Evolution' becomes a reality. A reality that has well justified the long periods of seclusion and isolation gone into verbal depiction of the chordates climaxing to Humans.

Dr. Pranveer Singh
Assistant Professor
Department of Zoology
Indira Gandhi National Tribal University (IGNTU)
Amarkantak - 484886 (MP), India
mailpranveer@gmail.com

Chapter 1

CHORDATES

1.1. HISTORY OF CHORDATES

The existence of such chordates (Chordata) was justified by well-known Russian zoologist SA Kovalevsky (1865), who, studying the development (ontogeny), tunicates (Tunicata) and bescherepnyh (Acrania), established the similarity of their organization with vertebrates. Although the name Chordata is attributed to **William Bateson** (1885), it was already in prevalent use by 1880. **Ernst Haeckel** described a taxon comprising tunicates, protochordates, and vertebrates in 1866. Though he used the German vernacular form, it is allowed under the **ICZN code** because of its subsequent latinization [Bateson, 1886; Berrill, 1955; Geneva, 2006; Benton, 2009].

The name of the type proposed by **Bell Chord in 1878** bescherepnye subtypes and tunicates is usually called the lower chordates and vertebrates – the higher chordates. For the chord are 43 thousand species distributed around the globe: they inhabit the seas and oceans, lakes and rivers, continents and islands. Type chord combines vtorichnorotyh dvustoronnesimmetrichnyh coelomic animals with metamerism, expressed predominantly in the early stages of embryonic development. Chordates are characterized by their unique characteristics: the presence of a lifetime or only one of the phases of chorda – the chords, playing the role of the internal skeleton, which has endodermal origin, and surrounded by a connective tissue sheath [Jenkinsen, 1913; Kingsley, 1923, 1927]. In primitive animals the notochord persists throughout life, while in the higher forms it is gradually replaced (superseded) by the vertebral column, composed of individual vertebrae, resulting in a connective tissue sheath of the chord; developing central nervous system, which has the shape of the tube (the inner cavity of said nevrotselem) and formed from the endoderm [Rychel et al., 2006]. Neural tube is located above the chord. In vertebrates, it is differentiated into two parts: the brain and spinal cord. In the walls of the pharynx – the anterior part of the digestive system – form the gill slits that act as digestive and respiratory organs. In aquatic animals in the septa between the gill slits are formed gills, terrestrial vertebrates have gill slits only in the embryo, then they are overgrown, and adults develop lungs - paired invaginations on the ventral side of the back of the throat. The digestive system is located under the chord. Heart in vertebrates is under the digestive system. Terrestrial ancestors of chordates remains were not preserved, so development of the chordates have on indirect evidence, namely, by comparing the structure of the adult forms and comparative study of embryonic development.

Most zoologists believe that the precursors of chordate animals were coelomic forms that have fallen to a sedentary life (or sessile), which led to a decrease in the number of segments of their bodies and the formation of the secondary cavity [Jenkinsen, 1913; Kingsley, 1923, 1927].

1.2. Comparative Anatomy, an Introduction

Comparative anatomy studies the descriptive and functional morphology of vertebrates. Basically it is a study of **evolution** and phylogeny. It elucidates the hidden secrets of origin of life, in particular the Chordates, which have swamped the 'Mother Earth' by its shear adaptability and diversity. It involves comparative study of the anatomy of an organ in different groups of vertebrates and to derive the evolutionary significance from it to understand as to why an organ evolved the way it is present now. Genetic and environmental forces are responsible for the development of an organ, on which then natural selection operates to determine its survival or elimination. Comparative vertebrate anatomy studies the structure, function of the structure, and range of variation in structure and function among vertebrates.

It helps in comprehending the structural basis of biology, to probe the ancient past and study of evolution, to develop appreciation of vertebrates and finally understanding man's place in the vertebrate world, without emphasis on man himself.

It will be very important to understand the contrasting terms like Morphology, which is the study of external features of an organism, and Anatomy, which studies the internal structures of an organism. **Morphology** studies structural evolution and **Anatomy** studies description of structures [Jenkinsen, 1913; Kingsley, 1923, 1927].

DESCRIPTIVE MORPHOLOGY *compares the* structures of representative vertebrate species while the **FUNCTIONAL MORPHOLOGY** *compares* the function of structures. It incorporates following (in random order)-

Zoology: animals
Physiology: function
Histology: tissues
Genetics: heredity, effects on variation and development
Embryology: from fertilization → parturition
Evolution: natural selection and adaptation (versus Lamarckian, intelligent design)
Paleontology: historical record of past anatomy (fossils)
Ecology: organism-environment interactions

1.3. Classification

All forms of life are classified into five major groups known as kingdoms. The generally recognized kingdoms are Monera (bacteria), Fungi, Protista (unicellular organisms), Plant and

Animals (multicellular animals). Within each kingdom, each group of organisms with similar characteristics is classified into a category known as phylum.

Phylum chordate was established in 1874. These are the taxonomic group that possess following four characters at some point of their development [Benton, 2009]. These **four common chordate** characters are (Figure 1 and 2):

1. Notochord
2. Post-anal tail
3. Dorsal hollow CNS/ Nerve cord
4. Pharyngeal gill slits (Kent's endostyle/ hypobranchial groove)

Other character of imminence might be cells that produce thyroxin.

These morphological characteristics may be transitory and may be present only during a particular stage of development, or they may be present throughout the animal's life.

Phylum chordate is divided into three subphyla:

1. Urochordata,
2. Cephalochordata
3. Vertebrata.

The Urochordata and Cephalochordata consists of small, non-vertebrate marine animals and are often referred collectively as **Protochordates**.

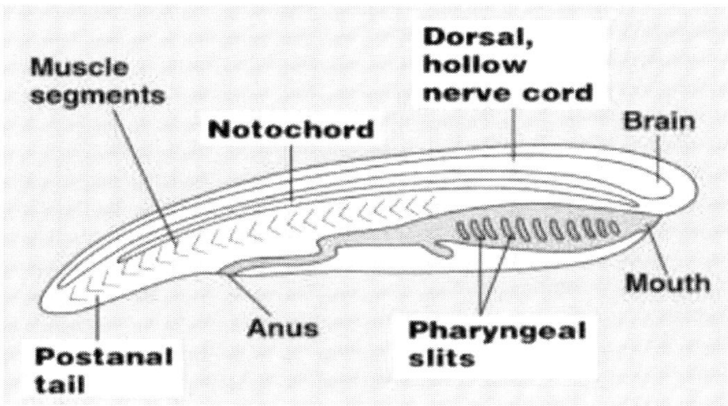

Figure 1. Four common chordate characters.

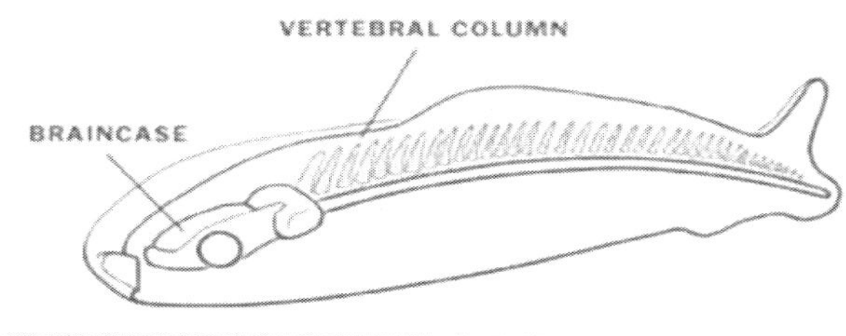

Figure 2. Vertebral column and Brain case.

Notochord is the rod of living cells ventral to central nervous system and dorsal to alimentary canal. It is a hydrostatic organ that acts as an endoskeleton – it maintains the elongated shape of the animal, preventing the body from collapsing on itself. Generally, the notochord is dorsal in position and runs from anterior to posterior. It lies immediately ventral to the spinal cord.

Notochord meets the following fate during development:

Head region - incorporated into floor of skull

Trunk and tail - surrounded by cartilaginous or bony vertebrate (except in Agnathans)

In adults notochord shows following fate -

Pisces (Fishes) and Amphibians - notochord persists the length of the trunk and tail but is constricted within the centrum of each vertebra

Reptiles, Aves (birds), and Mammals - notochord almost disappears during development (e.g., remains as a pulpy nucleus in the vertebrae of mammals)

Protochordates - notochord remains as the chief axial skeleton

Agnathans - lateral neural cartilages are located on notochord lateral to the spinal cord

Pharynx: The pharynx is the anterior end of the gut, immediately posterior to the mouth. In vertebrates, the pharyngeal slits become gill slits. In other forms, the slits are usedfor filter feeding, allowing water to pass through them, but being too narrow to let food through. Dorsal hollow nerve cord is also called the spinal cord and, fills the nervous function, connecting the brain with the rest of the nervous system. It is the region of alimentary canal exhibiting pharyngeal pouches in embryo; pouches may open to the exterior as slits:

Permanent slits - adults that live in water and breathe via gills

Temporary slits - adults live on land

Dorsal, hollow central nervous system - consists of brain and spinal cord and contains a central cavity called the neurocoel.

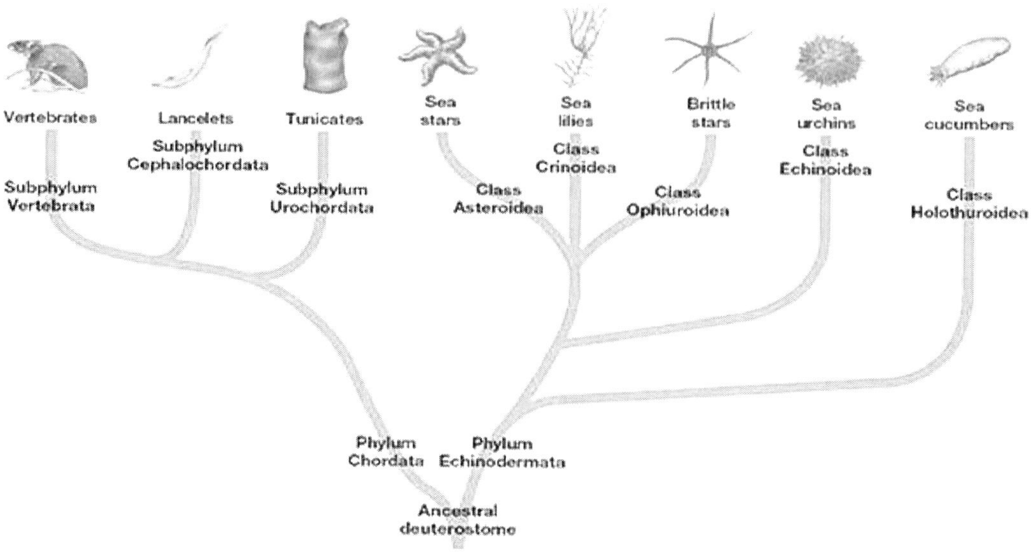

Figure 3. Detail classification scheme of Chordates.

1.4. Vertebrate Beginnings

Among the oldest, best known and well characterized are the Ostracoderms. They have the following identifying features:

- fishes that occurred in the late Cambrian period (see The Cambrian Explosion) through the Devonian (about 400 - 525 million years before present)
- had bony plates and scales (therefore, were easily fossilized)
- jawless vertebrates called 'armored fishes'

1.5. Ostracoderms

- *Myllokunmingia fengjiaoa* and *Haikouichthys ercaicunensis* - primitive fish that have many similarities to living hagfishes and are the oldest vertebrates 530 million year before (mybf) ever found.
- Before Vertebrates?
- *Cathaymyrus diadexus* (literally the 'Chinese eel of good fortune') is not the fossil of an eel. At just 5 cm long, but 535 m.y. old, it is the earliest known chordate. Researchers think that *Cathaymyrus* is a fossil relative of modern lancelets (Amphioxus).

Non-vertebrate chordates still alive today include tunicates (or sea squirts; Urochordates) and Amphioxus (or Branchiostoma), a Cephalochordates. A 530 million-year-old (although perhaps as old as 560 million years) creature, *Cheungkongella ancestralis*, probably a tunicate, found in the Chengjiang fauna in China's southwest Yunnan Province, might be the earliest known fossil evidence of primitive chordates [Zhang et al. 2001].

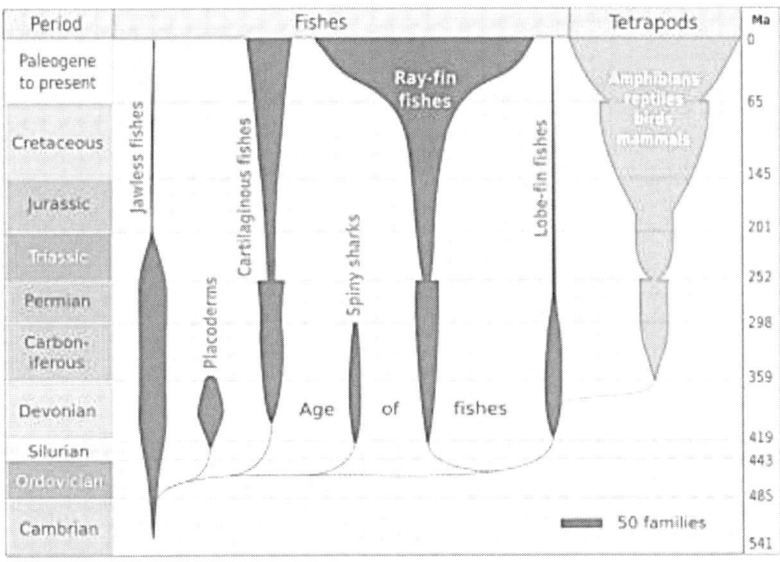

Figure 4. Evolution of jawless fishes.

Subphylum Urochordata [Ruppert et al., 2004; Delsuc et al., 2006] = Tunicates
Chordate 'ancestor' of vertebrates were:

- sessile (like adult tunicates)
- tail evolved as an adaptation in larvae to increase mobility
- 'higher forms' - came about by retention of tail (neoteny)

Tunicate larva - also called 'Sea squirt' had:

- Notochord, confined to the tail
- Notochord, lost during metamorphosis into sessile adult
- possess pharyngeal slits

Subphylum Cephalochordata [Chen et al., 1995] = Amphioxus (or Branchiostoma)

- Vertebrate features:
- Dorsal notochord,
- hollow nervous system
- pharyngeal gill slits'
- circulatory' system - vertebrate pattern with 'pumping vessels' (but no heart)

Hemichordates [Hyman, 1959] = Acorn worms
Bateson added acorn worms to the phylum Chordata in 1884 because they have:

- dorsal, hollow nervous system
- gill slits
- short diverticulum of the gut called the stomochord

Present consensus holds that stomochord is not homologous with the notochord and Hemichordates are placed in a separate phylum.
Possible invertebrate ancestors:

Hemichordates show *similarity* with Annelids in terms of

- bilateral symmetry
- segmented body organization and
- central nervous system with brain and longitudinal nerve cord.

However, it differs from annelids in having solid and ventral nerve cord (contrary to hollow and ventral nerve cord in chordates).

Echinodermata - chordate characteristics include:

- radial cleavage - blastomeres in adjacent tiers lie directly above one another (as opposed to spiral cleavage)

- anus forms near or at blastopore (deuterostomous)
- mesoderm arises as outpocketing of the gut wall
- indeterminate cleavage (i.e., fate of blastomeres isn't predetermined)

1.6. PHYLUM: CHORDATA

Subphylum:Vertebrata

Superclass 1: Pisces
Class I: Agnatha
Class II: Placodermii
Class III: Chondricthyes
Class IV:Acanthodii
Class V: Osteichthyes

Superclass 2: Tetrapoda
Class I: Amphibia
Class II: Reptilia
Class III: Aves
Class IV: Mammalia

Agnathans (Jawless vertebrates) versus Gnathostomes (Jawed vertebrates):

- **Semicircular canals**
  - agnathans have 1 or 2
  - gnathostomes have 3
- **Jointed, paired lateral appendages**
  - agnathans have none
  - gnathostomes do

Superclass 1: Pisces

Class I: Agnatha
Orders:

1. Osteostraci
2. Anaspida
3. Thelodonti
4. Galeaspida
5. Pituriaspida
6. Petromyzontia (Lampreys)
7. Myxinoidea (Hagfishes)

Ostracoderms [Maisey, 1996]:
(Osteostraci, Anspida, Heterostraci, and Coelolepid)

1. Extinct Paleozoic (Cambrian to Devonian) jawless fish with an external skeleton of bone ('bony armor')
2. oldest known vertebrates
3. many had flattened appearance (some may have been bottom-dwellers)

Cyclostomes: (Petromyzontia and Myxinoidea)

Lampreys - parasitic with horny, rasping teeth
Hagfishes - primarily scavengers

Myxinoidea (hagfish)

Myxini (hagfish) are eel-shaped slime-producing marine animals (occasionally called **slime eels**). They are the only known living animals that have a skull but not a vertebral column. Along with lampreys, hagfish are jawless and are living fossils; hagfish are basal to vertebrates, and living hagfish remain similar to hagfish 300 million years ago [Strahan, 1963; Martini, 1998]. The classification of hagfish has been controversial. The issue is whether the hagfish is itself a degenerate type of vertebrate-fish (most closely related to lampreys), or else may represent a stage which precedes the evolution of the vertebral column (as do lancelets). The original scheme, groups hagfish and lampreys together as cyclostomes (or historically, Agnatha), as the oldest surviving clade of vertebrates alongside gnathostomes (the now-ubiquitous jawed-vertebrates). An alternative scheme proposed that jawed-vertebrates are more closely related to lampreys than to hagfish (i.e., that vertebrates include lampreys but exclude hagfish), and introduces the category craniata to group vertebrates near hagfish. Recent DNA evidence has supported the original scheme [Strahan, 1963; Martini, 1998].

Hyperoartia (lamprey)

Hyperoartia is a disputed group of vertebrates that includes the modern lampreys and their fossil relatives. Examples of hyperoartians from early in their fossil record are *Endeiolepis* and *Euphanerops*, fish-like animals with hypocercal tails that lived during the Late Devonian Period. Some paleontologists still place these forms among the "ostracoderms" (jawless armored "fishes") of the class Anaspida, but this is increasingly considered an artificial arrangement based on ancestral traits. Placement of this group among the jawless vertebrates is a matter of dispute. While today enough fossil diversity is known to make a close relationship among the "ostracoderms" unlikely, this has muddied the issue of the Hyperoartia's closest relatives. Traditionally the group was placed in a superclass Cyclostomata together with the Myxini (hagfishes). More recently, it has been proposed that the Myxini are more basal among the skull-bearing chordates, while the Hyperoartia are retained among vertebrates. But even though this may be correct, the lampreys represent one of the oldest divergences of the vertebrate lineage, and whether they are better united with some "ostracoderms" in the Cephalaspidomorphi, or not closer to these than to e.g., to other "ostracoderms" of the Pteraspidomorphi, or even the long-extinct conodonts, is still to be resolved. Even the very existence of the Hyperoartia is disputed, with some analyses favoring a treatment of the "basal Hyperoartia" as a monophyleticlineage Jamoytiiformes that may in fact be very close to the ancestral jawed vertebrates [Hardisty et al., 1971; Gess et al., 2006].

Gnathostomes

Acanthodians:

1. Earliest known gnathostomes (Silurian; about 440 mybp)
2. *probably* related to modern bony fishes
3. small (less than 20 cm long) with large eyes
4. Acanthodians most likely died out because of the rapidly increasing number of ray-finned fishes and sharks during the Permian.

The relationships of acanthodians to other vertebrates has been the subject of considerable debate. Early researchers considered them to be most closely related to the ray-finned fishes, but most scientists during the mid-20th Century considered acanthodians to have a closer affinity to the sharks. Opinion has now generally swung back in favor of a closer relationship with ray-fins, but this is far from universally accepted [Forey et al., 2004; Janvier, 2008].

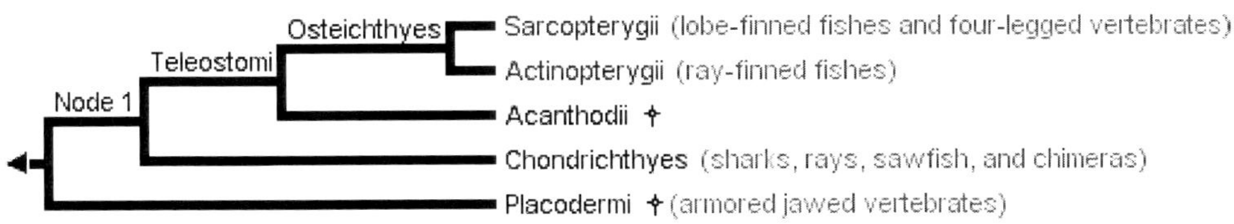

Inter-relationships of the Gnathostomata:

The main characteristics supporting the nodes of this phylogeny are:

Node 1: Superior oblique muscle of eye attached anteriorly to eyeball. Braincase includes nasal capsules. Jaw muscle external to mandibular arch.

Teleostomi: Mouth terminal in position, narrow-based braincase, 3 otoliths in ear.

Osteichthyes: Endochondral bone, lepidotrichs on fins, jaws lined by dentary, premaxillary and maxillary.

Introduction

The Gnathostomata, or gnathostomes, are the dominant group of the Middle Devonian (-380 million years ago). They differ from all other craniates or vertebrates in having a vertically biting device, the jaws, which consist of an endoskeletal mandibular arch and a variety of exoskeletal grasping, crushing, or shearing organs, i.e., the teeth, and jaw bones.

Among recent vertebrates, the gnathostomes include sharks, rays, chimaeras, ray-finned fishes, lobe-finned fishes and land vertebrates [Janvier, 1996].

Extant gnathostomes fall into two major clades, the Chondrichthyes and Osteichthyes. In addition, there are two extinct major gnathostome clades, the Placodermi (Early Silurian-Late Devonian) and the Acanthodii (Latest Ordovician or Earliest Silurian - Early Permian). There may be other fossil gnathostome taxa which fall outside of these four taxa. This could be the case for the Mongolepida, only known from isolated scales from the Early Silurian, and which are provisionally assigned to the chondrichthyans, yet with great reservations [Young, 1986; Smith and Johanson, 2003].

The Chondrichthyes are characterized by a special type of hard tissue lining the cartilages of the endoskeleton: the prismatic calcified cartilage. Another chondrichthyan characteristic is the pelvic clasper, as special copulatory organ derived from the metapterygium, i.e., the posterior part of the pelvic fin. A pelvic clasper may, however, be present in the fossil Placodermi. Chondrichthyans include two major extant clades, the Elasmobranchii and the Holocephali, and a number of fossil clades (Cladoselachidae, Symmoriida, Xenacanthiformes, Iniopterygia, Eugeneodontida) which may fall outside these two clades [Goudjet, 1984, 2001].

The Osteichthyes are characterized by endochondral ("spongy") bone in the endoskeleton, dermal fin rays made up by lepidotrichiae (modified, tile-shaped scales), and three pairs of tooth-bearing dermal bones lining the jaws (dentary, premaxillary and maxillary). The Osteichthyes include two major clades, the Actinopterygii and the Sarcopterygii [Goudjet, 1984, 2001].

The Placodermi are characterized by a dermal armor consisting of head armor and a thoracic armor. In the thoracic armor, the foremost dermal plates form a complete "ring" around the body and always include at least one median dorsal plate [Goudjet, 1984, 2001].

The Acanthodii are characterized by dermal spines inserted in front of all fins but the caudal one. They also possess minute, growing scales which have a special onion-like structure, i.e., the crown consists of overlying layers of dentine or mesodentine [Goudjet, 1984, 2001].

Characteristics

Gnathostomes are characterized by:

- A vertically biting device called jaws, and which is primitively made by two endoskeletal elements, the palatoquadrate and Meckelian cartilage, and a number of dermal elements called teeth, sometimes attached to large dermal bones.
- **Pelvic fins**. These are the paired fins or limbs situated just in front of the anus.
- **Interventrals and basiventrals in the backbone**. These are the elements of the backbone which lie under the notochord, and match the basidorsals and interdorsals respectively.
- Gill arches which lie internally to the gills and branchial blood vessels, contrary to the gill arches of all jawless craniates, which are external to the gills and blood vessels.
- A horizontal semicircular canal in the inner ear [Johanson, 2003].

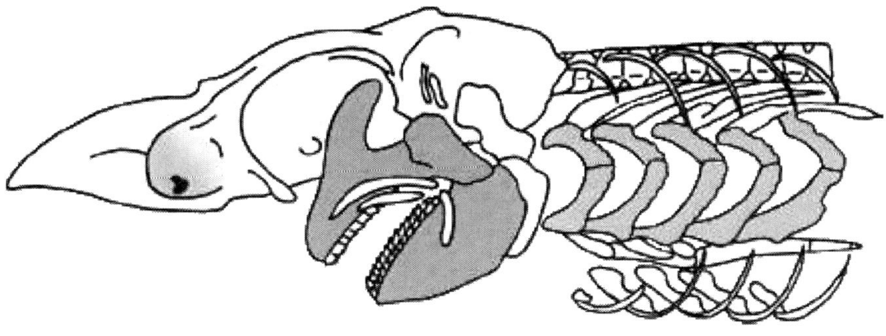

Figure 5. The skull of a gnathostome, or jawed vertebrates (here a shark), are characterized by vertically biting jaws (red) consisting of the palatoquadrate dorsally and the Meckelian cartilage ventrally. The gill arches (green) are situated internally to the gill filaments, and the nasal capsules (blue) open to the exterior by means of paired nostrils [Janvier 1996].

- Paired nasal sacs which are independent from the hypophysial tube. In all extant and fossil jawless craniates, the nasal sacs, which contain the olfactory organs, open into a median duct, the nasohypophysial duct, which takes part in the formation of the pituitary gland and either leads postriorly to the pharynx (e.g., in hagfish and galeaspids) or ends as a blind pouch (e.g., in lampreys and osteostracans). In the gnathostomes, this pouch remains as a thin canal in the palate, the buccohypophysial canal, whereas the nasal sacs open separately to the exterior by external nostrils (Figure 5) [Johanson, 2003].

There are numerous other characteristics of the soft anatomy and physiology (e.g., myelinated nerve fibres, sperms passing through urinary ducts, etc.), which are unique to the gnathostomes among extant craniates, but cannot by observed in fossils.

Discussion of Phylogenetic Relationship

Gnathostomes have long been placed as the sister-group of the ensemble of the Agnatha, or jawless vertebrates, regarded as a clade [Long, 1995; Long et al., 1997]. Current phylogenies, however, suggest that the Agnatha are not a clade, and that, among recent craniates, the gnathostomes are the sister-group of the Hyperoartia (lampreys) only. Among fossil craniates, the Osteostraci share the largest number of uniquely shared derived characteristics with the gnathostomes [Goudjet, 1984, 2001].

The controversies about gnathostome inter-relationships mainly bear on the position of the two major fossil groups, the Placodermi and Acanthodii. Placoderms have long been regarded as the sister-group of chondrichthyans, with which they share an eye-stalk (a small cartilage linking the eyeball to the braincase) and possibly a pelvic clasper. It has also been suggested that they could be the sister-group of the osteichthyans, because they share with them large dermal plates and a median dermal bone in the palate, the parasphenoid. The current consensus is that placoderms are the sister-group to all other known gnathostomes [Long, 1995; Long et al., 1997]. All other gnathostomes share a special arrangement of the eye muscles (the superior oblique eye muscle being attached anteriorly in the orbit), whereas that of placoderms is similar to that in the Osteostraci (the superior oblique eye muscle being attached posteriorly in the orbit).

Also, the braincase of placoderms shows a separate ossification containing the nasal capsules, whereas the latter are always included in the braincase in all other gnathostomes [Goudjet, 1984, 2001].

The Acanthodii have long been regarded as closely related to either chondrichthyans or placoderms, but they are now almost unanimously regarded as the sister-group of the Osteichthyes. They share with them otoliths in the inner ear, lepidotrich-like scales on fins, and a narrow-based braincase with a transverse ventrale fissure. Acanthodians, however, are rarely known from articulated specimens, and their internal anatomy remains poorly known. It is possible that Acanthodii is paraphyletic [Goudjet, 1984, 2001].

Class II: Placodermii (plated skin):

1 - Silurian (about 420 million years before present)
2 - probably off the main line of vertebrate evolution
3 - many had bony dermal shields
4 - some were probably predators (with large, sharp 'tooth plates')

Placoderms were armored jawed fishes that first appeared about 420 million years ago (MYA) during the Silurian Period. Placoderms were named for their heavy armor of dermal bone, which formed large shields on the head and thorax. The rest of their bodies were covered with small bony scales or was without dental armor. The head and trunk shields of most placoderms were articulated by bony joints.

This joint apparently allowed the forward part of the skull to tilt up, increasing the gape. Placoderms lacked teeth, but biting or grinding structures are often being found in the dermal bones lining their mouths [Young, 1980, 1986; Goudjet, 1984, 2001].

Placoderms evolved into a variety of body forms in a relatively short time. Many were torpedo-shaped, but there were notable expections, including the flatten Phyllolepida and the bottom feeding Antiarchi. Most placoderms were less than 30 cm (2 feet) in length, but some members of the dinichthyids (= terrible fish) reached or exceeded 6 m (20 ft), making them the first giants of the vertebrate lineage.

They had diversified dramatically by the beginning of the Devonian and came to dominate most marine and freshwater ecosystems before becoming extinct at the end of that period (355 MYA).

About 200 genera of placoderms have been discovered, with most of these occurring during the Devonian radiations. The rapid evolution and diversity of placoderms have made them useful in dating Devonian rocks [Young, 1980, 1986; Goudjet, 1984, 2001].

Class III: Chondrichthyes

The cartilaginous fishes

1. ancestors had bony skeletons so cartilaginous skeleton is specialized
2. pelvic fins of males are modified as claspers
3. placoid scales
4. numerous today but more abundant in the past

***Subclass A: Elasmobranchii** – most common cartilaginous fishes*

Order Cladoselachii - primitive sharks (300-400 mybp)

Order Selachii - 'modern' sharks
Order Batoidea - Rays and Skates

Elasmobranchs

1. 1st pharyngeal slit modified as a spiracle
2. naked gill slits (no operculum)
3. mouth located ventrally

Subclass B: Holocephali
Order. Chimaeriformes

- marine
- gill slits have a fleshy operculum & the spiracle is closed
- few scales
- common ancestor with sharks but an independent line

Class III: Osteichthyes - Bony fishes

1. skeleton is partly or chiefly bone
2. gill slits are covered by a bony operculum
3. skin has scales with, typically, little bone
4. most have a swim bladder
5. ray-finned or lobe-finned

Subclass A: Actinopterygii - Ray-fins
Superorder: Chondrostei

- most primitive ray-fins
- chiefly Paleozoic (300-400 mybp)
- include present-day Sturgeons & Paddlefish

Superorder: Neopterygii
Order: Semionotiformes

- dominant Mesozoic fishes
- possess ganoid scales

two extant genera:

Lepidosteus - predatory; includes present-day gars
Amia - includes present-day bowfins (or dogfish)
Division Teleostei: **modern ray-finned fishes**

- recent bony fishes
- 95% of all living fish
- about 40 living orders

- well-ossified skeleton
- cycloid and ctenoid scales (flexible & overlapping)
- pelvic fins often located far forward
- no spiracle

Subclass B: Sarcopterygii: lobe-finned fishes

Order Crossopterygii - chiefly Paleozoic except *Latimeria*

1. resemble early amphibians
2. skeleton of fin lobe corresponds closely to proximal skeletal elements of early tetrapod limbs
3. skull similar to that of early amphibians
4. had swim bladders that may have been used as lungs

Order Dipnoi - lungfish (3 living genera; Africa, Australia and South America)

African and South American species have inefficient gills and will drown if held under water. Australian species (*Neoceratodus* spp.) relies on gills unless oxygen content of water is too low.

Superclass 2: Tetrapoda

Class I: Amphibia

Subclass A: Labyrinthodontia (Oldest known class)

Fish-like features:

1. small bony scales in the skin
2. fin-rays in the tail (for swimming)
3. Skull similar to that of some Crossopterygians
4. Sensory canal system (like the lateral line system) that indicates a primarily aquatic existence

Labyrinthodonts are distinguished by deeply folded structure of enamel and dentine layers in the teeth that look like an intricate labyrinth in the cross section, hence the name of this group. Labyrinthodonts were probably similar to fishes in their mode of living [Milner, 1993; Jarvik, 1996]. Labyrinthodonts, like fishes and most modern amphibians, laid eggs in the water, where their larvae developed into mature animals. All labyrinthodonts had special sense organs in the skin that formed a system for perception of water fluctuations. Moreover, some of them possessed well developed gills. In contrast, many labyrinthodonts seemingly had primitive lungs. They could breathe atmospheric air that was a great advantage for residents of warm shoals with low oxygen levels in the water. The air was inflated into the lungs by contractions of a special throat sac. Primitive members of all labyrinthodont groups were probably true water predators, and only advanced forms that arose independently in different groups and times, gained an amphibious, semi-aquatic mode of living. Mature individuals of advanced labyrinthodonts could live on land, feeding mainly on insects and other small invertebrates. Well ossified robust skeletons in some Late Carboniferous and

Early Permian labyrinthodonts prove their adaptation to the terrestrial mode of life. It suggests that amphibians had successfully 'organized' actual terrestrial assemblages prior to the wide expansion of reptiles [Milner, 1993; Jarvik, 1996].

The most diverse group of the labyrinthodonts was the 'batrachomorphs' ('similar to a frog'). Though these animals looked more like crocodiles, they most probably gave rise to the order Anura, the amphibians without tails, which include, in particular, the modern frogs. Batrachomorphs appeared in the Late Devonian, but they had worldwide distribution in the continental shallow basins of the Permian (Platyoposaurus, Melosaurus) and Triassic Periods (Thoosuchus, Benthosuchus, Eryosuchus). Some batrachomorphs existed until the end of the Cretaceous [Panchen and Smithson, 1988; Ahlberg and Milner, 1994].

Subclass B: Lepospondyli

- ancestry uncertain due to lack of fossil evidence
- probably on a 'side branch' of vertebrate evolution

Subclass C: Lissamphibia (modern amphibians)

Order Anura - frogs and toads
Order Urodela - tailed amphibians
Order Gymnophiona (apodans) - wormlike, burrowing amphibians

Modern amphibian characteristics:

1. aquatic larval stage with external gills
2. middle ear cavity with ear ossicle (columella)
3. no bony scales (except apodans)

Class II: Reptilia - the first amniotes

1. scaly
2. clawed
3. large, yolk-laden, shell-covered eggs laid on land

Stem reptiles = Cotylosaurs (about 300mybp)

Subclasses A: Anapsida

Cotylosauria - stem reptiles
Chelonia - turtles and tortoises

- unchanged for about 175 million years
- identified by bony dermal plates to which ribs and trunk vertebrae are fused

Subclass B: Lepidosauria

Rhynchocephalia (Sphenodonta) - only living representative is the Tuatara
Squamata - lizards, geckos, and snakes

Subclass C: Archosauria

Thecodontia - stem archosaurs Pterosauria (check this short video and this one)

Saurischia - 2 major groups: sauropods and theropods

Ornithischia (like Iguanodon)

Crocodilia

Saurischia (sawr-RIS-kee-ah) and **Ornithischia** are the two orders of dinosaurs,

Subclass D: Euryapsida - marine reptiles, includes the plesiosaurs and ichthyosaurs

Subclass E: Synapsida

Order Pelycosauria - first stage in evolution to mammals

Order Therapsida

With the division based on the shape of the pelvic bone. The saurischian pubis (left) juts forward, and its ischium points backward. The ornithischian pubis and ischium (right) both point backward (Figure 6). The ornithischians were all herbivorous, and included some of the most interesting-looking dinosaurs [Frazetta, 1968].

Ornithischian dinosaurs include three suborders:

- Ornithopoda,
- Marginocephalia
- Thyreophora

The famous carnivorous dinosaurs were from the saurischian order, as were the largest herbivorous dinosaurs. The saurischian dinosaurs include two suborders:

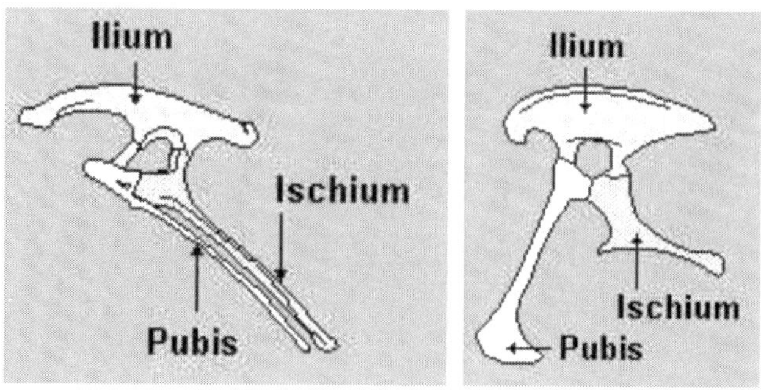

Figure 6. Pelvic bone with its component.

- Theropoda
- Sauropodomorpha

Pterosaurs, first vertebrates to evolve true flight, the flying archosaurian reptiles (Figure 7). After the discovery of pterosaur fossils in the 18th century, it was thought that pterosaurs were a failed experiment in flight; a humorous mishap; or that they were simply gliders, too weak to fly [Naish and Martil, 2003; Templin et al., 2004]. More recent studies have revealed that pterosaurs were definitely proficient flyers, and were no evolutionary failure; as a group they lasted about 140 million years (about as long as birds have)! Pterosaurs are thought to be derived from a bipedal, cursorial (running) archosaur in the late Triassic period (about 225 million years ago) [Arnold et al., 2004; Barrett et al., 2008]. No other phylogenetic hypothesis has withstood examination; however, the early history of pterosaurs is not yet fully understood because of their poor fossil record in the Triassic period [Wang, 2004; Hone and Benton, 2007; Andres, 2010]. We can infer that the origin of flight in pterosaurs fits the "ground up" evolutionary scenario, supported by the fact that pterosaurs had no evident arboreal adaptations.

The pterosaur wing was supported by an elongated fourth digit. Pterosaurs had other morphological adaptations for flight as a keeled sternum for the attachment of flight muscles, a short and stout humerus (the first arm bone), and hollow but strong limb and skull bones. Pterosaurs also had modified scales that were wing-supporting fibers, and that possibly formed hair like structures to provide insulation -- bird feathers are analogous to the wing fibers of pterosaurs, and both are thought to possibly have been evolved originally for the primary purpose of thermoregulation (which implies, but does not prove, that both pterosaurs and the earliest birds were endothermic).

Early pterosaurs (such as Dimorphodon) had long tails that assisted balance, but later pterosaurs had no tails, and may have been more adept flyers [Padian, 1983, 2003; Buffetaut et al., 2004; Dyke et al., 2006; Grellet et al., 2007; Elgin et al., 2011]. The most derived pterosaurs, such as Pteranodon and Quetzalcoatlus, were so large that soaring was the only feasible option; **these were the largest flyers ever to cast a shadow on the Earth's surface**.

Reptile subclasses - classified in part according to presence or absence of temporal openings

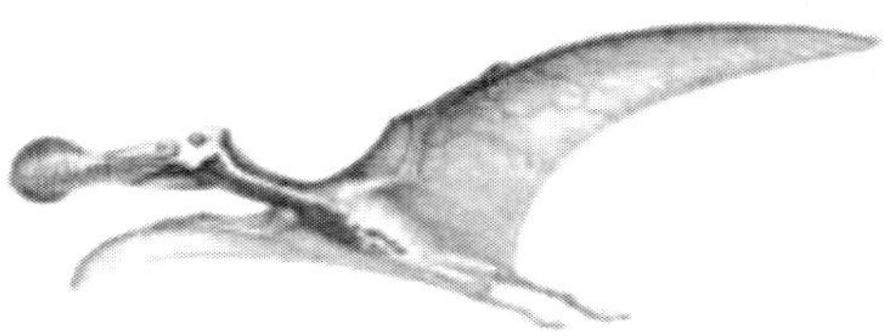

Figure 7. Archosaurian reptiles.

- Synapsid type = mammal-like reptiles
- Anapsid type = stem reptiles and turtles
- Diapsid type = rhynchocephalians, lizards, and snakes
- Euryapsid type = extinct plesiosaurs.

Class III: Aves - birds

1. May have arisen from an archosaurian reptile (Figure 6), perhaps a small bipedal dinosaur
2. Lost several dinosaur characteristics (e.g., long tail and teeth) but retained others (e.g., claws, scales, diapsid skull, single occipital condyle and, perhaps, feathers.

Subclass A: Archaeornithes

Genera: Archaeopteryx and Archaeornis

Characteristics:

1. solid bones
2. weakly developed keel and, probably, weakly developed flight muscles

Subclass B: Neornithes

Superorder: Odontognathae

- extinct
- many features of modern birds (e.g., hollow bones and short tail)

Superorder: Paleognathae

- ratites
- small wings but powerful leg muscles

Superorder: Neognathae - birds adapted for sustained flight

Modifications to reduce weight include:

- loss of some bones
- pneumatic bones
- reduced tail
- loss of teeth
- loss of urinary bladder

Class IV: Mammalia

Characteristics:

1. hair
2. mammary glands
3. 3 middle ear bones
4. muscular diaphragm
5. sweat glands
6. marrow within bones
7. 2 sets of teeth
8. biconcave, enucleate red blood cells
9. well-developed cerebral cortex

Subclass A: Prototheria - egg-laying mammals

Order Monotremata - platypus and spiny anteaters

1. lay eggs
2. testes within the abdominal cavity
3. no pinna
4. no corpus callosum
5. less stable body temperature

Subclass B: Theria

Infraclass: Metatheria

Marsupialia - pouched mammals; young born alive, but at a very immature stage

Infraclass: Eutheria - placental mammals

System Type	Chordate's System
Muscular-Skeletal	Chordate has an inner skeleton with a backbone. It generally has four appendages that are in the forms of legs, arms, wings or fins.
Digestion	Chordate takes in food through the mouth supplied with tongue. Mouth may or may bear teeth. It has a digestive system with stomach and intestines. Chordates can be Herbi-carni-omnivorous.
Respiration	Chordate takes in oxygen and gets rid of carbon dioxide through lungs or gills.
Circulation	Chordate can be warm or cold blooded.
Nervous	Chordate has a brain and nervous system. Most chordates have two eyes and at least a minimal system of hearing as fish do underwater. Many have excellent hearing. Some chordates are literally blind as the bat, but most have good vision. The chordates have the most developed brains of all the phyla and complex nervous systems
Reproduction	All chordates reproduces sexually
Excretion	Chordate has complex skin that is many times covered with feathers, scales or hair. They get rid of wastes through kidneys and intestines
Symmetry	All chordates has bilateral symmetry
Coloration	Chordates can be almost any color in nature

REFERENCES

Ahlberg, P. E. & Milner, A. R. (1994). The origin and early diversification of tetrapods. *Nature*, *368*, 507–514.

Andres, B. B. (2010). A review of pterosaur phylogeny. In Martill, D., Unwin, D., & Loveridge, R. F. (eds). *The Pterosauria*. Cambridge University Press.

Arnold, C. & Caple, L. A. (2004). Pterodactyl. *Pterosaurs: rulers of the skies in the dinosaur age*. Houghton Mifflin Harcourt. p. 23.

Barrett, P. M., Butler, R. J., Edwards, N. P. & Milner, A. R. (2008). Pterosaur distribution in time and space: an atlas. *Zitteliana*, 61-107.

Bateson, W. (1886). The ancestry of the chordata. *Q J Microsc Sci*, *26*, 535–571

Benton, M. (2009). *Vertebrate Palaeontology*. Edition 3, page 44, John Wiley & Sons.

Berrill, N. J. (1955). The origin of vertebrates. Oxford: Oxford University Press

Buffetaut, E., Martill, D. & Escuillié, F. (2004). Pterosaurs as part of a spinosaur diet. *Nature*, *430*, 33.

Chen, J. Y., Dzik, J., Edgecombe, G. D., Ramsköld, L. & Zhou, G. Q. (1995). A possible Early Cambrian chordate. *Nature*, *377*, 720-722.

Chen, J. Y., Huang, D. Y., Peng, Q. Q., Chi, H. M., Wang, X. Q. & Feng, M. (2003). The first tunicate from the Early Cambrian of South China. *Proc Natl Acad Sci*, *100*, 8314–8318.

Coates, M. I. & Clack, J. A. (1990). Polydactyly in the earliest known tetrapod limbs. *Nature* 347, 66–67.

Delsuc, F., Brinkmann, H., Chourrout, D. & Philippe, H. (2006). Tunicates and not cephalochordates are the closest living relatives of vertebrates. *Nature*, *439*, 965–968.

Dyke, G. J., Nudds, R. L. & Rayner, J. M. (2006). Limb disparity and wing shape in pterosaurs. *J Evol Biol*, *19*, 1339–42.

Elgin, R. A., Hone, D. W. E. & Frey, E. (2011). The Extent of the Pterosaur Flight Membrane. *Acta Palaeontol Pol*, *56*, 99–111.

Forey, P. & Janvier, P. (2000). Agnathans and the origin of jawed vertebrates. In Gee, Henry. *Shaking the tree: readings from Nature in the history of life*. USA: University of Chicago Press; Nature/Macmillan Magazines. 251–266.

Frazetta, T. H. (1968). Adaptive problems and possibilities in the temporal fenestration of tetrapod skulls. *J Morphol*, *125*, 145–158.

Geneva, ghorevere. (2006). "The authorship of higher chordate taxa". *Zoologica Scripta*, *41*, 435–436.

Gess, R. W., Coates, M. I. & Rubidge, B. S. (2006). A lamprey from the Devonian period of South Africa. *Nature*, *443*, 981–984.

Goujet, D. F. (2001). Placoderms and basal gnathostome apomorphies, Ahlberg P.E In Major events in early vertebrate evolution: palaeontology, phylogeny and development, vol. *61* Systematics association series. London, UK:Systematics Association, 209-222.

Goujet, D. F. (1984). Placoderm interrelationships: a new interpretation, with a short review of placoderm classifications. *Proc Linn Soc New South Wales*, *107*, 211–243.

Grellet-Tinner, G., Wroe, S., Thompson, M. B. & Ji, Q. (2007). A note on pterosaur nesting behavior . *Histor Biol*, *19*, 273–7.

Hardisty, M. W. & Potter, I. C. (1971). Hardisty, M. W.; Potter, I. C., eds. *The Biology of Lampreys* (1 ed.). Academic Press.

Hone, D. W. E. & Benton M. J. (2007). An evaluation of the phylogenetic relationships of the pterosaurs to the archosauromorph reptiles. *J Syst Palaeontol*, *5*, 465–469.

Hyman, L. H. (1959). The enterocoelous coelomates — the Hemichordata. In The Invertebrates, Volume *5*, McGraw-Hill; 72-207.

Janvier, P. (1996). Early Vertebrates, Oxford, 393 pp.

Janvier, P. (2008). "arly Jawless Vertebrates and Cyclostome Origins. *Zool Sci*, *25*, 1045–1056.

Jarvik, E. (1996). The Devonian tetrapod Ichthyostega. *Fossils & Strata* 40, 1–213.

Jenkins, F. A. Jr., Shubin, N. H., Gatesy, S. M. & Warren, A. (2008). *Gerrothorax pulcherrimus* from the Upper Triassic Fleming Fjord Formation of East Greenland and a reassessment of head lifting in temnospondyl feeding. *J Vert Paleontol*, *28*, 935–950.

Jenkinsen, J. W. (1913). Vertebrate embryology. Oxford.

Johanson, Z. (2003). Placederm branchial and hypobranchial muscles and origin in jawed vertebrates. *J Vert Paleontol*, *23*, 735-749

Kingsley, J. S. (1923). The vertebrate skelton. London

Kingsley, J. S. (1927). Comparative anatomy of vertebrates. London

Kowalevsky, A. O. (1865). Embryological studies in Wurmern und Arthropoden. Mem. Akad Sc St Petersburg, Ser VII, 16

Liem, K. F., William, E. B., Warren F. W. Jr. & Lance, G. (2001). *Functional Anatomy of the Vertebrates*. The United States of America: Thomson: Brooks/Cole. p. 50.

Long, J. A. (1995). The Rise of Fishes: 500 Million Years of Evolution. Johns Hopkins Univ. Press, Baltimore & London, 223 pp.

Long, J. A., Anderson, M. E., Gess, R. & Hiller, N. (1997). New placoderm fishes from the Late Devonian of South Africa. *J Vert Paleontol*, *17*, 253-268.

Maisey, J. G. (1996). *Discovering Fossil Fishes* (illustrated ed.). New York: Henry Holt & Company. p. 37.

Martini, F. H. (1998). The ecology of hagfishes. In J. M. Jorgensen, J. P. Lomholt, R. E. Weber and H. Malte. *The Biology of Hagfishes*. London: Chapman and Hall. 57–77.

Milner, A. R. (1993). Amphibian-grade Tetrapoda. In *The Fossil Record*, vol *2*, (ed.Benton, M. J.) Chapman & Hall, London, 665–679

Naish, D. & Martill, D. M. (2003). Pterosaurs — a successful invasion of prehistoric skies. *Biologist*, *50*, 213–6.

Padian, K. (1983). A Functional Analysis of Flying and Walking in Pterosaurs. *Paleobiology*, *9*, 218–39.

Padian, K. (2003). Pterosaur Stance and Gait and the Interpretation of Trackways. *Ichnos*, *10*, 115–126.

Panchen, A. L. & Smithson, T. R. (1988). The relationships of the earliest tetrapods. In Benton, M. J. (ed.) *The Phylogeny and Classification of the Tetrapods*, Volume *1*, Amphibians, Reptiles, Birds, 1–32. Oxford: Clarendon Press.

Ruppert, E. E., Fox, R. S. & Barnes, R. D. (2004). *Invertebrate Zoology, 7th edition*. Cengage Learning. 940–956.

Rychel, A. L., Smith, S. E., Shimamoto, H. T. & Swalla, B. J. (2006). "Evolution and Development of the Chordates: Collagen and Pharyngeal Cartilage". *Molecular Biology and Evolution*, *23*, 541–549.

Smith, M. M. & Johanson, Z. (2003). Separate Evolutionary Origins of Teeth from Evidence in Fossil Jawed Vertebrates. *Science*, *299*, 1235-1236.

Strahan, R. (1963). The behavior of myxinoids. *Acta Zool*, *44*, 73–102.

Templin, R. J. & Chatterjee, S. (2004). *Posture, locomotion, and paleoecology of pterosaurs*. Boulder, Colo: Geological Society of America. p. 60.

Wang, X. & Zhou, Z. (2004). Palaeontology: pterosaur embryo from the Early Cretaceous. *Nature*, *429*, 621.

Young, G. C. (1980). A new Early Devonian placoderm from New South Wales, Australia, with a discussion of placoderm phylogeny: *Palaeontographica*, *167A*, 10–76.

Young, G. C. (1986). The relationships of placoderm fishes. *Zool J Linn Soc*, *88*, 1-57.

Chapter 2

CHORDATES ORIGIN AND AFFINITIES

2.1. CHORDATES ORIGIN

Chordates, members of the phylum **Chordata**, are deuterostome animals possessing a notochord, a hollow dorsal nerve cord, pharyngeal slits, an endostyle, and a post-anal tail for at least some period of their life cycles.

Chordates evolved sometime during Cambrian period, 500 million years ago during Cambrian explosion, almost at the same time when invertebrates were beginning to evolve. They may have evolved from some freshwater forms as all modern chordates possess glomerular kidneys that are designed to remove excess water from body. However, early fossils of chordates have all been recovered from marine sediments and even modern protochordates are all marine forms. Also, glomerular kidneys are found in some marine forms such as myxinoids and sharks. That makes the marine origin of chordates more plausible [Bateson, 1886; Benton, 2000; Cowen, 2000; Rychel et al., 2006].

2.2. OVERVIEW OF AFFINITIES

Tunicate larvae have both notochord and a nerve cord which were lost in adulthood. Cephalochordates have a notochord and a nerve cord (but no brain or specialist sensory organs) and a very simple circulatory system. Craniates are the only subphylum whose members have skulls. In all craniates except for hagfish, the dorsal hollow nerve cord is surrounded with cartilaginous or bony vertebrae and the notochord is generally reduced; hence, hagfish are not universally regarded as vertebrates, though recent DNA comparisons suggest that they are in fact vertebrates. The chordates and two sister phyla, the Hemichordata and the Echinodermata, make up the deuterostomes, one of the two superphyla that encompass all fairly complex animals. An article by Nakano et al. demonstrates that the Xenoturbellida are not deuterostomes [Bateson, 1886; Cowen, 2000].

Attempts to work out the evolutionary relationships of the chordates have produced several hypotheses. The current consensus is that chordates are monophyletic, meaning that the Chordata include all and only the descendants of a single common ancestor *which is itself a chordate*, and that craniates' nearest relatives are cephalochordates. All of the earliest chordate fossils have been found in the Early Cambrian Chengjiang fauna, and include two species that are regarded as fish, which implies they are vertebrates. Because the fossil record of chordates is poor, only molecular phylogenetics offers a reasonable prospect of dating their emergence. However, the use of molecular phylogenetics for dating evolutionary transitions is controversial [Bateson, 1886; Benton, 2000; Cowen, 2000].

It has also proved difficult to produce a detailed classification within the living chordates. Attempts to produce evolutionary "family trees" give results that differ from traditional classes because several of those classes are not monophyletic. As a result, vertebrate classification is in a state of flux.

2.3. Origin of Name

Although the name Chordata is often attributed to William Bateson (1886), it was already in prevalent use in by 1880. Ernst Haeckel described a taxon comprising tunicates, cephalochordates, and vertebrates in 1866. Though he used the German vernacular form, it is allowed under the ICZN code because of its subsequent latinization.

Chordates form a phylum of creatures that are based on a bilateral body plan, and is defined by having at some stage in their lives all of the following [Cameron et al., 2000]:

1. **A notochord**, in other words a fairly stiff rod of cartilage that extends along the inside of the body. Among the vertebrate sub-group of chordates the notochord develops into the spine, and in wholly aquatic species this helps the animal to swim by flexing its tail.
2. **A dorsal neural tube**. In fish and other vertebrates this develops into the spinal cord, the main communications trunk of the nervous system.
3. **Pharyngeal slits**. The pharynx is the part of the throat immediately behind the mouth. In fish, the slits are modified to form gills, but in some other chordates they are part of a filter-feeding system that extracts particles of food from the water in which the animals live.
4. **Post-anal tail**. A muscular tail that extends backwards behind the anus.
5. **Endostyle**. This is a groove in the ventral wall of the pharynx. In filter-feeding species, it produces mucus to gather food particles, which helps in transporting food to the esophagus. It also stores iodine, and may be a precursor of the vertebrate thyroid gland.

2.4. Closest Non-chordates Relatives

Hemichordates

Enteropneust hemichordate: Balanoglossus

Hemichordates ("half (½) chordates") have some features similar to those of chordates: branchial openings that open into the pharynx and look rather like gill slits; stomochords, similar in composition to notochords, but running in a circle round the "collar", which is ahead of the mouth; and a dorsal nerve cord — but also a smaller ventral nerve cord [Horst, 1939; Hyman, 1959; Ruppert, 2005].

There are two living groups of hemichordates. The solitary enteropneusts, commonly known as "acorn worms", have long proboscises and worm-like bodies with up to 200 branchial slits, are up to 2.5 metres (8.2 ft) long, and burrow though seafloor sediments. Pterobranchs are colonial animals, often less than 1 millimetre (0.039 in) long individually, whose dwellings are interconnected [Nu¨bler-Jung and Arendt, 1996]. Each filter feeds by means of a pair of branched tentacles, and has a short, shield-shaped proboscis. The extinct graptolites, colonial animals whose fossils look like tiny hacksaw blades, lived in tubes similar to those of pterobranchs [Nu¨bler-Jung and Arendt, 1996; Lowe et al., 2003].

Echinoderms differ from chordates and their other relatives in three conspicuous ways: they possess bilateral symmetry as larvae and in adulthood they have radial symmetry, meaning their body pattern is shaped like a wheel; they have tube feet; and their bodies are supported by skeletons made of calcite, a material not used by chordates [Richards et al., 2010]. Their hard, calcified shells keep their bodies well protected from the environment, and these skeletons enclose their bodies, but are also covered by thin skins. The feet are powered by another unique feature of echinoderms, a water vascular system of canals that also functions as a "lung" and are surrounded by muscles that act as pumps. Crinoids look rather like flowers, and use their feather-like arms to filter food particles out of the water; most lie anchored to rocks, but a few can move very slowly. Other echinoderms are mobile and take a variety of body shapes, for example starfish, sea urchins and sea cucumbers [Jefferies et al., 1995; Smith, 2005].

2.5. ORIGINS

The majority of animals more complex than jellyfish and other Cnidarians are split into two groups, the protostomes and deuterostomes, the latter of which contains chordates. It seems very likely the 555 million-year-old *Kimberella* was a member of the protostomes. If so, this means the protostome and deuterostome lineages must have split some time before *Kimberella* appeared — at least 558 million years ago, and hence well before the start of the Cambrian 541 million years ago. The Ediacaran fossil *Ernietta*, from about 549 to 543 million years ago, may represent a deuterostome animal [Blair, and Hedges, 2005; Cameron et al., 2005].

Figure 1. Starfish.

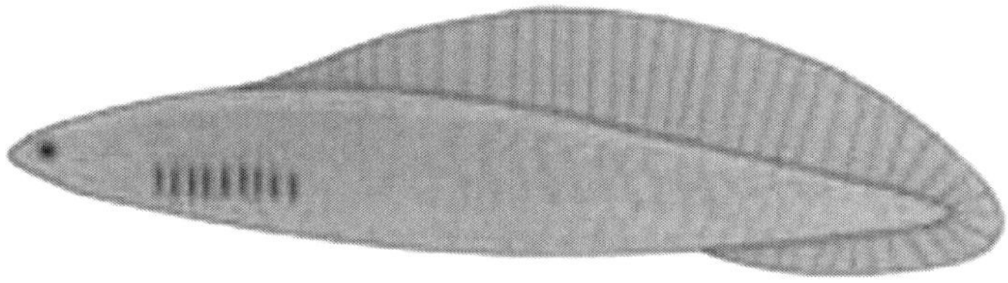

Figure 2. Haikouichthys.

Haikouichtyhs (Figure 2) from about 518 million years ago in China may be the earliest known fish. Fossils of one major deuterostome group, the echinoderms (whose modern members include starfish, sea urchins and crinoids), are quite common from the start of the Cambrian, 542 million years ago. The Mid Cambrian fossil *Rhabdotubus johanssoni* has been interpreted as a pterobranch hemichordate. Opinions differ about whether the Chengjiang fauna fossil *Yunnanozoon*, from the earlier Cambrian, was a hemichordate or chordate. Another fossil, *Haikouella lanceolata*, also from the Chengjiang fauna, is interpreted as a chordate and possibly a craniate, as it shows signs of a heart, arteries, gill filaments, a tail, a neural chord with a brain at the front end, and possibly eyes — although it also had short tentacles round its mouth. *Haikouichthys* and*Myllokunmingia*, also from the Chengjiang fauna, are regarded as fish. *Pikaia*, discovered much earlier (1911) but from the Mid Cambrian Burgess Shale (505 Ma), is also regarded as a primitive chordate. On the other hand fossils of early chordates are very rare, since non-vertebrate chordates have no bones or teeth, and only one has been reported for the rest of the Cambrian [Blair and Hedges, 2005; Cameron et al., 2005].

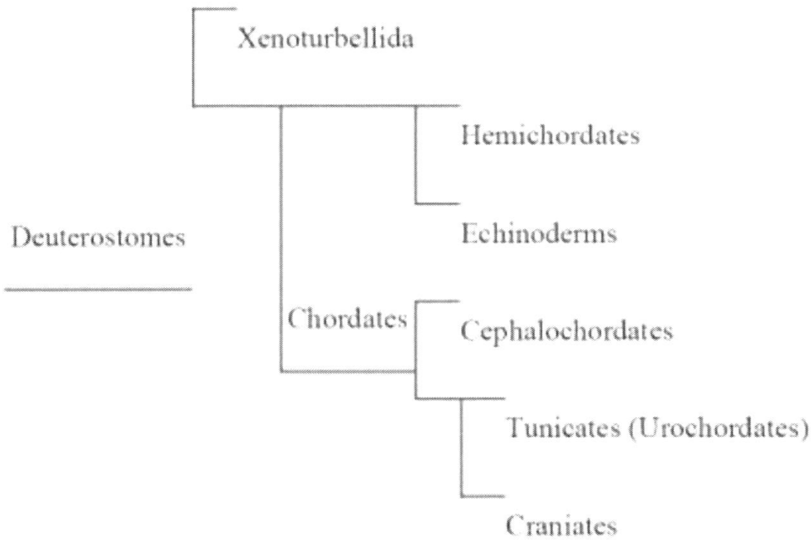

Figure 3. Evolutionary relationship between chordates groups and deuterostome relatives.

2.6. A CONSENSUS FAMILY TREE OF THE CHORDATE

The evolutionary relationships between the chordate groups and between chordates as a whole and their closest deuterostome relatives have been debated since 1890. Studies based on anatomical, embryological, and paleontological data have produced different "family trees". Combining such analyses with data from a small set of ribosomal RNA genes eliminated some older ideas, but open the possibility that tunicates (urochordates) are "basal deuterostomes", surviving members of the group from which echinoderms, hemichordates and chordates evolved. Some researchers believe that, within the chordates, craniates are most closely related to cephalochordates, but there are also reasons for regarding tunicates (urochordates) as craniates' closest relatives. One other phylum, Xenoturbellida, has been thought for a while to be basal within the deuterostomes, closer to the original deuterostomes than to the chordates, echinoderms and hemichordates. But an article by Nakano et al. (2013) demonstrates that the Xenoturbellida are not deuterostomes.

Since chordates have left a poor fossil record, attempts have been made to calculate the key dates in their evolution by molecular phylogenetics techniques - by analyzing biochemical differences, mainly in RNA. One such study suggested deuterostomes arose before 900 million years ago and the earliest chordates around 896 million years ago. However, molecular estimates of dates often disagree with each other and with the fossil record, and their assumption that the molecular clock runs at a known constant rate has been challenged (Figure 3).

Chordates evolved from some deuterostome ancestor (echinoderms, hemichordates, pogonophorans etc.) as they have similarities in embryonic development, type of coelom and larval stages. Fossils of the earliest vertebrates are known from the Silurian-Devonian period, about 400 million years ago. The following theories have been given to explain the origin of chordates [Gee, 1996]:

1. **Coelenterate Theory-Origin of Chordates:** According to this theory, chordates were developed from coelenterates. Radial symmetry of coelenteron, cnidoblasts etc, were first and advanced characters developed to give rise to chordates. This theory infers that chordates might have acquired higher characters independently. It is not correct and hence this theory is not acceptable [Gastrang, 1894; De Robertis and Sasai, 1996].
2. **Annelid Theory-Origin of Chordates:** This theory suggests that the chordates have evolved from an annelid stock. The annelids show bilateral symmetry, metamerism, head, lateral coelome complete digestive tract, closed circulatory system, hemoglobin, etc., like chordates. The resemblance is enhanced if; an annelid is turned upside down. But the mouth would be dorsal which is unlike that of chordates. Metamerism and appendages of annelids differ in nature from those of the chordates. Bilateral symmetry, head and complete digestive tract occur in other non-chordate phyla also. Coelome is schizocoelic in annelids and enterocoelic in lower chordates. Haemoglobin is dissolved in the plasma in annelids but it is present in the red blood corpuscles in chordates. Annelid nerve cord is double, and, ventral in contrast to single, hollow, dorsal nerve cord of chordates (Figure 4). Some striking differences exist between the annelids and the chordates in their embryology. Hence it is difficult to accept this theory.

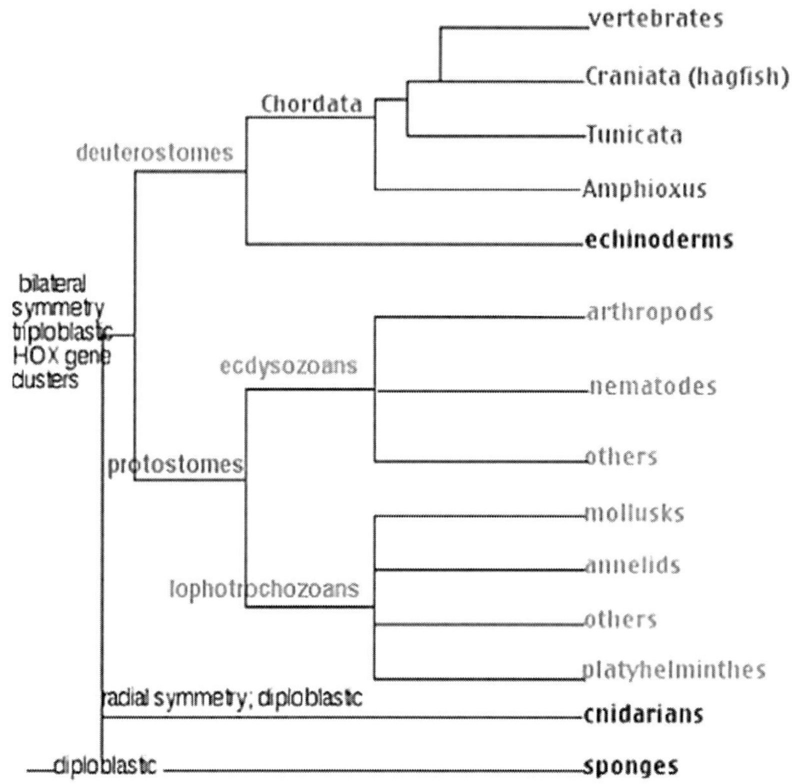

Figure 4. Evolutionary relationships of chordates and its origin.

1. **Echinoderm Origin.** The theory was given by Johannes Muller (1860) and is based on the comparative studies of larval stages of echinoderms and hemichordates. Tornaria larva of hemichordates resembles echinoderm larvae such as Bipinnaria, Auricularia, Dipleurula and Doliolaria, which all possess ciliary bands and apical tuft of cilia. Johannes Muller, W. Garstang and DeBeers proposed that echinoderm larvae gave rise to chordates by neoteny. Also, like chordates, echinoderms are also deuterostomes and possess mesodermal skeletal elements (Figure 4).

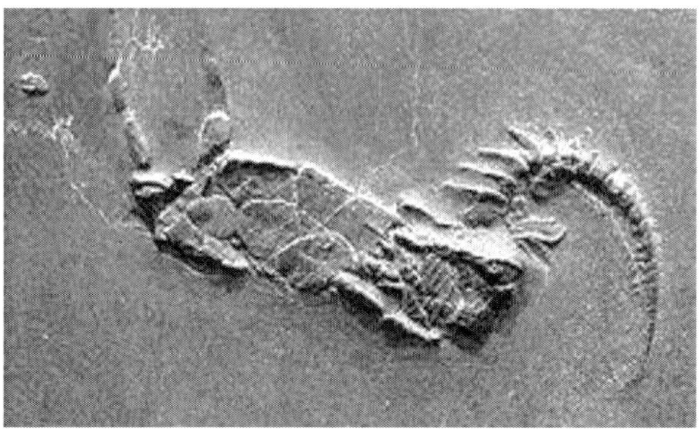

Figure 5. Calcichordata.

The discovery of fossil echinoderms called **Calcichordata** [Peterson, 1995] (Figure 5) from Ordovician period (450 mya) further confirms echinoderm ancestry of chordates. Calcichordates were asymmetrical animals which demonstrate affinities with both echinoderms and chordates but their skeleton is made of $CaCO_3$ whereas in vertebrates the bones are made of hydrated Ca and phosphate. They had large pharynx with a series of gill slits, each covered with flaps for filter feeding, a small segmented body and a postanal tail. A perforated pharynx for filter feeding appears to have evolved in diverse groups of animals during Cambrian-Orodovician periods when planktons were abundant in water [Peterson, 1995].

a) **Embryological Evidence -Origin of Chordates:** Both echinoderms and chordates have enterocoelic coelome, mesoderm and deuterostomous mouth. There is resemblance between the bipinnaria larva of certain echinoderms and the tornaria larva of hemichordates. In echinoderm chordates the central nervous system develops from a dorsal strip of ectoderm [De Beer, 1937].
b) **Serogical Evidence-Origin of Chordates:** A close similarity exists between the proteins of the body-fluid of chordates and echinoderms. Hence, the chordates are more related to echinoderms.

The radial symmetry of adult echinoderms will disprove their relationship with the bilaterally symmetrical chordates. In echinoderms, radial symmetry is secondarily developed from a basically bilateral symmetry. Both the primitive and the early echinoderm larvae show bilateral symmetry.

2. **Hemichordate Origin.** Romer (1967) suggested that ancestral deuterostomes were sedentary tentacle feeders whose mucous-laden ciliated tentacles served to trap planktons as they were waved in water as do the modern lophophorates and pterobranch hemichordates, *Cephalodiscus* and *Rhabdopleura.* By some mutation pharyngeal gill slits evolved in these ancestors, who made the pharynx sieve-like to trap planktons as the water current passed through it. Extant pterobranchs possess both ciliated arms and pharyngeal gill slits. Tornaria larva of hemichordates shows phylogenetic relationship with echinoderm larvae and hemichordates also show affinities with chordates.
3. **Urochordate Origin.** Garstang (1928) and Berrill (1955) gave importance to the tadpole-like larva of urochordates which carries typical chordate characters, namely, a notochord in tail along with segmented myotomes, dorsal hollow nerve cord, sense organs and pharyngeal gill slits. Garstang (1928) suggested that chordates evolved from some sessile filter feeding urochordate by the larval stage evolving into adult by neoteny and by losing the sedentary adult stage.
4. **Cephalochordate Origin.** Chamberlain (1900) studied the primitive and advanced characters of cephalochordates and proposed that while extant cephalochordates possess all chordate characters in typical state, they also show some primitive features of non-chordates, such as, absence of heart, head, sense organs, respiratory pigment, filter-feeding mode of food capture and excretion by solenocytes. Fossils of 60 specimens from mid-Cambrian of the earliest chordate, *Pikaia gracilens* have been discovered from Burgess Shale in British Columbia, Canada. The Amphioxus-like fossils show streamlined, ribbon-shaped, 5 cm long body having notochord in the posterior two-third of body and myomeres. It has a small head with two tentacles and gill slits in the neck region. Other chordate-like fossils are: *Cathaymyrus* from

early Cambrian sediments in China and *Palaeobranchiostomata* from early Permian from South Africa that appears to be more similar to Amphioxus.

5. **Combined theory.** Barrington (1965) combined all the above theories and proposed that the common ancestor of echinoderms and chordates was a sessile ciliary arm feeder that lived in the plankton-rich environment of the Cambrian. Modern Crinoidea (Echinodermata), Pogonophora and Pterobranch hemichordates evolved from a similar ancestor by retaining the original mode of feeding, perhaps because they continued to inhabit the same environment as occurred in ancestral days. However, pharyngotremy (perforation of pharynx with gill slits) must have evolved in a large number of groups at that time, which must have been much more superior method of food gathering by filtering water through pharynx as compared to ciliated arm feeding. Hence, the sedentary Protoascidians of that time lost ciliated arm feeding and adopted pharyngeal filter feeding as the only method of food gathering. Sometime later, when the plankton population in water declined, free-swimming tailed larva of these urochordates did not metamorphose and became a neotenic adult, since free-swimming mode was superior in food searching at a time of food scarcity. Cephalochordate-like ancestors evolved by perfection and expansion of chordate characters that were already present in the ascidian tadpole larva.

Echinoderm -Hemichordate Theory-Origin of Chordates: This theory infers origin of chordates, hemichordates and echinoderms from a common ancestor. This theory is based on the following evidences:

The generalised chordate ancestry is as follows: This was suggested by N.J.Berrill, (1953) in his book “The origin of Vertebrates”. ECHINODERM - AURICULARIA - HEMICHORDATA- TORNARIA -PROTO CHORDATE ASCIDIAN TADPOLE - FREE SWIMMING CHORDATE.

Hyman (1959) and others concluded that all the three groups have a common ancestor, Echinoderms and hemichordates branched off from a common evolutionary line which ended in the chordate group.

chordate characteristics	notochord, dorsal hollow nerve cord, pharyngeal perforations, endostyle or thyroid gland, muscular post anal tail
coelom	main body cavity containing viscera
protostomes	nonchordate lineage, blastopore becomes mouth, spiral cleavage during early development, schizocoelic coelom (forms by pouches in mesoderm tissue), ectodermal skeleton
deuterostomes	blastopore becomes anus; radial cleavage; enterocoelic coelom (outpockets in gut); mesodermal calcified skeleton; chordates
spiral cleavage	cells reproduce in offset spiral form; protostomes
radial cleavage	radiate from center; deuterostomes
muscular post anal tail	evolve to aid swimming
notochord	cartilaginous, flexing, support, firm
dorsal hollow nerve cord	precursor to spinal cord
endostyle	in invertebrates, metabolic processes, iodine
thyroid gland	in vertebrates, metabolic processes, iodine
pharyngeal slits	food/water into mouth, passes out perforations and out of atriopore

amphioxus	represents hypothetical chordate
pelagic	floating young
benthic	bottom of ocean
sessile	not moving
protochordates	all marine, filter feeders, pelagic young, benthic/sessile adults; juvenile metamorph into sexual adult; growth - reproductive form; phylogenetically between vertebrates and all other invertebrates; some or all chordate characteristics; Phylum Hemichordata, Phylum Chordata, Subphylum Cephalochordata, Urochordata, NOT vertebrata
dorsoventral pattern	top/bottom; back/front
pattern inversion	Dorsal/Ventral flips; body plan flips
metamirism	blocks of muscle
cephalized brain	regions in brain/head
phylum hemichordata	sister group to phylum chordata; link between chordates and echinoderms; chordate: deuterostomes, pharyngeal slits; echinoderm: tornaria larva, geometric, molecular
enteropneusta	hemichordate example: acorn worm, enteropneusts
hemichordates	acorn worms, pterobranchs
subphylum cephalochordata	all chordate characters present, amphioxus
subphylum urochordata	all chordate characteristics present as larvae; tunicates (covering): larvae, thaliacea, ascidacea; filter feeders; quick/dramatic metamorph
metamorphic	larvacae, ascidacea, thaliacea
planktonic larvae	proposed phylogenetic link between chordates and echinoderms
extant protochordates	filter food in pharynx by passing water in mouth and out slits, notochord and post anal tail for movement (better than cilia)
chordates origin	controversial; annelids/arthropods, echinoderms?
evidence for chordates from annelids/arthropods	body segmentation, brain regionalization, dorsal nerve cord and ventral gut when flipped (opposite body plan), movement of some larval crustaceans
evidence against chordates from annelids/arthropods	analogous characters, protostomes versus deuterostomes (basic developmental differences), body segmentation (exoskeleton versus myotomal segmentation), solid nerve cord
Garstang's hypothesis	echinoderms as chordate ancestors; echinoderm larvae that failed to reproduce; larval echinoderm structures precursor to chordate structures (circumoral/adoral band); deuterostomes; pelagic larva; bilateral symmetry in larva; larval similarities; hemi/chordate similarities; allele recognition
allele recognition	recognition of self versus others (organ transplant); only in echinoderms, vertebrates, and random chordates, rare
paedomorphosis	sexual adult that resembles larval form: progenesis and neoteny
progenesis	sexual maturity accelerated; gonads develop faster than body, body stops (mudpuppy)
neoteny	halitng of development of particular traits; resemble some juvenile characteristics (human face)
evidence against Garstang	an echinoderm gave rise to the protochordate -- echoniderms and chordates share a common ancestor
size increase advantages	advantages: predators; better stability/mobility in ocean current
size increase	requires alternate way of moving; cilia not effective; develop notochord, segmented, muscle, elongated body; increased musculature
size increase favors:	change in feeding apparatus (increased efficiency in developing pharyngeal sits versus cilia around mouth); development of organ systems (circulatory, respiratory, digestive, etc)

2.7. CHORDATES AFFINITY

The main lines of invertebrate animals show the closest affinity with chordate. Many of the supposed linking similarities between chordates and annelids or arthropods result from ANALOGY rather than HOMOLOGY. Some scientists believe that Annelids and arthropods have the closest similarity because of following reasons [see review by Holland, 2005]:

a) **Metamerism or Metameric Segmentaion:**
Both group's shows metamerism but this similarity is superficial.
REASON: The segmentation of this annulates is almost complete division of entire body into rings and units, all organ systems are affected by segmentation. However, in Chordates, only dorsal myotomal region is segmented whereas even the ventral part of mesoderm is not segmented [Holland, 1988].

b) **Orientation of the Chordates:**
The two group's namely Chordates and annulates shows difference in this. The NERVE CORD is dorsal to the duct whereas in annulates it is below the gut and the brain above. Similarly the blood circulation in different directions in two groups and the limbs are also based on different plans. It is thought that the Eurypterids show some superficial resemblance to the Agnatha of Devonian making the scientists believe that, the type of an annulate could be the possible ancestors of chordates [Holland , 1988].

c) **Cleavage Pattern:**
Development of modern annulates is completely different from chordates. In annulates, the cleavage by which the fertilized eggs gets divided into blastomeres follows a SPIRAL PATTERN in which each blastomere arises a regular way and the future fate of each is exactly defined. In contrast the cleavage in chordates is RADIAL AND IRREGULAR [Holland , 1991; McGinnis al, 1992].

d) **Gastrulation:** The next stage of development i.e., gastrulation is also different in two groups. In chordates, it occurs by invagination of one side of the ball of cells to form an archenteric cavity communicating with exterior. However, in annulates, the cells forming guts migrate inward all around the sphere and later get arranged into a tube that opens to the outside [McGinnis et al., 1992] (Table 1).

e) **Fate of Blastofore:**
It is another point of comparison in chordates, protochordates and in Echinoderm blastopore becomes the anus while the mouth is a new opening hence these animals come under group DEUTEROSTOMIA. Distinct from non-chordata that belongs to PROTOSTOMIA in which blastopore becomes the mouth.

f) The method of mesoderm and coelom formation are also important i.e., in DEUTEROSTOMIA this 3rd layer is produced by separation of endoderm so that coelom is at first continuous with the archenteron and it is said t be an ENTEROCOEL. However, in the annulates, the cells separate in various ways to form a mesoderm and a coelom than arises within this solid mass as a SCHOZOCOEL (Table 1).

Thus in all the above points the chordates differ from all the annulates but resembles Echinoderms and their allies [Holland , 1991].

Further features also support this relationship. For e.g.: Both have a larva with longitudinal ciliated bands, very different from the TROCHOPHORE LARVA OF ANNULATES which has transverse cilia.

Similarly NERVOUS SYSTEM of annulates consists of a set of ganglionated cords whereas in Echinoderm-like animals it is a diffused sheet of fibres and cells below the epidermis. The nerve chord of the chordate can be easily derived from Echinoderm nervous system and not from the annulates type of nervous system [Gans and Northcutt, 1983].

A MESODERMAL SKELETON is found in both chordates and echinoderms but absent in annulates [Bengton, 2004].

The mechanism for providing energy liberation also shows systematic differences. For e.g., regeneration of ATP is brought about by Phosphoarginine and its Kinase in the arthropods and molluscs, while Phosphocreatinine and its Kinase bring about the same reaction in Amphioxus and chordate while both this systems operate in echinoderms and urochordates. This may have the advantage; a rich energy source like Creatinine which could be easily synthesized from glycine which could be kept aside for locomotion whereas the rare Arginine could be utilized only for the important histone formation. Early on during evolution it was found creatinine was characteristic of early larval life but in neoteny, creatinine became the only substrate for ATP generation.

Table 1. Comparative plans of Protostomes and Deuterostomes

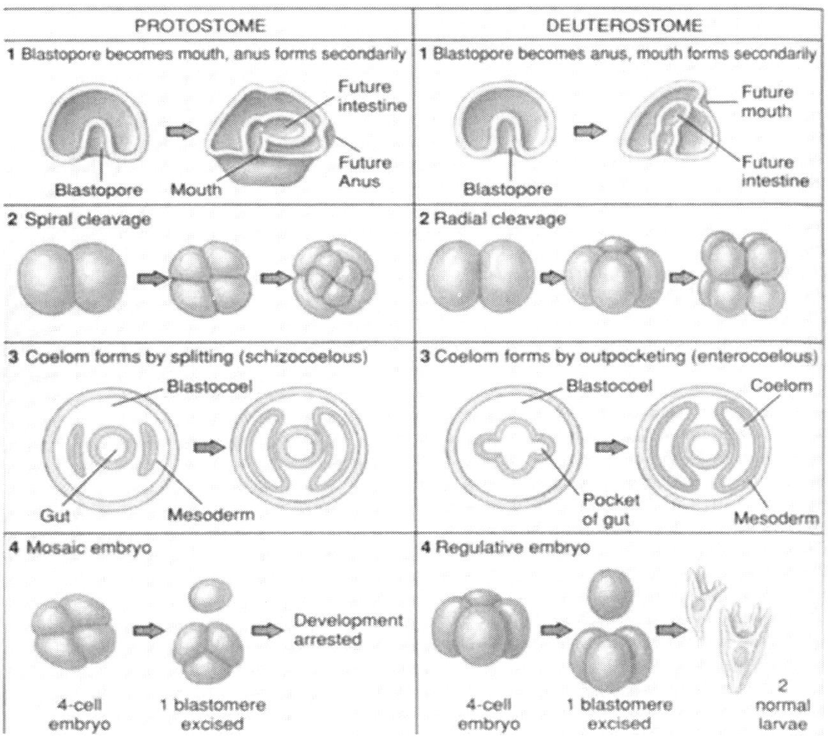

2.8. THE ORIGIN OF TETRAPODS

The word "tetrapod" means "four feet" and includes all species alive today that have four feet — but this group also includes many animals that don't have four feet [Long and Gordon, 2004]. That's because the group includes all the organisms (living and extinct) that descended from the last common ancestor of amphibians, reptiles, and mammals (Figure 7). So, for example, the ichthyosaur, an extinct swimming reptile, is a tetrapod even though it did not use its limbs to walk on land. So is the snake, even though it has no limbs. And birds and humans are tetrapods even though they only walk on two legs. All these animals are tetrapods because they descend from the tetrapod ancestor described above, even if they have secondarily lost their four-feet [Jennifer, 1997; Laurin, 2010].

The invasion of the land had to be slow because finned fishes of any sort are poorly adapted to life on land. The physical properties of water require significantly different appendages and support structures in an organism than those required on land [Laurin, 2010]. Fluids, including both air and water, exert forces evenly in all directions on anything in it. However, critically, water is about 1,000 times denser than air, and thus is much more "buoyant", i.e., it supports things much better. Since organisms are primarily composed of water (and thus not much denser in it), organisms will not sink very readily in water and experience the illusion of weightlessness. The immediate problem this presents to the invasion onto land is that , this means that most fishes lack load-bearing structures — that is, are physically incapable of supporting their own bodies in such a lightly buoyant medium like air, and are thus unable to move. Finally, the exaptation of primitive swim bladder/lung structures to be the primary source of oxygen intake would enable actual excursions onto the land [Canoville, 2010].

For any of these novel structures or uses of existing structures to become evolutionarily favorable, it is important to recognize that there must be a benefit to the organism. The first precursor structures that would ultimately be adapted to land living can be found in the sarcopterygian line of fishes [Kardong, 1998; Haaramo, 2003; Barley, 2010; Niedźwiedzki, 2010; Retallac, 2011; Lu et al., 2012].

The invasion of land by vertebrates, the so-called "sea-land transition", was an incredibly important event, but also very gradual, occurring over several million years. Contrary to many popular depictions, there was no clear point when fins became limbs, at which point the creature hauled itself onto the land; instead, the ancestry of **tetrapods** and the invasion of the land can be traced to the gradual alteration of fin structures seen in **sarcopterygians**, **actinopterygian** line approximately 400 million years ago [Kardong, 1998; Niedźwiedzki, 2010; Retallac, 2011; Lu et al., 2012]. As can be seen in Figure 6, as the sarcopterygian-tetrapod line became more derived, the radius and ulna became more suitable for load-bearing and propulsive movement against a rigid medium, and the **distal elements** became less suited for propulsive action in a dense fluid (fins) and more adapted toward providing traction on a rigid substrate (more leg-like) [Canoville et al., 2010].

Additionally, there were further adaptations associated with transitioning to a land-based existence that are somewhat more subtle. While the evolution of more robust limbs are associated with moving into shallower, brackish waters in which fins are no longer useful, there are other associated adaptations that are evolutionarily favorable under the same circumstances. Shallower water gives organisms less room to maneuver, which means if the head is attached such that it has a definite orientation with respect to the body, there can be positions where it is difficult to eat food, to submerge gills, or to breathe air. All three of these issues are addressed if the shoulder structures decouples from the head (permitting the head some motion with respect to the body trunk), and a neck that permits independent head movement allowing more movement still. As these organisms became capable of moving into shallower waters, buoyant forces on the body decrease, forcing modifications of the rib cage to provide better support structures for the internal organs. Similarly, less buoyant support required the limbs to be more firmly attached to the backbone for locomotor purposes [Panchen, 1967; Hildebran and Goslow, 2001], as shown in Figure 6.

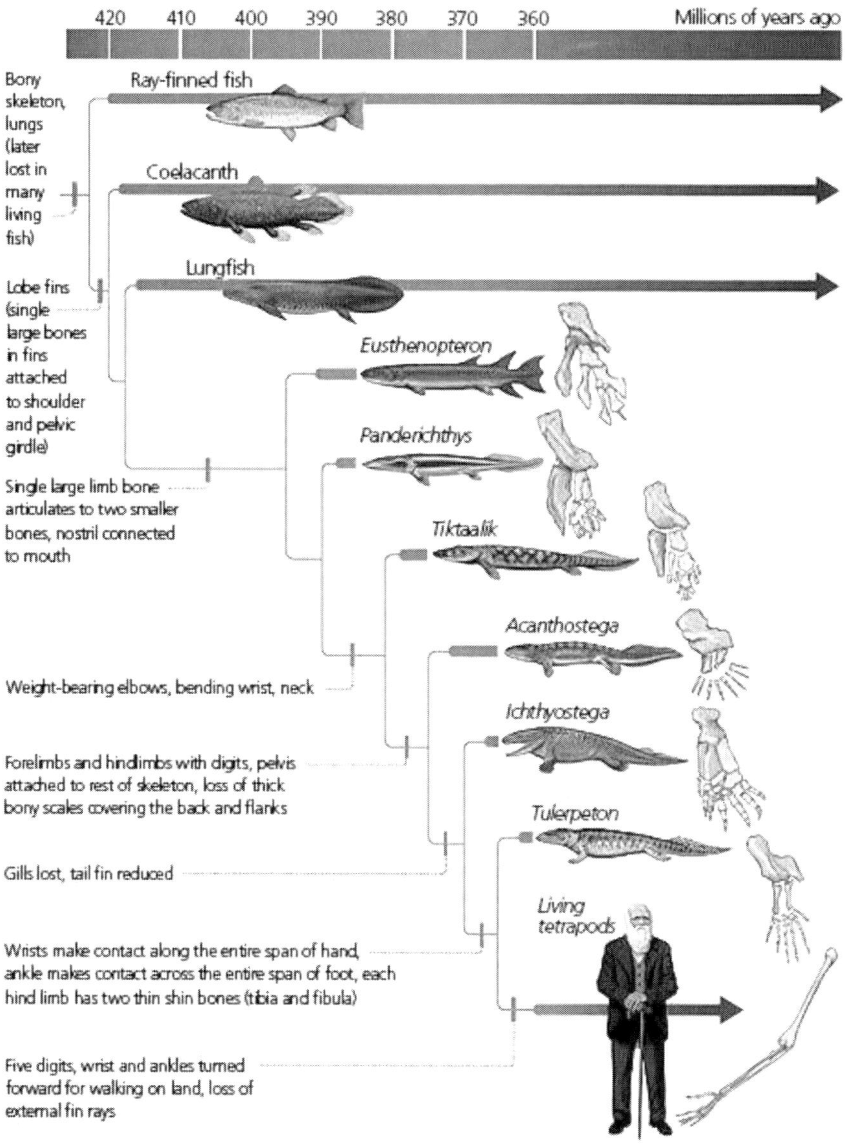

(Source: http://evolution.berkeley.edu/).

Figure 6. Evogram showing Evolution of Tetrapods.

While the physiological changes associated with the land-sea transition are locked in the fossil record, the behavioral changes prompting or prompted by these changes are, necessarily, harder to tease from the rocks. Current theories postulate that these physiological changes were driven by the progressive movement of these creatures into shallower, more brackish water. In turn, this progression was prompted by the availability of an unfilled niche — there was simply little or no competition for food at the water's edge, as nothing had yet moved to occupy that position. Further, the actual excursions beyond the water would be prompted by a lack of competition for consuming invertebrates on land, a diet still retained by **extant** amphibians. While a complete invasion of the land was not yet possible, due to desiccation problems of both the animals and their eggs, these were the critical issues which enabled movement about the land. The final innovations which enabled complete colonization were the evolution of scales to prevent desiccation, and the shell-covered amniotic egg to allow eggs to be laid away from water [Sahney et al., 2010].

[1] This is more accurately called the water-land transition, as it is not yet sure if the transition occured in saltwater or in freshwater

[2] Contrary to popular belief, an organism is not truly weightless in water. If life were made of lead, it would have had to overcome similar structural issues in water that life had to overcome moving onto land.

[3] More accurately, it must promote the spread of a piece of its genome in the population. Sometimes this distinction is important, though for the purposes of this article treating the organism as the beneficiary is sufficient.

[4] The reasoning behind the water generally being dirty and brackish is that it lowers the oxygen content of the water, thus prompting a greater reliance on atmospheric oxygen.

[5] The invertebrates were found on land due to a lack of competition themselves, both for preying on each other and for the plants which had by then widely colonized the land.

[6] Again, a trait still found in modern amphibians which are tied to water or moist environments.

Tetrapods evolved from a finned organism that lived in the water. However, this ancestor was not like most of the fish we are familiar with today. Most animals we call fishes today are ray-finned fishes, the group nearest the root of this evogram. Ray-finned fishes comprise some 25,000 living species, far more than all the other vertebrates combined. They have fin rays — that is, a system of often branching bony rays (called lepidotrichia) that emanate from the base of the fin [Coates, 2002; Zhang et al., 2010; Stephanie et al., 2012] (Figure 7).

In contrast, the other animals in the evogram as shown in figure 7 — coelacanths, lungfishes, all the other extinct animals, plus tetrapods (represented by Charles Darwin) — have what we call "fleshy fins" or "lobe fins." That is, their limbs are covered by muscle and skin. Some, such as coelacanths, retain lepidotrichia at the ends of these fleshy limbs, but in most fleshy-finned animals these have been lost [Zhang et al., 2010; Stephanie et al., 2012].

The common ancestor of all those different organisms (ray-fins, coelacanths, lungfishes, tetrapods, etc.) was neither a lobe-fin nor a ray-fin. This ancient vertebrate lineage had fins (with lepidotrichia), scales, gills, and lived in the water. Yet they also had air bladders (air-filled sacs) connected to the back of their throats that could be used for breathing air (i.e., as lungs) or for buoyancy control. The air bladders of many ray-fins no longer connect to their throats, and so they are not able to breathe air. In these ray-fins, the air bladder is used mainly for buoyancy control and is known as a swim bladder [Romer, 1949; Kardong, 1998]. By contrast, tetrapod have taken an alternative route: they have lost the buoyancy control function of their air bladders, and instead this organ been elaborated to form the lungs that we all use to get around on land.

When we get past coelacanths and lungfishes on the evogram, we find a series of fossil forms that lived between about 390 and 360 million years ago during the Devonian Period. During this interval, this lineage of fleshy-finned organisms moved from the water to the land. Many parts of the skeleton changed as new innovations that permitted life on land evolved (Figure 7). For example, the ancestors at the base of this evogram lived fully in the water and had skulls that were tall and narrow, with eyes facing sideways and forwards. This allowed them to look around in their watery environments for predators and prey. However, as ancestors of the first tetrapods began to live in shallower waters, their skulls evolved to be flatter, with eyes on the tops of their heads. This probably allowed them to look up to spot food. Then, as tetrapods finally moved fully onto land and away from the water, many lineages once again evolved skulls that were tall and narrow, with eyes facing sideways and forwards, allowing them to look around their terrestrial environments for predators and prey [Romer, 1949; Kardong, 1998].

As lineages moved into shallower water and onto land, the vertebral column gradually evolved as well. You may have noticed that fishes have no necks. Their heads are simply connected to their shoulders, and their individual vertebrae look quite similar to one another, all the way down the body. Mobile necks allow land animals to look down to see the things on the ground that they might want to eat. In shallow water dwellers and land dwellers, the first neck vertebra evolved different shapes, which allowed the animals to move their heads up and down. Eventually, the second neck vertebra evolved as well, allowing them to move their heads left and right. Later tetrapods evolved necks with seven or more vertebrae, some long and some short, permitting even more mobility [Hildebran and Goslow, 2001; Coates et al., 2002; Zhang et al., 2010; Stephanie et al., 2012].

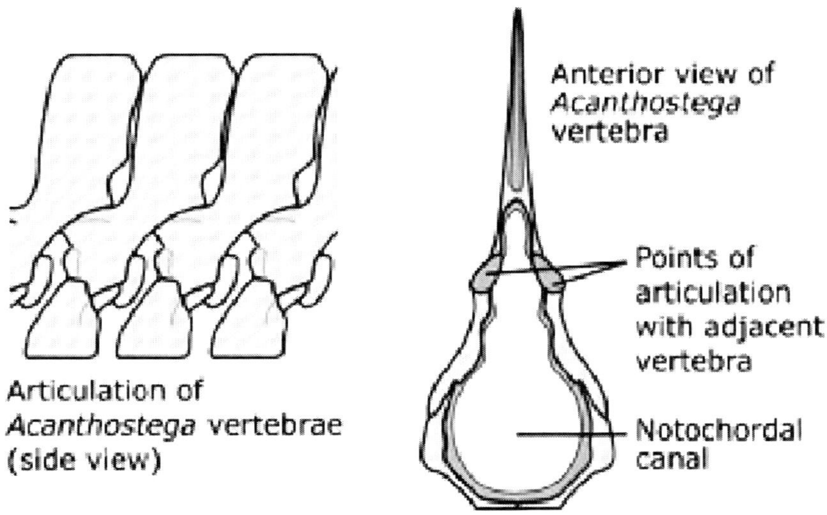

Figure 7. Vertebra shape and connections in the early tetrapod *Acanthostega*.

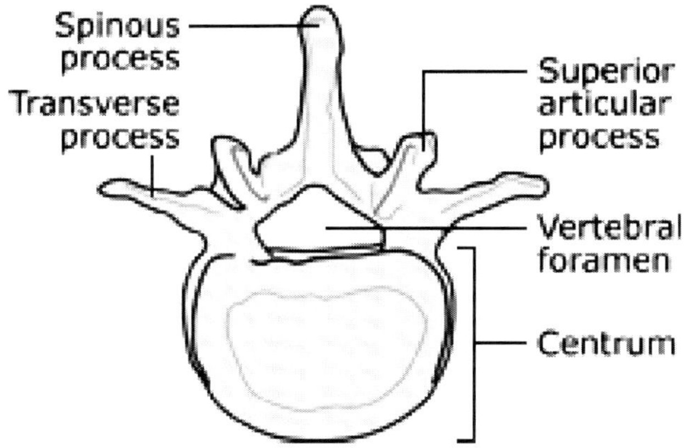

Figure 8. One element of the human vertebral column.

A typical vertebrae consist of a spool-like centrum, which connects in front and back with other centra. On top of the centra are vertebral spines and arches to which muscle segments attach, and lateral to the centra are the ribs; these anchor muscles that flex as the animals move. Fishes swim with simple lateral motions, so their arches are relatively straight and needle-like, and so are their ribs. Because fishes live in the water, gravity is not a big problem for them. But on land, a quadruped with a backbone between forelimbs and hindlimbs faces the same problems As the fleshy-finned organisms began to venture onto land, they evolved a series of interlocking articulations on each vertebra, which helped them overcome sag and hold the backbone straight with minimal muscular effort [Hildebran and Goslow, 2001; Coates et al., 2002; Zhang et al., 2010; Stephanie et al., 2012].

The connection between the pelvis and hindlimbs (sacrum) in early tetrapods is a prime example of **exaptation**. It is extremely useful for terrestrial organisms because it allows them to use their hindlimbs efficiently for locomotion on land. Since the aquatic ancestors of fishes and tetrapods had no such connection, one might guess that this feature first evolved serving the function of enabling terrestrial locomotion (Figure 7 and 8). However, the earliest form of this connection (as seen in Acanthostega) evolved while these tetrapod precursors were still living in the water. Based on current evidence, Acanthostega appears to have been fully aquatic, so this connection likely evolved to function in something other than terrestrial locomotion. Only later, as tetrapod ancestors moved onto land, was this trait co-opted for terrestrial support — and as it was, additional vertebrae were fused in the same way, providing further support [Hildebran and Goslow, 2001; Coates et al., 2002; Zhang et al., 2010; Stephanie et al., 2012].

As the limbs and their connections to the rest of the skeleton evolved, limb bones took on distinct roles and many bones were lost. The humerus and the femur were already connected to two outer bones (the radius and ulna in the forelimb, the tibia and fibula in the hindlimb) (Figure 7 and 8). This is something that evolved about 30 million years before vertebrates came onto land. However, muscular connections between these bones began to change on the road to land and allowed the limbs to be used for terrestrial locomotion. The ankle was originally composed of many small bones arranged in two rows, but gradually many of these small bones were lost. The first animals to get close to walking on land had eight digits on each limb. Over time, some of these digits were lost, leading to animals with seven digits, then six, and then five, which is the common condition now seen in living tetrapods [Hildebran and Goslow, 2001; Coates et al., 2002; Zhang et al., 2010; Stephanie et al., 2012].

As these animals evolved to live on land, other changes in the rest of their bodies evolved. Many would eventually lose their gills, which only work well for getting oxygen when wet and their tail fins got smaller. Similarly they lost the lateral line system, a network of vibration-sensitive canals along the skull and jaw, which doesn't work out of water.

The environments of the animals shown in this evogram (Figure 6) also changed through time. In fact, if you were to venture back to Arizona at the beginning of the "Age of Dinosaurs" in the Triassic Period, some 225 million years ago, you would find ray-fins, coelacanths, and lungfishes living in the marshes, streams, and temporary ponds of that day, along with freshwater sharks. So the habitats that these animals occupy today are not necessarily the ones in which they have always lived, or in which they originally evolved. It is still unclear exactly where the transition from water to land took place ecologically. Paleontologists have discovered fossils involved in this transition preserved from freshwater, brackish, and marine habitats. Regardless of where the transition occurred, eventually early ancestors of the first tetrapods came up onto land — although not all stayed. Some, like the whales, made the transition back into the water [Sahney et al., 2010].

REFERENCES

Barrington, E. J. W. (1965). The Biology of Hemichordates and Protochordates. Oliver and Boyd. Edinburg and London.

Bateson, W. (1886). The ancestry of the chordata. *Quart J Microscop Sci*, *26*, 535-571.

Bengtson, S. (2004). *Early skeletal fossils*. In Lipps, J.H., and Waggoner, B.M. Neoproterozoic-Cambrian Biological Revolutions. *Paleontol Soc Pap*, *10*, 67–78.

Benton, M. J. (2000). *Vertebrate Palaeontology: Biology and Evolution*. Blackwell Publishing. 12–13.

Berrill, N. J. (1955). The origin of vertebrates. Oxford: Oxford University Press

Blair, J. E., Blair, S. & Hedges, S. B. (2005). Molecular Phylogeny and Divergence Times of Deuterostome Animals. *Mol Biol Evol*, *22*, 2275–2284.

Cameron, C. B., Garey, J. R. & Swalla, B. J. (2000). Evolution of the chordate body plan: new insights from phylogenetic analyses of deuterostome phyla. *Proc Natl Acad Sci USA*, *97*, 4469-4474.

Canoville, A. & Laurin, M. (2010). Evolution of humeral microanatomy and lifestyle in amniotes, and some comments on paleobiological inferences. *Biol J Linn Soc*, *100*, 384–406.

Chamberlain, T. C. (1900). On the habitat of the early vertebrates. *J Geol*, *8*, 400.

Clack, J. A. (1997). Devonian tetrapod trackways and trackmakers; a review of the fossils and footprints. *Palaeogeo, Palaeoclim, Palaeoecol*, *130*, 227–250.

Coates, M. I., Jeffery, J. E. & Ruta, M. (2002). Fins to limbs: what the fossils say. *Evol Dev*, *4*, 390–401.

Cowen, R. (2000). *History of Life* (3rd ed.). Blackwell Science. p. 412.

De Beer, G. R. (1937). The vertebrate skull. Oxford University Press.

De Robertis, E. M. & Sasai, Y. (1996). A common plan for dorsoventral patterning in bilateria. *Nature, 380*, 37-40.

Gans, C. (1883). Northcutt RG. Neural crest and the origin of the vertebrates: a new head. *Science, 220*, 268

Garstang, W. (1894). Preliminary note on a new theory of the phylogeny of the Chordata. *Zool Anz, 17*, 122–125.

Garstang, W. (1928a). The morphology of the tunicata, and its bearingson the phylogeny of the chordata. *Q J Microsc Sci, 72*, 51–187.

Garstang, W. (1928b). The origin and evolution of larval forms. *Rep Brit Ass Adv Sci*, 77-98.

Gee, H. (1996). Before the backbone: views on the origin of the vertebrates.London: Chapman & Hall

Grzegorz, N., Piotr, S., Katarzyna, N., Marek, N. & Per, E. A. (2010). Tetrapod trackways from the early Middle Devonian period of Poland. *Nature, 463*, 43–48.

Haaramo, M. (2003). Sarcopterygii. *in Mikko's Phylogeny Archive*.

Hildebran, M. & Goslow, G. (2001). *Analysis of Vertebrate Structure* (5th ed.). New York: John Wiley.

Hiroaki, N., Kennet, L., Sarah, J., Bourlat, M. J., Telford, P. F., Jens, R., Nyengaard, M. O. & Michael, C. T. (2013). *Xenoturbella bocki* exhibits direct development with similarities to Acoelomorpha. *Nat Comm, 4*, 1537.

Holland, N. D. (1988). The meaning of developmental asymmetry for echinoderm evolution: a new interpretation. In C. R. C. Paul and A. B. Smith (eds.). Echinoderm Phylogeny and Evolutionary Biology. Oxford Science Publications and Liverpool Geological Society, Clarendon Press, Oxford, 13–25.

Holland, N. D. (1991). Echinodermata: Crinoidea. In A. C. Geise, J. S. Pearse, and V. B. Pearse (eds.). Reproduction of Marine Invertebrates: Echinoderms and Lophophorates. The Boxwood Press, California, 247–292.

Holland, N. D. (2005). Chordates. *Curr Biol, 15*, 911-914.

Hyman, L. H. (1959). The enterocoelous coelomates — the Hemichordata. In The Invertebrates, Volume 5, McGraw-Hill:72-207.

Hyman, L. H. (1959). Phylum Hemichordata. In: The invertebrates: Smaller coelomate groups. New York: McGraw-Hill. 72–207.

Jefferies, R.P.S., Brown, N.A.& Daley, P.E.J.(1996). The early phylogeny of chordates and echinoderms and the origin of chordate left right asymmetry and bilateral symmetry. *Acta Zool (Stockh), 77*, 101-22.

Jing, L., Min, Z., John, A. L., Wenjin, Z., Tim, J., Senden, L. J. & Tuo, Q. (2012). The earliest known stem-tetrapod from the Lower Devonian of China. *Nat Comm, 3*, 1160.

Kardong, K. V. (1998). Vertebrates: Comparative Anatomy, Function, Evolution, second edition, USA: McGraw-Hill, 747 pp.

Laurin, M. (2010). *How Vertebrates Left the Water*. Berkeley, California, USA.: University of California Press.

Long, J. A. & Gordon, M. S. (2004). The greatest step in vertebrate history: a paleobiological review of the fish-tetrapod transition. *Physiol Biochem Zool, 77*, 700–19.

Lowe, C. J., Wu, M., Salic, A., Evans, L., Lander, E., Stange-Thomann, N., Gruber, C. E., Gerhart, J. & Kirschner, M. (2003). Anteroposterior patterning in hemichordates and the origins of the chordate nervous system. *Cell, 113*, 853-865.

McGinnis, W. & Krumlauf, R. (1992). Homeobox genes and axial patterning. *Cell*, *68*, 283-302.

Nu¨bler-Jung, K. & Arendt, D. (1996). Enteropneusts and chordate evolution. *Curr Biol* 6, 352-353.

Panchen, A. L. (1967). The nostrils of choanate fishes and early tetrapods. *Biol Rev*, *42*, 374–419.

Peterson, K. J. (1995). A phylogenetic test of the calcichordate scenario. *Lethaia*, *28*, 25–38.

Retallack, G. (2011). Woodland Hypothesis for Devonian Tetrapod Evolution. *J Geol* (University of Chicago Press), *119*, 235–258.

Romer, A. (1967). Major steps in vertebrate evolution. *Science* 158, 1629–1637.

Romer, A. S. (1949). *The Vertebrate Body*. Philadelphia: W.B. Saunders. (2nd ed. 1955; 3rd ed. 1962; 4th ed. 1970)

Ruppert, E. (2005). Key characters uniting hemichordates and chordates: homologies or homoplasies? *Can J Zool*, *83*, 8–23.

Rychel, A. L., Smith, S. E., Shimamoto, H. T., & Swalla, B. J. (2006). Evolution and Development of the Chordates: Collagen and Pharyngeal Cartilage. *Mol Biol Evol*, *23*, 541–549. doi:10.1093/molbev/msj055. PMID 16280542

Sahney, S., Benton, M. J. & Falcon-Lang, H. J. (2010). Rainforest collapse triggered Pennsylvanian tetrapod diversification in Euramerica. *Geology*, *38*, 1079–1082.

Shanta, B. (2010). Oldest footprints of a four-legged vertebrate discovered. New Scientist.

Smith, A.B. (2005). The pre-radial history of echinoderms. *Geol J*, *40*, 255-280.

Stephanie, E. P., Jennifer, A. C. & John, R. H. (2012). Three-dimensional limb joint mobility in the early tetrapod *Ichthyostega*. *Nature*, *486*, 524–527.

Van der Horst, C. J. (1939). Hemichordata. In Klassen und Ordnungen des Tierreichs, Volume 4, Abt 4, Buch 2, Teil 2. Edited by HG Bronn:1-739.

Zhang, J., Wagh, P., Guay, D., Sanchez-Pulido, L., Padhi, B. K., Korzh, V., Andrade-Navarro, M. A. & Akimenko, M. A. (2010). Loss of fish actinotrichia proteins and the fin-to-limb transition. *Nature*, *466*, 234–237.

Chapter 3

ORIGIN, EVOLUTION AND AFFINITIES OF REPTILES

3.1. ADAPTIVE RADIATION OF REPTILES

In evolutionary biology, **adaptive radiation** is a process in which organisms diversify rapidly into a multitude of new forms, particularly when a change in the environment makes new resources available and opens environmental niches [Stanley, 1979; Schluter, 2000] . Starting with a recent single ancestor, this process results in the speciation and phenotypic adaptation of an array of species exhibiting different morphological and physiological traits with which they can exploit a range of divergent environments [Stanley ; 1979].

Adaptive radiation, a characteristic example of cladogenesis, can be graphically illustrated as a "bush", or clade, of coexisting species (on the tree of life)

Four features can be used to identify an adaptive radiation:

1. **A common ancestry of component species:** specifically a *recent* ancestry. This is not the same as a monophyly in which *all* descendants of a common ancestor are included.
2. **A phenotype-environment correlation:** a *significant* association between environments and the morphological and physiological traits used to exploit those environments.
3. **Trait utility:** the performance or fitness advantages of trait values in their corresponding environments.
4. **Rapid speciation:** presence of one or more *bursts* in the emergence of new species around the time that ecological and phenotypic divergence is underway.

Causes

Innovation

The evolution of a novel feature may permit a clade to diversify by making new areas of morpho space accessible. A classic example is the evolution of a fourth cusp in the mammalian tooth. This trait permits a vast increase in the range of foodstuffs which can be fed on. Evolution of this character has thus increased the number of ecological niches available to mammals. The trait arose a number of times in different groups during the

Cenozoic, and in each instance was immediately followed by an adaptive radiation [Mayr, 2001]. Birds find other ways to provide for each other, i.e., the evolution of flight opened new avenues for evolution to explore, initiating an adaptive radiation [Simpson, 1944]. Other examples include placental gestation (for eutherian mammals), or bipedal locomotion (in hominins).

Opportunity

Adaptive radiations often occur as a result of an organism arising in an environment with unoccupied niches, such as a newly formed lake or isolated island chain. The colonizing population may diversify rapidly to take advantage of all possible niches.

In Lake Victoria, an isolated lake which formed recently in the African rift valley, over 300 species of cichlid fish adaptively radiated from one parent species in just 15,000 years.

Adaptive radiations commonly follow mass extinctions: following extinction, many niches are left vacant. A classic example of this is the replacement of the non-avian dinosaurs with mammals at the end of the Cretaceous, and of brachiopods by bivalves at the Permo-Triassic boundary (Figure 1).

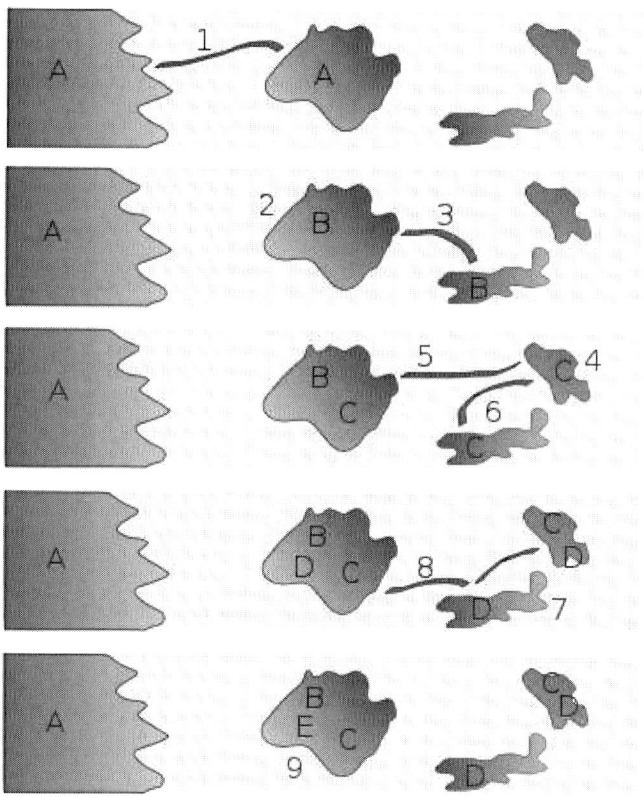

Figure 1. 1. Species A migrates from the mainland to the first island. 2. Isolated from the mainland, species A evolves to species B. 3. Species B migrates to the second island. 4. Species B evolves in species C. 5. Species C recolonizes the first islands, but is now unable to reproduce with species B. 6. Species C migrates to the third island. 7. Species C evolves into species D. 8. Species D migrates to the first and second island. 9. Species D evolves to species E. This process might go on indefinitely until a large diversity is reached.

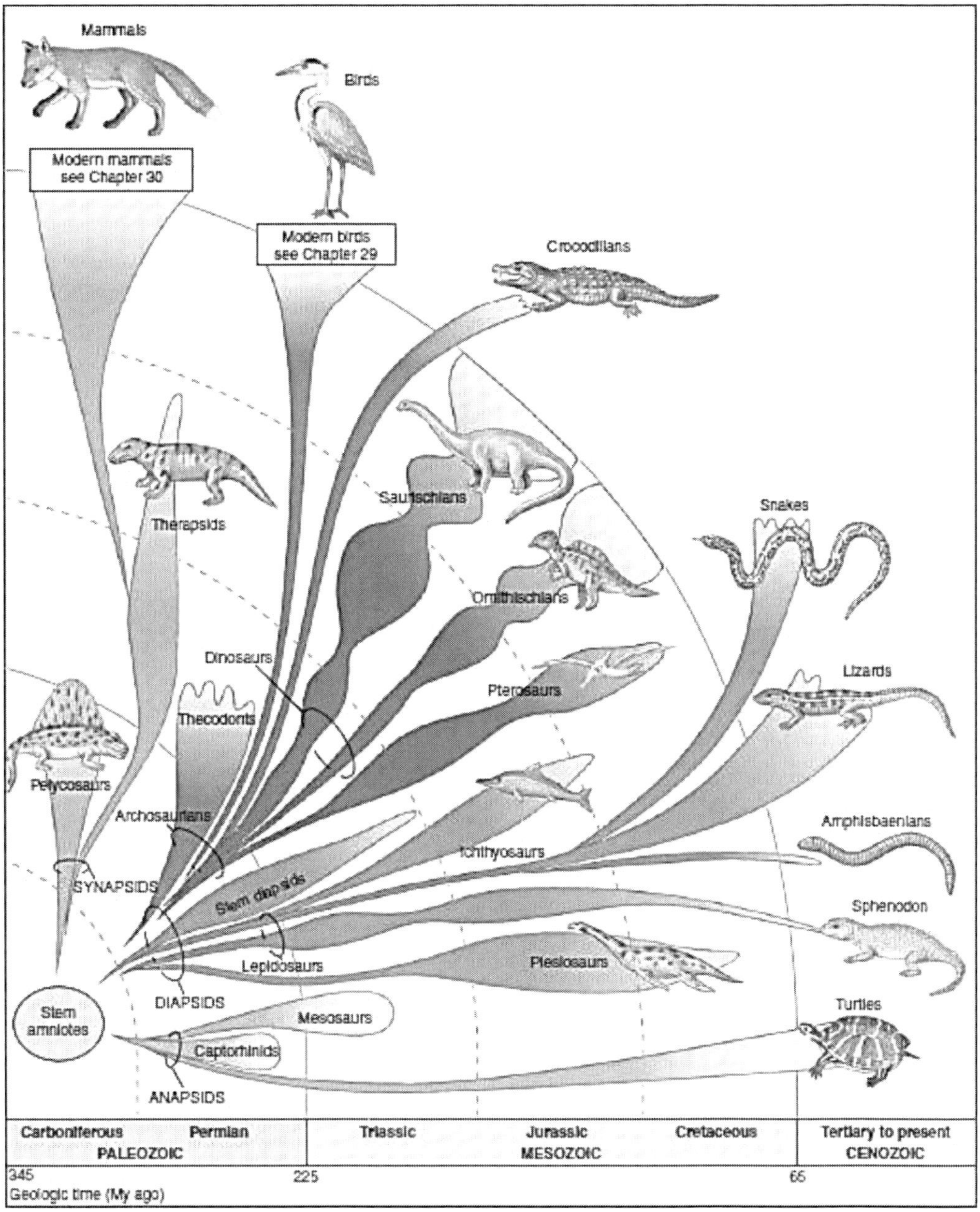

Figure 2. Evolution of the amniotes. The evolutionary origin of amniotes occurred by the evolution of an amniotic egg that made reproduction on land possible, although this egg may well have developed before the earliest amniotes had ventured far on land. The amniote assemblage, which includes the reptiles, birds, and mammals, evolved from a lineage of small, lizard like forms known as captorhinids that retained the skull pattern of the early tetrapods. First to diverge from the primitive stock were the mammal-like reptiles, characterized by a skull pattern termed the synapsid condition. All other amniotes, including the birds and all living reptiles except the turtles, have a skull pattern known as diapsid. The turtles have a skull pattern known as anapsid. The great Mesozoic radiation of reptiles may have resulted partly from the increased variety of ecological habitats into which the amniotes could move.

Origin and Adaptive Radiation of Reptilian Groups As mentioned in the prologue to this section, amniotes are a monophyletic group that evolved in the late Paleozoic. Most paleontologists agree that the amniotes arose from a group of amphibian-like tetrapods, the anthracosaurs, during the early Carboniferous period of the Paleozoic. By the late Carboniferous (approximately 300 million years ago), amniotes had separated into three lineages. The first lineage, the **anapsids** (Gr. *an*, without, + *apsis*, arch), is characterized by a skull having no temporal opening behind the orbits, the skull behind the orbits being completely roofed with dermal bone (Figure 3). This group is represented today only by the turtles. Their morphology is an odd mix of ancestral and derived characters that has scarcely changed at all since turtles first appeared in the fossil record in the Triassic some 200 million years ago.

The second lineage, the ***diapsids*** (Gr. *di*, double, + *apsis*, arch), gave rise to all other reptilian groups and to the birds (Figure 2). The diapsid skull was characterized by the presence of two temporal openings: one pair located low on the cheeks, and a second pair positioned above the lower pair and separated from them by a bony arch (Figure 3). Three subgroups of diapsids appeared. The **lepidosaurs** include extinct marine ichthyosaurs and all modern reptiles except the turtles and **crocodilians**. The more derived **archosaurs** comprised dinosaurs and their relatives, and living crocodilians and birds. A third, smaller subgroup, the **sauropterygians** included several extinct aquatic groups, most conspicuous of which were the large, long-necked plesiosaurs.

The third lineage was the **synapsids** (Gr. *syn*, together, + *apsis*, arch), the mammal-like reptiles. The synapsid skull had a single pair of temporal openings located low on the cheeks and bordered by a bony arch (Figure 3). The synapsids were the first amniote group to diversify, giving rise first to pelycosaurs, later to therapsids, and finally to mammals (Figure 2).

3.2. Changes in Traditional Classification of Reptilian Groups

With increasing use of cladistic methodology in zoology, and its insistence on hierarchical arrangement of monophyletic groups, important changes have been made in the traditional classification of reptiles. The class Reptilia is no longer recognized by cladists as a valid taxon because it is not monophyletic. As customarily defined, class Reptilia excludes birds, which descend from the most recent common ancestor of the reptiles. Consequently, reptiles are a paraphyletic group because they do not include all descendants of their most recent common ancestor. Reptiles can be identified only as amniotes that are not birds. This is clearly shown in the phylogenetic tree of the amniotes (Figure 3).

An example of this problem is the shared ancestry of birds and crocodilians. Based solely on shared derived characteristics, crocodilians and birds are sister groups; they are more recently descended from a common ancestor than either is from any other living reptilian lineage. In other words, birds and crocodilians belong to a monophyletic group apart from other reptiles and, according to the rules of cladism, should be assigned to a clade that separates them from the remaining reptiles. This clade is in fact recognized; it is the Archosauria (Figures 2 and 3), a grouping that also includes the extinct dinosaurs. Therefore, birds should be classified as reptiles. Archosaurs plus their sister group, the lepidosaurs (tuataras, lizards, snakes, and amphisbaenids), comprise a monophyletic group that some taxonomists call the Reptilia. The term "Reptilia" is thereby redefined to include birds in contrast to its traditional usage. However, evolutionary taxonomists argue that birds represent a novel adaptive zone and grade of organization whereas crocodilians remain within the traditionally recognized reptilian adaptive zone and grade. In this view, the morphological and ecological novelty of birds has been recognized by maintaining the traditional classification that places crocodilians in class Reptilia and birds in class Aves. Such conflicts of opinion between proponents of the two major competing schools of taxonomy (cladistics and evolutionary taxonomy) have had the healthy effect of forcing zoologists to reevaluate their views of amniote genealogy and how vertebrate classifications should represent genealogy and degrees of divergence. In our treatment, "reptilian group," "reptile," and "reptilian" refer to members of four monophyletic groups (turtles, crocodilians, squamates, tuataras) that formerly were combined into the paraphyletic class Reptilia.

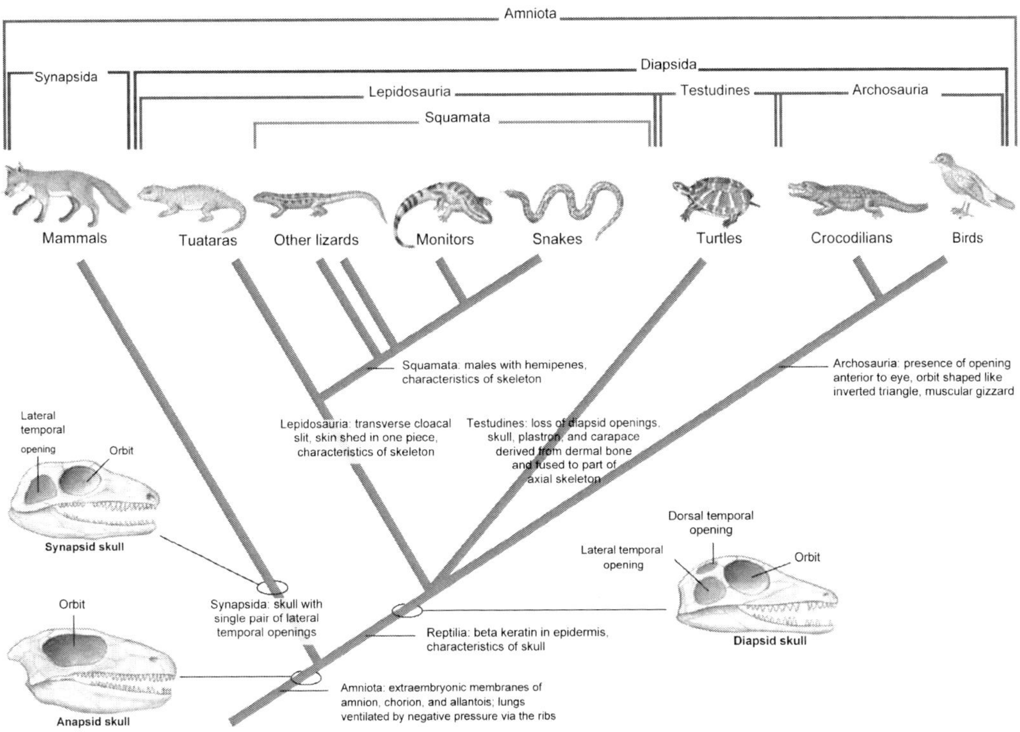

Figure 3. Cladogram of the living amniota showing monophyletic groups. Some of the shared derived characters (Synanomorphies) that are diagnostic for the lineages are given. The skull represents the ancestral conditions of the three groups. The skull of modern diapsids and synapsids are highly modified by the loss or fusion of skull bones that obscures the ancestral condition. Representative skull for anapsids are Nyctiphruetus of the upper Permian; for diapsids Youngina of upper Permian; for synapids, aerosaur, a pelycosaur of the lower Permian. The relationhip expressed in the cladogram is tentative, especially that between birds and mammals. Contrary to the view shown here, in which mammals are the outgroup, some authorities support a sister-group relationship between birds and mammals based on several kind of physiological and molecular evidences.

3.3. Genotypic Sex Determination Enabled Adaptive Radiations of Extinct Marine Reptiles

Adaptive radiations often follow the evolution of key traits, such as the origin of the amniotic egg and the subsequent radiation of terrestrial vertebrates. The mechanism by which a species determines the sex of its offspring has been linked to critical ecological and life-history traits [Stably, 1979; Schluter, 2000] but not to major adaptive radiations, in part because sex-determining mechanisms do not fossilize. Research from several groups were used to establish a previously unknown coevolutionary relationship in 94 amniote species between sex-determining mechanism and whether a species bears live young or lays eggs. This relationship was used to predict the sex-determining mechanism in three independent lineages of extinct Mesozoic marine reptiles (mosasaurs, sauropterygians and ichthyosaurs), [Simpson, 1944, 1953; Mayr, 2001; Carroll, 1997], each of which is known from fossils to have evolved live birth. Our results indicate that each lineage evolved genotypic sex determination before acquiring live birth. This enabled their pelagic radiations, where the relatively stable temperatures of the open ocean constrain temperature-dependent sex determination in amniote species. Freed from the need to move and nest on land, extreme physical adaptations to a pelagic lifestyle evolved in each group, such as the fluked tails, dorsal fins and wing-shaped limbs of ichthyosaurs. With the inclusion of ichthyosaurs, mosasaurs and sauropterygians, genotypic sex determination is present in all known fully pelagic amniote groups (sea snakes, sirenians and cetaceans), suggesting that this mode of sex determination and the subsequent evolution of live birth are key traits required for marine adaptive radiations in amniote lineages [Benton, 1993].

Tuatara

- No intromittent organ

hemipenes - pair of intromittent organs / works like penis / barbed to anchor male

Dinosaurs

- terrestrial
- many kinds
- dominated the mesozoic (age of reptiles)
- some had parental care
- massive extinction about the end of the Cretaceous

Pterosaurs

- flying reptiles

Ichthyosaurs

- Shark shaped
- Marine Plesiosaurs
- long neck
- paddle feet
- marine

Diapsids

- Two temporal openings
- all living reptiles except turtles
- Dinosaurs
- Gave rise to birds

Synapsids

- One temporal opening
- no living reptiles
- gave rise to mammals

Are reptile scales homologus or analogous with fish scales?

Homologus to feathers but not comparable to hair because they are formed from keratin in the epidermis, which gives the skin protection from physical trauma and limit water loss. **Analogous** to fish scales because reptile scales are made of keratin and fish scales are made of bone but both are used for osmoregulation or abrasion resistance functions [Barlow, 2002].

Amniotic eggs consist of

1. Embryo
2. Extraembryonic membranes = Amnion membrane / Yolk sac/ Allantosis/ Chorion
3. Albumen
4. Shell

Amnion membrane

- membrane covering embryo
- Provides the embryo with its own aquatic environment

Yolk Sac

- contains nutrients

Allantosis

- stores wastes
- used in respiration

Chorion

- respiration membrane in Amnionic egg [Laurin and Reisz,1995; Paton et al., 1999].

3.4. Origin and Evolution of Reptiles

Reptiles evolved from amphibians of Carboniferous period, which depended on water bodies for laying eggs and development of larval stages and hence could not exploit arid habitats far away from water bodies. They invented a large yolk-laden shelled egg that could be laid on land and in which an amniotic sac contained fluid in which embryo could develop to an advanced stage, capable of fending for itself when hatched. The following anatomical changes transformed the ancestral amphibians into land adapted reptiles [Monastersky,1999; Falcon-Lang et al., 2007]:

1. Body developed a covering of epidermal scales to prevent loss of body moisture, and skin glands were lost.
2. Skull became monocondylic for better movement and flexibility. Atlas and axis vertebrae together permitted skull movement in all directions.
3. Limb bones and girdles became stronger but limbs were attached on the sides of body, and belly touched the ground during creeping mode of locomotion.
4. Sacral region involved two strong and fused vertebrae to support the body weight on hind legs.
5. Pentadactyle limbs developed claws that helped in climbing on rocks and trees. Lung respiration became more efficient.
6. As a water conservation strategy, metanephros kidneys excreted uric acid, which did not require water for excretion.
7. Reptiles continued to be ectothermal since ventricle was not completely partitioned by a septum and blood mixed in heart.
8. Internal fertilization evolved as a large cleioid shelled egg was laid on land.
9. Embryonic membranes amnion, allantois and yolk sac evolved to enable embryonic development in arid conditions.

3.5. Ancestors of Reptiles

The Cotylosaurs

They were the most primitive stem reptiles that evolved from the labyrithodont amphibians (Embolomeri) in Carboniferous period.

Seymoria was a lizard-like animal, with pentadactyle limbs and a short tail. It had homodont labyrinthine teeth on the jaw bones as well as on vomer and palatine bones. Presence of lateral line indicates its amphibious habits. Skull was monocondylic for better movement of head. *Seymoria* indicates gradual transition from labyrinthodont amphibians to reptiles. Another 5 foot long cotylosaur fossil, *Limnoscelis* was found in Mexico that had large premaxillary teeth and long tail [Pawely, 2006].

The Parapsids

They possessed superior temporal vacuity in the skull and were adapted for aquatic mode of life.

Plesiosaurus was marine long-necked, fish-eating animal with 15 metre long fusiform body, short tail and paddle-like limbs modified for swimming. The skull was euryapsid type with a superior temporal vacuity. The fossils are from lower Jurassic (about 180 million years) and they are believed to have become extinct in end-Cretaceous mass extinction [Ruta et al., 2003].

Ichthyosaurus had fish-like body with fore limbs modified into paddle-like fins and hind limbs disappeared. There was a fleshy dorsal fin too. Caudal fin was large and bilobed. Jaws projected into an elongated snout and teeth were homodont, an adaptation for fish-catching. Skull was parapsid type with additional postfrontal and supratemporal bones behind the eye orbit. Vertebral column became secondarily simplified with amphicoelous vertebrae [Ruta et al., 2003].

The Synapsids

Synapsids split off from the primitive reptilian stock very early in evolution, perhaps in the middle carboniferous period. Synapsids had started developing mammalian characteristics that enabled them to be fleet-footed and active predators. Their legs commenced to move under the body. Heterodont dentition and false palate started developing in pelycosaurs and had been completely formed in therapsids. Two types of synapsids occurred from carboniferous to Permian, namely, the primitive Pelycosaurs and advanced therapsids [Linda and Müller, 2009].

Pelycosaurs are represented by *Dimetrodon* whose fossils were discovered from North America and Russia from the late Carboniferous to Permian periods. They were primitive reptile-like animals in which limbs had moved under the body but not completely and each limb had 5 digits with claws. Neural spines on the back were excessively long, stretching highly vascularized skin between them that formed a fin-like or sail-like structure. They had heterodont dentition with incisors, canines and molars clearly defined but the false palate had not been completely formed [Van Tuninen and Hadly, 2004].

Therapsids were more advanced and active synapsids which were perhaps endothermic animals with high rate of metabolism.

Heterodont dentition with false palate allowed these animals to chew and grind food for quick digestion in the gut so that high metabolic demand of the body could be fulfilled. Jaw muscles were attached to zygomatic arch to make chewing effective. Carnivore therapsids were called Cynodonts (ex. *Cynognathus)* and herbivores were Dicynodonts [Van Tuninen and Hadly, 2004].

The Thecodonts

They evolved from the sauropsid Archosauria, a group of insignificant lizard-like reptiles that survived the Triassic mass extinction. They evolved into bipedal and highly agile predators.

Euperkeria and **Ornithosuchus** fossils were unearthed from South Africa and Europe. They were about 2 ft long bipedal lizard-like animals with small head but very long tail for balancing while they chased flying insects by rapid running. Endothermy must have evolved in thecodonts to meet the extraordinary energy demands of their predatory life style [Van Tuninen and Hadly, 2004].

The Saurischians

They were dinosaurs with lizard-like pelvic girdle in which ischium and pubis bones radiated away from each other. They were both bipedal and quadrupedal and carnivores as well as herbivores.

The Ornithischians

They were dinosaurs with bird-like pelvic girdle in which ischium and pubic bones were directed towards posterior as found in modern birds. These were also highly diversified carnivores as well as herbivores and both bipedal and quadruped.

The Pterosauria

They were flying or gliding dinosaurs of Mesozoic that varied in size from sparrow-sized to some species, like *Pteranodon,* having a wing span of 8 meters. They had pneumatic bones. Last digit of the fore limb was extraordinarily long and served to attach the membranous patagium between fore limb, hind limb and the body. Hind limbs were used for clinging on to the rocks and cliffs and 3 digits of fore limbs also had curved claws, an adaptation for clinging. Their jaws were modified into beak that possessed homodont dentition but *Pteranodon* did not have teeth [Van Tuninen and Hadly, 2004].

Though not a reptile, *Gephyrostegus bohemicus* is where reptile evolution begins. It was an Upper Carboniferous (~310 mya) lizard-like tetrapod, or reptilomorph, from Nyran in the Czech Republic.

Since some reptile fossils, like *Westlothiana* (~350 mya), are older than *Gephyrostegus*, it's clear that oldest gephyrostegid were not found yet. Carroll (1970) thought that reptiles originated as tiny tetrapods.

Cephalerpeton, the most primitive reptile was smaller than this specimen of *Gephyrostegus*. Even smaller was another specimen of *Gephyrostegus*.

The reptile family tree begins with *Gephyrostegus* because, by definition, **the Reptilia** includes turtles, lizards, snakes, crocodilians and birds, their last common ancestor and all of its descendants. That last common ancestor (*Cephalerpeton*, according to the present study) was close to *Gephyrostegus*. In the present study, distinct from all prior studies, all known reptiles nest in two major branches with turtles and lizards on one branch and crocodilians, birds and mammals on the other.

Gephyrostegus bohemicus itself was probably too large to produce embryos enclosed within an amniotic membrane, the defining character of **the Amniota** (= Reptilia in this study). However, *Gephyrostegus watsoni* is a more likely candidate.

Shortly after a very early sister to *Cephalerpeton*, members of the reptilia split into two clades represented at their bases by *Westlothiana* and *Cephalerpeton*. The new **Lepidosauromorpha**, are all those sharing a more recent common ancestor with living lizards, the Lepidosauria, than with Archosauria [Gauthier 1986]. The new **Archosauromorpha** share more traits in common with living archosaurs (birds and crocs) than with Lepidosauria. This primitive division has not been reported before because all prior studies were more limited in scope, without the benefit of an umbrella study like this one [Van Tuninen and Hadly, 2004].

Pederpes finneyae was an early tetrapod and the first to have only five toes. It lived for 350 million years and grew to 1m long. It is also distinguished by its short skull, large eyes and compact torso with relatively short neural spines (Figure 4).

The Microsauria

Several specimens that have been labeled as basal reptiles are widely recognized as reptile like amphibians known as microsaurs. This is a segment from the larger tree to show where microsaurs nest.

Earlier, with just two microsaurs included, the large study found Utaherpeton and Tuditanus to nest with reptilian with Westlothiana. The addition of several other microsaurs clarified the situation and moved all microsaurs out of the reptilian.

Ichthyostega is at the base of this tree. Reptilomorphs are tetrapod texa leading toward reptiles (not including Eryops and kin). Silvanerpeton and Gephyrostegus are reptile sister taxa with the latter considered as protoreptile. In other studies Diadectes, Chroniosuchus and Solenodonsaurus were also considered reptilomorphs. That is an error resulting from a too small taxon list. Here, those taxa nest deep within the reptiles.

A similar study of large scope looked at pre-amniote tetrapods [Ruta, et al. 2003]. It was not accurate near and within the reptilian (= amniota) due to the short and restricted amniote taxon list employed.

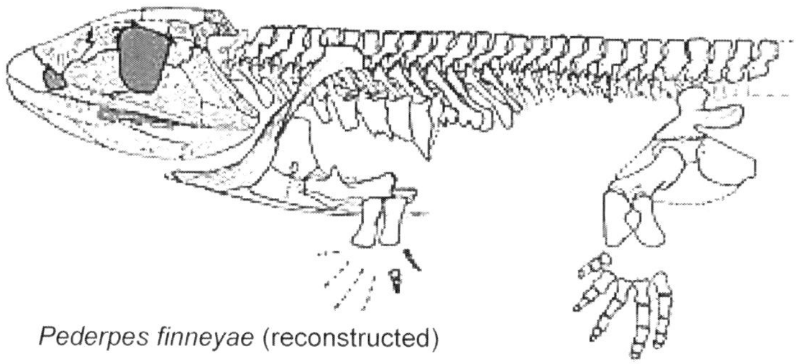

Figure 4. Reconstructed image of *Pederpes finneyae.*

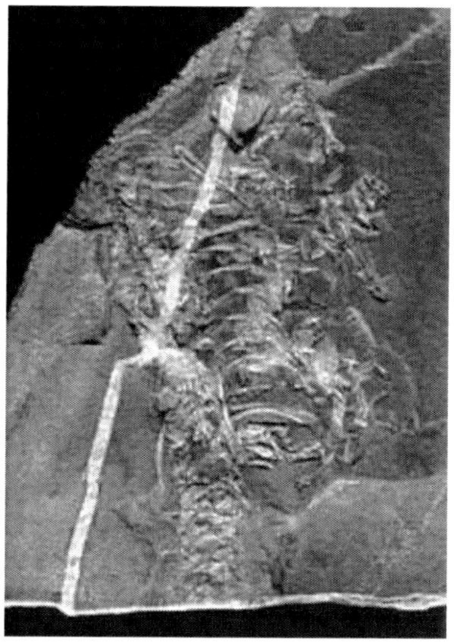

Figure 5. Fossilized specimen of an early reptile.

3.6. Evolution of Reptile

Reptiles arose about 310-320 million years ago during the **Carboniferous** period. Reptiles, in the traditional sense of the term, are defined as animals that have **scales** or **scutes**, lay land-based hard-shelled eggs, and possess **ectothermic** metabolisms. So defined, the group is **paraphyletic**, excluding **endothermic** animals like birds and mammals that are descended from early reptiles. A definition in accordance with **phylogenetic nomenclature**, which rejects paraphyletic groups, includes birds while excluding mammals and their **mammal-like reptile** ancestors. So defined, Reptilia is identical to **Sauropsida** [Van Tuninen and Hadly, 2004].Though few reptiles today are **apex predators**, many examples of apex reptiles have existed in the past. Reptiles have an extremely diverse evolutionary history that has led to biological successes such as **dinosaurs**, **pterosaurs**, **plesiosaurs**, **mosasaurs**, and **ichthyosaurs** (Figure 5).

First Reptiles (Rise from Water)

Reptiles first arose from **amphibians** in the swamps of the late **Carboniferous**. Increasing evolutionary pressure and the vast untouched niches of the land powered the evolutionary changes in amphibians to gradually become more and more land based. Environmental selection propelled the development of certain traits, such as a stronger skeletal structure, muscles, and more protective coating (scales) became more favorable; the basic foundation of reptiles were founded. The evolution of lungs and legs are the main transitional steps towards reptiles, but the development of **hard-shelled external eggs** replacing the amphibious water bound eggs is the defining feature of the class Reptilia and is what allowed these amphibians to fully leave water. Another major difference from amphibians is the increased brain size, more specifically, the enlarged **cerebrum** and **cerebellum**. Although their brain size is small when compared to birds and mammals, these enhancements prove vital in hunting strategies of reptiles. The increased size of these two regions of the brain allowed for improved motor skills and an increase in sensory development [Benton, 1993].

The origin of the reptiles lies about 320–310 million years ago, in the steaming swamps of the late Carboniferous period, when the first reptiles evolved from advanced reptiliomorph labyrinthodonts. The oldest known animal that may have been an amniote, a reptile rather than an amphibian [Juan et al., 2004], is *Casineria* (though it has also been argued to be a temnospondyl amphibian) [Gauthier, 1994; Sean Modesto et al., 2011].

A series of footprints from the fossil strata of Nova Scotia, dated to 315 million years, show typical reptilian toes and imprints of scales. The tracks are attributed to *Hylonomus*, the oldest unquestionable reptile known [Simpson, 1953]. It was a small, lizard-like animal, about 20 to 30 cm (8–12 in) long, with numerous sharp teeth indicating an insectivorous diet (Figure 6). Other examples include *Westlothiana* (for the moment considered a reptiliomorph amphibian rather than a true amniote) and *Paleothyris*, both of similar build and presumably similar habit. One of the best known early reptiles is *Mesosaurus*, a genus from the early Permian that had returned to water, feeding on fish [Paton et al., 1999].

Figure 6. An early reptile *Hylonomus*.

Figure 7. Mesozoic scene showing typical reptilian megafauna: the dinosaurs. *Europasaurus holgeri* and *Iguanodon*, and the early bird *Archaeopteryx* perched on the foreground tree stump.

The earliest reptiles were largely overshadowed by bigger labyrinthodont amphibians, such as *Cochleosaurus*, and remained a small, inconspicuous part of the fauna until after the small ice age at the end of the Carboniferous (Figure 7).

Anapsids, Synapsids, Diapsids and Sauropsids

It was traditionally assumed that first reptiles were anapsids, having a solid skull with holes only for the nose, eyes, spinal cord, etc.; the discoveries of synapsid-like openings in the skull roof of the skulls of several members of Parareptilia, including lanthanosuchoids, millerettids, bolosaurids, some nycteroleterids, some procolophonoids and at least some mesosaurs made it more ambiguous and it's currently uncertain whether the ancestral reptile had an anapsid-like or synapsid-like skull. Very soon after the first reptiles appeared, they split into two branches. One branch, **Synapsida** (including both "mammal-like reptiles" and modern, extant mammals), had one opening in the skull roof behind each eye [Osborn, 1903].

The other branch, **Sauropsida**, is itself divided into two main groups [Goodrich, 1916; Watson, 1957; Pawley, 2006]. One of them, the aforementioned Parareptilia, contained taxa with anapsid-like skull, as well as taxa with one opening behind each eye. Members of the other group, Diapsida, possessed a hole in their skulls behind each eye, along with a second hole located higher on the skull. The function of the holes in both synapsids and diapsids was to lighten the skull and give room for the jaw muscles to move, allowing for a more powerful bite (Figure 8).

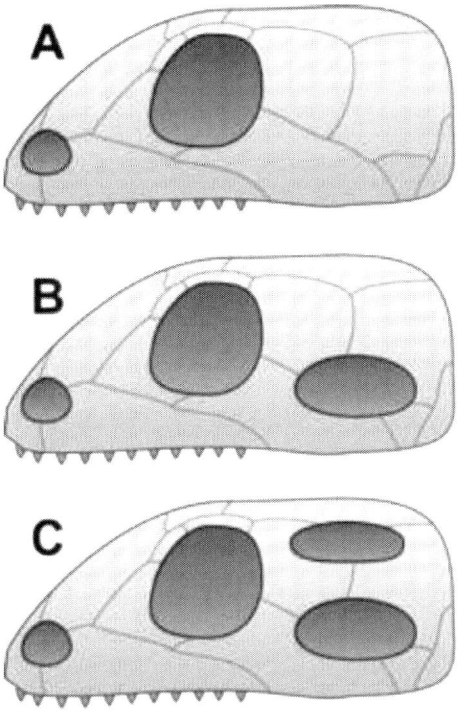

Figure 8. Cavities in reptile skull, A = Anapsid, B = Synapsid, C = Diapsid.

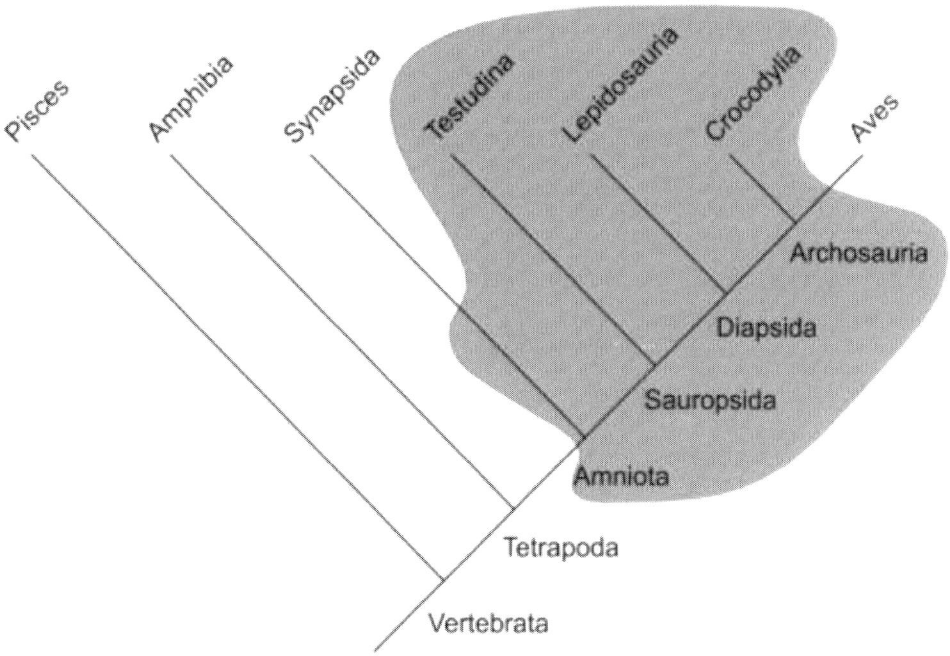

Figure 9. Cladogram showing reptilian affinities.

Turtles have been traditionally believed to be surviving anapsids, on the basis of their skull structure. The rationale for this classification was disputed, with some arguing that turtles are diapsids that reverted to this primitive state in order to improve their armor (see Parareptilia). Later morphological phylogenetic studies with this placed turtles firmly within Diapsida. All molecular studies have strongly upheld the placement of turtles within diapsids, most commonly as a sister group to extant archosaurs [Goodrich, 1916; Pawley, 2006].

Reptiles in the traditional sense (green field) are a paraphyletic group comprising all non-avian and non-mammalian amniotes (Figure 9).

Permian Reptiles

Near the end of the Carboniferous, while the terrestrial reptiliomorph labyrinthodonts were still present, the synapsids evolved the first fully terrestrial large vertebrates, the pelycosaurs such as *Edaphosaurus*. In the mid-Permian period, the climate turned dryer, resulting in a change of fauna: The primitive pelycosaurs were replaced by the more advanced therapsids [Modesto et al., 2011].

The anapsid reptiles, whose massive skull roofs had no postorbital holes, continued and flourished throughout the Permian. The pareiasaurs reached giant proportions in the late Permian, eventually disappearing at the close of the period.

Early in the period, the diapsid reptiles split into two main lineages, the **archosaurs** (forefathers of crocodiles and dinosaurs) and the **lepidosaurs** (predecessors of modern tuataras, lizards, and snakes). Both groups remained lizard-like and relatively small and inconspicuous during the Permian [Modesto et al., 2011].

3.7. The Mesozoic Era, the "Age of Reptiles"

The close of the Permian saw the greatest mass extinction known. Most of the earlier anapsid/synapsid megafauna disappeared, being replaced by the archosauromorph diapsids [Raup, 1995; Dean, 1999]. The archosaurs were characterized by elongated hind legs and an erect pose, the early forms looking somewhat like long-legged crocodiles. The archosaurs became the dominant group during the Triassic period, developing into the well-known dinosaurs and pterosaurs, as well as the pseudosuchians. The Mesozoic is often called the "Age of Reptiles", a phrase coined by the early 19th-century paleontologist who recognized the dinosaurs and the ancestors of the crocodilians as the dominant land vertebrates. Some of the dinosaurs were the largest land animals ever to have lived while some of the smaller theropods gave rise to the first birds [Raup, 1995; Haines, 2000].

The sister group to Archosauromorpha is Lepidosauromorpha, containing squamates and rhynchocephalians, as well as their fossil relatives. Lepidosauromorpha contained at least one major group of the Mesozoic sea reptiles: the mosasaurs, which emerged during the Cretaceous period. The phylogenetic placement of other main groups of fossil sea reptiles – the sauropterygians and the ichthyosaurs, which evolved in the early Triassic and in the Middle Triassic respectively – is more controversial. Different authors linked these groups either to lepidosauromorphs or to archosauromorphs, and ichthyosaurs were also argued to be diapsids that did not belong to the least inclusive clade containing lepidosauromorphs and archosauromorphs [Jacobs, 1997; Pope et al., 1997].

The therapsids came under increasing pressure from the dinosaurs in the Jurassic; the first mammals and the tritylodontids were the only survivors of the line by the end of the period [Langer et al., 2010].

Demise of the Dinosaurs

Cretaceous–Paleogene Extinction Event

The close of the Cretaceous period saw the demise of the Mesozoic era reptilian megafauna. Along with massive amount of volcanic activity at the time, the meteor impact that created the Cretaceous–Paleogene boundary is accepted as the main cause for this mass extinction event. Of the large marine reptiles, only sea turtles are left, and, of the dinosaurs, only the small feathered theropods survived in the form of birds [Pope et al., 1997]. The end of the "Age of Reptiles" led to the "Age of Mammals". Despite the change in phrasing, reptile diversification continued throughout the Cenozoic, with squamates undergoing a greater radiation than they did in the Mesozoic [Pope et al., 1997]. Today, squamates make up the majority of extant reptiles (over 90%). There are approximately 9,547 extant species of reptiles, compared with 5,400 species of mammals, so the number of reptilian species (without birds) is nearly twice the number of mammals [Wang and Dodson, 2006].

Role Reversal

After the Cretaceous–Paleogene extinction event wiped out all of the non-avian dinosaurs (birds are generally regarded as the surviving dinosaurs) and several mammalian groups, placental and marsupial mammals diversified into many new forms and ecological niches throughout the Paleogene and Neogene eras. Some reached enormous sizes and almost as wide a variation as the dinosaurs once did. Nevertheless, mammalian megafauna never quite reached the skyscraper heights of some sauropods.

3.8. The Four Orders of Reptilia

Testudines

Testudines, or turtles, may have evolved from anaspids, but their exact origin is unknown and heavily debated. Fossils date back to around 220 million years ago and share remarkably similar characteristics. These first turtles retain the same body plan as do all modern testudines and are mostly herbivorous, with some feeding exclusively on small marine organisms. The trade-mark shell is believed to have evolved from extensions from their backbone and widened ribs that fused together. This is supported by the fossil of *Odontochelys semitestacea*, which has an incomplete shell originating from the ribs and back bone. This species also had teeth with its beak, giving more support to it being a transitional fossil, although this claim is still controversial. This shell evolved to protect itself from predators, but also slows down the land based species by a great amount. This has caused many species to go extinct in recent times. Because of alien species out-competing them for food and the inability to escape from humans, there are many endangered species in this order [Rieppel, 1996; Bernard et al., 2010].

Sphenodontia

Sphenodontians arose in the mid Triassic and now consists of a single genus, tuatara, which comprises two endangered species that live on New Zealand and some of its minor surrounding islands. Their evolutionary history is filled with many species. Recent paleogenetic discoveries show that tuataras are prone to quick speciation.

Squamata

The most recent order of reptiles, squamates, is recognized by having a movable quadrate bone (giving them upper-jaw movement), possessing horny scales and hemipenes. They originate from the early Jurassic and are made up of the three suborders Lacertilia (paraphyletic), Serpentes, and Amphisbaenia. Although they are the most recent order, squamates contain more species than any of the other reptilian orders. Squamates are a monophyletic group included, with the Sphenodontia (e.g., tuataras) in the Lepidosauria. The latter superorder, together with some extinct animals like the plesiosaurs, constitute the Lepidosauromorpha, the sister infraclass to the group, the Archosauromorpha, that contains crocodiles, turtles, and birds. Although squamate fossils first appear in the early Jurassic, mitochondrial phylogenetics suggests that they evolved in the late Permian. Most evolutionary relationships within the squamates are not yet completely worked out, with the relationship of snakes to other groups being most problematic. From morphological data, Iguanid lizards have been thought to have diverged from other squamates very early, but recent molecular phylogenies, both from mitochondrial and nuclear DNA, do not support this early divergence. Because snakes have a faster molecular clock than other squamates, and there are few early snake and snake ancestor fossils, it is difficult to resolve the relationship between snakes and other squamate groups [Huey, 1982].

Crocodilia

The first organisms that showed similar characteristics of Crocodilians were the Crurotarsi, which appeared during the early Triassic 250 million years ago. This quickly gave rise to the Eusuchia clade 220 million years ago, which would eventually lead to the order of Crocodilians, the first of which arose about 85 million years ago during the late Cretaceous. The earliest fossil evidence of eusuchians is of the genus *Isisfordia*. Early species mainly feed on fish and vegetation. They were land-based; most having long legs (when compared to modern crocodiles) and many were bipedal. As diversification increased, many apex predators arose, all of which are now extinct. Modern Crocodilia arose through specific evolutionary traits. The complete loss of bipedalism was traded for a generally low quadrupedal stance for an easy and less noticeable entrance to bodies of water. The shape of the skull/jaw changed to allow more grasp along with upward pointing nostrils and eyes. Mimicry is evident, as the backs of all crocodilia resemble some type of floating log and their general color scheme of brown and green mimics moss or wood. Their tail also took on a paddle shape to increase swimming speed. The only remaining groups of this order are the alligator, crocodile, caiman, and gharial [Bernard et al., 2007; Erickson et al., 2012].

3.9. Origin of Reptiles

Creatures such as dinosaurs, lizards, tortoises and crocodiles all belong to the reptile family. Reptiles possess unique characteristics: They are all covered in scales. They are cold-blooded and cannot produce their own body heat, for which reason they need to warm their bodies in the sun every day. Their young hatch from eggs [Gauthier et al., 1988; Modesto, 1999].

Evolution theory faces another major dilemma in explaining the origin of these creatures. Darwinists claim that reptiles evolved from amphibians, yet no concrete evidence supports this. On the contrary, a comparison of reptiles and amphibians shows enormous physiological differences between the two groups, and that any half-reptile, half-amphibian would stand no chance of survival [Modesto, 1999].

One reason concerns the two groups' different egg structures. Amphibians deposit their eggs in water, and they are ideally suited to aqueous development. These eggs are highly porous, they have a transparent membrane and a jelly-like consistency. Reptiles, on the other hand, lay their eggs on land, and their eggs are ideally suited to dry conditions. The hard shell of the reptile egg, known as an amniotic egg, contains an air sac, and is impermeable to water. The water needed by the young is contained inside the egg until it hatches [Watson, 1957; Juan et al., 2004].

Were amphibians to deposit their eggs on land, they would soon dry out and the embryos inside would die. This is a point that the evolution theory, which maintains that reptiles gradually evolved from amphibians, cannot explain. **To begin life on dry land, the amphibian's egg would have to become an amniotic one within a single generation. This could hardly occur through mutation and natural selection, the two suggested mechanisms of evolution**.

On the other hand, the fossil record also removes the possibility of an evolutionary origin for reptiles. In an article, "**Problems of the Origin of Reptiles**," the well-known evolutionist paleontologist Lewis L. Carroll admits as much:

> "Unfortunately, not a single specimen of an appropriate reptilian ancestor is known prior to the appearance of true reptiles. The absence of such ancestral forms leaves many problems of the amphibian-reptilian transition unanswered".

Robert L. Carroll, regarded as an authority on vertebrate paleontology, accepts that "*The early amniotes are sufficiently distinct from all Paleozoic amphibians that their specific ancestry has not been established.*" Carroll also makes these comments in his book:

> When they first appear in the fossil record, both frogs and salamanders appear essentially modern in their skeletal anatomy. . . Despite these similarities, frogs, salamanders, and caecilians are very different from one another in skeletal structure and ways of life, both now and throughout their known fossil record . . . we have found no fossil evidence of any possible antecedents that possessed the specialized features common to all three modern orders.

The most important creature that evolutionists have so far sought to portray as the "ancestor of reptiles" is the fossil amphibian Seymouria. Yet the discovery that reptiles were alive 30 million years before Seymouria's appearance revealed that this cannot be an intermediate species.

It is of course impossible for "the forerunner of reptiles" to have first appeared long after them. **The scientific facts show that reptiles emerged on Earth not through the gradual change maintained by the theory of evolution, but suddenly, with no antecedents**.

3.10. Problems of the Origin of Reptiles

The fossil records of the four living reptilian orders can be traced into the Triassic. The earlier ancestry of the turtles has not been established. Squamates and rhyncho-cephalians evolved from the Late Permian eosuchians; crocodiles from the thecodonts. The ancestry of the eosuchians and thecodonts is to be found in the central stock of Permo-Carboniferous reptiles, the captorhinomorphs. The earliest captorhino-morphs, from the Lower Pennsylvanian, are already fully developed reptiles. The limnoscelids and solenodonsaurids are more primitive forms, retaining features typical of anthracosaurian amphibians [Romer and Parsons, 1977].

Neither reptiles nor any appropriate ancestors are known prior to the Lower Pennsylvanian. Because of the absence of any true ancestors, the nature of the amphibian-reptilian transition must be studied on the basis of amphibians contemporary with the early reptiles. The Permian seymouriamorphs have long been accepted as relicts of the group which gave rise to reptiles, although Seymouria itself is specialized in many features of its anatomy. The Middle Pennsylvanian genus Gephyrostegus appears to resemble much more closely the anatomy expected in the ancestors of reptiles [Romer and Parsons, 1977]. This genus forms the basis for consideration of the anatomical, physiological and behavioural changes which culminated in the origin of reptiles.

Study of the earliest known reptiles and their closest relatives among contemporary amphibians indicates that the initial adaptation leading to the emergence of the class was assumption of a terrestrial habit, with accompanying small body size. The small body size of the immediate ancestors of reptiles would have made it possible for them to produce sufficiently small eggs that they could develop in damp places on land without initially being supported and protected by extraembryonic membranes. The rapid increase in body size in all lineages of Pennsylvanian reptiles indicates the prior development of an amniotic egg [Tuninen and Hadly, 2004].

Fundamental to the emergence of reptiles was modification in the jaw mechanism from the kinetic inertial system of amphibians to a static pressure system. The latter was presumably developed in order for the developing reptiles to utilize more active terrestrial prey. This change in the jaw mechanism is reflected in the reorganization of the palate which serves as a morphological basis for denning the establishment of reptilian status. At approximately the same stage as the change in palatal structure, the definitive reptilian vertebral pattern was developed [Piñeiro et al., 2012].

The apparent closure of the otic notch and the probable reorientation of the stapes in the amphibian-reptilian transition presumably resulted from the decrease in relative skull size and do not appear to be related to any change in hearing ability. The tympanum probably maintained the same relative relationship with the squamosal and supratemporal throughout this transition.

On the basis of the present fossil record, all adequately known Palaeozoic reptiles appear to have had a common ancestry among the predecessors of the known **gephyro-stegids**. The family Diadectidae is the only important group whose specific relationships cannot be established.

On the basis of this study, the following taxonomic changes are suggested: the family Limnoscelidae should not be included among the captorhinomorphs. The seymouria-morph concept should be restricted to forms having the specializations of Seymouria, the discosauriscids and kotlassids. Gephyrostegids should be specifically excluded from the Seymouriamorpha and should be included in a separate taxon among the anthra-cosaurs of equal rank with embolomeres and seymouriamorphs [Benton, 2000].

3.11. Seymouria: An Early Small Reptile

In 1904, the paleontologist Ferdinand Broli found in Texas reptile fossils belonging to Seymouria. According to studies, it has lived in the Lower Permian, North America and Europe (approximately 280 to 270 million years ago). It was a small reptile, of 60 centimeters and prefered plain water areas. Its spine was robust and they had strong and long legs [Carroll, 1991].

Seymouria was a reptile-like labyrinthodont. It was small, only 2 ft (60 cm) long. Regarding the skin, it is believed that it was dry and allowed the opportunity to keep the water in the body. The scientists believe that it was capable of removing the salt from the body through a gland, located on the nose. Due to Seymouria' specific features this species was able to live in difficult climate conditions, reason for which it could live both in water and land [Laurin, 1996].

This reptile's skull is similar to that of reptiles and amphibians and had a rounded snout. Their teeth appear both on the jaw and on the palate, helping it to feed on insects and small animals. Its nostrils were near the muzzle. It may also be equipped with sense organs, allowing it to feel if there were also other animals or objects around. There were very thin hair follicles on the lateral parts, with sensory cells.

Seymouria was well adapted to life on land, with many reptilian features—so many, in fact, that it was first thought to be a primitive reptile (Figure 10).

The dry climate of the Permian suited reptiles better than amphibians, but *Seymouria* had many reptilian features that helped it in this harsh environment. It had long and muscular legs, and may had dry skin and the ability to conserve water. It may have been able to excrete excess salt from its blood through a gland in its nose, like modern reptiles. All of this meant that *Seymouria*, unlike amphibians and other early tetrapods, might have lived for extended periods of time away from water. If so, this would have allowed it to move about the landscape in search of insects, small amphibians, and other possible preys, such as the eggs of reptiles [Laurin, 1996].

Male *Seymouria* had thick skulls that may have been used to batter rivals in mating contests. After mating, the females would have had to return to water to lay their eggs. As in amphibians, the larvae would develop in water, hunting for worms and insects until they were strong enough to live on land. While no larvae are known from *Seymouria* itself, fossil larvae of its relative *Discosauriscus* and other species in the order Seymouriamorpha have been found, with impressions of external gill structures as in modern amphibians [Laurin, 1996].

Fossils of *Seymouria* were first found in Seymour, Baylor County, Texas (hence the name of the type species, *Seymouria baylorensis*, or "Baylor County Seymour one"). Over the years, many well preserved fossils have been found in North America and Germany, including the "Tambach Lovers", two individuals of *S. sanjuanensis* fossilized lying next to each other (though of course it cannot be determined whether they really were a couple killed during mating). *Seymouria agilis* [Olson, 1980] that is known from only one specimen (holotype UCMP 143 277), a nearly complete skeleton from the Chickasha Formation of Oklahoma, was reassigned by Michel Laurin and Robert R. Reisz as a Macroleter (Figure 11).

Figure 10. Seymouria, Images, fossilized, reconstructed and in actual habitat.

Figure 11. Fossil remains of Seymouria.

THE SEYMOURIA MISTAKE

Evolutionists at one time claimed that the Seymouria fossil on the left was a transitional form between amphibians and reptiles. According to this scenario, Seymouria was "the primitive ancestor of reptiles." However, subsequent fossil discoveries showed that reptiles were living on earth some 30 million years before Seymouria. In the light of this, evolutionists had to put an end to their comments regarding Seymouria.

No fossil amphibian seems clearly ancestral to the lineage of fully terrestrial vertebrates (reptiles, birds, and mammals).

So far, the most important animal put forward the "ancestor of reptiles" has been Seymouria, a species of amphibian. However, the fact that Seymouria cannot be a transitional form was revealed by the discovery that reptiles existed on earth some 30 million years before Seymouria first appeared on it. The oldest Seymouria fossils are found in the Lower Permian layer, or 280 million years ago. Yet the oldest known reptile species, Hylonomus and Paleothyris, were found in lower Pennsylvanian layers, making them some 315-330 million years old. It is surely implausible, to say the least, that the "ancestor of reptiles" lived much later than the first reptiles. In brief, contrary to the evolutionist claim that living being evolved gradually, scientific facts reveal that they appeared on earth suddenly and fully formed [Laurin, 1996].

Relationships of *Seymouria*, *diadectes*, and Chelonia

Evidence now available supports the concept of close relationship between the seymouriamorphs and diadectids, with the latter a specialized offshoot of the former. Diadectes itself clearly is not a progenitor of Chelonia. The seymouriamorph diadectid complex may be a sterile line, except for the gephyrostegids which, originating very early among seymouriamorphs, may have led to true reptiles.

There is some evidence, mostly highly tentative, that procolophons and their probable descendants, the pareiasaurs, may have arisen from the non-gephyrostegid seymouriamorph stock. Chelonia could have found their ancestery within the procolophons or pareiasaurs. If this is the case, and only additional study can confirm or deny it, then the concept of the Parareptilia, as proposed earlier [Olson, 1947], may retain much of its original meaning, with the Diadectidae, however, far removed from the ancestry of the Chelonia.

3.12. SPHENODON (ORDER RHYNCHOCEPHALIA)

Features and Affinities

Albert Günther (1867) of the British Museum proposed the order name **Rhynchocephalia**, meaning "***beak headed***" for the tuatara and its fossil relatives (Figure 12-15). The common name "**tuatara**" is derived from the Maori language, which means "*spines on the back*". Tuatara have been referred to as **living fossils**, which is a term used for any living species of organisms that appears to be the same as a species only known from fossils and which has no close living relatives. *Sphenodon* or **hatteria** is the last survivor of the reptilian order Rhynchocephalia, which evolved in the early Mesozoic era, about 200 million years ago, but is now survived by only two species, which are found on the islands off the New Zealand coast and in Karori Wildlife Sanctuary, Wellington, New Zealand. Since 1895, the country of New Zealand awarded the tuatara strict legal protection. Sphenodon are of great interest in the study of the evolution of lizardsand snakes, and for the reconstruction of the appearance and habits of the earliest diapsids (the group that also includes birds and crocodiles) [Russell, 1998; Musico, 1999; Kristina et al., 2007].

Rhynchocephalia, an Order of Class – Reptilia is considered as an aberrant group of reptiles which appeared first in Triassic and passed smoothly through the Mesozoic. It is a true diapsid reptile now represented today by one very conservative living species *Sphenodon punctata*. It is called by the natives of New Zealand as 'tuatara' (tua = back, tara = spine).

It is a lizard like reptile which grows to about 2 feet long and has well developed pentadactyle limbs adapted for walking. Gadow refers it as "the last living witness of the by-gone ages". This primitive reptile is called the "living fossil" [Russell, 1998; Jones, 2009]. It is a very lazy creature. It lives in the holes on the slopes of the sand hills of the shore with the petrels. This animal is rigidly protected by law. It emits a soft frog like croak. It feeds on insects, crustaceans and worms. Vertebrae are amphicoelous with remnants of notochord in the cavities between the centra. In addition to atlas and axis there is another small median bone, the **proatlas** in the occipital region of the skull. Ribs are single headed and some of them possess uncinate process. There is a median sternum with coracoid and interclavicle attached at the anterior end. There is a T-shaped intercalvicle and a clavicle. There are eleven carpal bones; four, including a pisiform, in the proximal row, two centrals and five in the distal row. In the pelvis, pubes are united in symphysis, in front of which is a cartilaginous epipubis. A cartilaginous hypoischium is attached to the ischia behind [Colenso, 1885; Jones, 2008].

Biology

It takes between 10 and 20 years for a tuatara to reach sexual maturity [Gand et al., 1984; Cree et al., 1992, 1995]. The female lays 5-18 eggs only once every 4 years. Mating occurs from mid-summer to early autumn and the eggs are laid the following spring or early summer. Incubation takes from 12 to 15 months, with the development of the embryo stopped during the winter months. It has an average lifespan of 60 years and reaches sexual maturity in a decade. It lays eggs in burrows that it shares with seabirds, and has the longest incubation period of all reptiles (up to 15 months). The sex of the offspring is determined by the temperature: warm soil temperatures produce males, while cooler soil produces females. Adult tuatara is terrestrial and nocturnal reptile, though it often basks in the sun to warm its body and hibernates during winter months. Their long life is the product of having an extremely low metabolic rate, and they can tolerate extremely cold weather. Indeed the activity levels of the tuatara peak at body temperatures of 12 to 17 degrees Celsius, which is the lowest for any reptile. This is probably why they have been able to survive in New Zealand's temperate climate for so long. Tuatara has the slowest growth rate among reptiles and continues to grow larger for the first 35 years of their lives. The average lifespan is about 60 years but some can live up to 100 years (Figure 12-15) [Thompson and Daugherty, 1998].

Systematics

Sphenodonts, and their sister group Squamata (which includes lizards, snakes and amphisbaenians), belong to the superorder Lepidosauria, the only surviving taxon within Lepidosauromorpha. Squamates and sphenodonts both show caudal autotomy (loss of the tail-tip when threatened), and have transverse cloacal slits. The origin of the sphenodonts probably lies close to the split between the Lepidosauromorpha and the Archosauromorpha. Though they resemble lizards, the similarity is superficial, because the group has several characteristics unique among reptiles. The typical lizard shape is very common for the early amniotes; the oldest known fossil of a reptile, the *Hylonomus*, resembles a modern lizard [Wu, 1994].

Tuatara was originally classified as lizards in 1831 when the British Museum received a skull. The genus remained misclassified until 1867, when Albert Günther of the British Museum noted features similar to birds, turtles, and crocodiles. He proposed the order Rhynchocephalia (meaning "beak head") for the tuatara and its fossil relatives. Many disparately related species were subsequently added to the Rhynchocephalia, resulting in what taxonomists call a "wastebasket taxon". Williston proposed the Sphenodontia to include only tuatara and their closest fossil relatives in 1925. *Sphenodon* is derived from the Greek for "wedge" and "tooth". However, today Rhynchocephalia is used to include *Gephyrosaurus* and Sphenodontia, while Sphenodontia excludes the former [Wu, 1994; Hay et al., 2010].

Habitat. The geographic range of tuatara are 32 offshore islands which are generally cliff-bound, frequently exposed to strong winds and which support a natural, often stunted, salt and wind tolerant vegetation. Most islands are also home to several species of sea birds, whose nutrient-rich guano helps support the island's ecosystem. Tuataras usually inhabit the breeding burrows of small petrels. They feed on small insects. The total population of tuatara of all species and subspecies is estimated to be between 60,000 and 100,000. A rodent species called kiore (Polynesian rat, *Rattus exulans*) has recently established on several islands and has threatened the survival of tuatara [Kristina et al., 2007; Hay et al., 2010].

Distribution

Sphenodon was formerly known on the main islands of New Zealand, but is now restricted to some small islands in the Bay of Plenty and appears to be on the verge of extinction [Musico, 1999; Hay et al., 2010].

***Sphenodon punctatus*. The Northern Tuatara** colour ranges from olive green through grey to dark pink or brick red, often mottled and always with white spots. Males measure 61 cm in length and weigh up to 1 kg and females are 45 cm long (Figure 13) [Musico, 1999].

***Sphenodon guntheri*. Brothers Island tuatara** is confined to North Brother Island in Cook Strait, with a population of approximately 400. It is smaller, weighing up to 660 grams and has olive brown skin with yellowish patches. It inhabits low forest, scrub areas and rock stacks on Brothers Island and is found up to 300 meters above sea level where the climate is cool. The island is also occupied by birds such as petrel and shearwater, which provide the tuatara with many benefits.

General Anatomical Features

1. The lizard like body of Sphenodon is about 2 feet long and the tail measures about 1/3 of the whole length of the body- The animal is nocturnal.
2. The body and tail are laterally compressed and a series of spines are dorsally present (along the middle line).
3. Limbs are pentadactyle, legs are short and primitive in nature. Sometimes 10-11 carpels are present.
4. Hind limbs are typically plantigrade.
5. The upper surface of the body is covered with small granular scales and the lower surface is covered with transverse rows of large squarish scales.
6. The head is large and there is prominent ridge over the eye.
7. There is a foramen above the outer and one above the inner condyle of the humerus.
8. There are eleven carpal elements in all, four in the proximal row, two centrals and five in the distal row.
9. In the tarsus, the tibial and fibular elements are quite distinct, though firmly united. The intermedium and the centrale are firmly fixed to the tibiale. There are three tarsal bones.
10. Teeth are pleurodont and homodont like reptiles but they are monophyodont. The teeth are pointed, triangular, laterally compressed and arranged in two parallel rows, one along the maxilla, and the other along the palatine. The teeth of the lower jaw bite in between the two upper rows, all the rows becoming worn down in such a way as to form continuous ridges. Each premaxilla bears a prominent, chisel-shaped incisor, represented in the young animal by two pointed teeth. In the young Sphenodont a tooth has been found on each vomer — a condition unusual among reptiles [Colenso, 1885; Jones, 2008].
11. The pubes are united in a Symphysis, infront of which is a cartilaginous epipubis. A large oval foramen intervenes between the ischium, and the pubis. A cartilaginous hypoischium is attached to the ischia behind.
12. In the skull, there is a complete lower temporal arch. The quadrate is immovably fixed, wedged in by the quadrate- jugal, squamosal, and pterygoid. The premaxilla are not fused together but separated by a suture. There is a broad palate, formed by the plate-like vomer, palatine and pterygoids. The pterygoids meet in the middle line and extend forwards to the vomers. There is an epipterygoid extending from the parietal to the pterygoid and quadrate [Colenso, 1885; Jones, 2008].
13. Sphenodon retains the large parietal foramen, in which, is the non-functional median pineal eye.
14. There is no antorbital foramen.
15. The anal opening of Sphenodon is transverse.
16. There is no copulatory organ (Penis) in the male.
17. The vertebrae are amphicoelous and fully ossified and intercentra are present.

18. The ribs are single headed and have uncinate processes. There is a sternum and abdominal ribs.
19. Chevron bones occur in the caudal region of the vertebrae.
20. There is a median bone, the Pro-atlas, intercalated between the atlas and the occipital region of the skull.
21. Caudal ribs become fused with the vertebrae.
22. The lateral temporal fossa is bound below, by an inferior temporal arch composed of jugal and quadrato-jugal (Palato-quadrate).
23. Jacobson's organ is present in a rather primitive form.
24. A urinary bladder is present.
25. There is a T-shaped interclavicle.
26. Coracoid is without fenestra.
27. It lays 10-13, leathery shelled eggs which are buried few inches below the surface of the soil. The young ones are hatched 13 months later.

Causes of Its Long Survival

There are three important causes of its long survival. These are —

a. It was free from its enemies because the placentals and other large terrestrial animals were introduced in New- Zealand quite lately.
b. Slow rate of metabolism and conservation of sufficient energy.
c. Lethargic habit and long period of incubation of the fertilized eggs.

Why Called a Living Fossil

The living fossil is an animal that has survived beyond its era. It is the one species or such a group of species that has continued existence when the other closely related contemporary species have become extinct [Jones, 2009; Jones et al., 2009]. Sphenodon is the sole survival of the order Rhynchocephalia. Its anatomical features indicate several primitive characters such as-

1. The skull bones are shaped and disposed in the manner of extinct group of reptiles.
2. Teeth are fused to the jaw bones.
3. Presence of pineal foramen.
4. There are several characters which resemble that of Dinosaurs and other extinct reptiles.
5. It resembles with the ancient Homaeosaurs in several features except the presence of uncinate process of the ribs in *Sphenodon*.

Thus, the term 'living fossil' is justified for *Sphenodon*. This animal which has been called the "Voice of the Past" has been given absolute protection by law in New Zealand.

Affinities

Sphenodon (Order – Rhynchocephalia) possesses several characters which are present in different groups of animals. So, actual position of Sphenodon in the animal kingdom was somewhat controversial for some time. Its affinities with different groups have been discussed below:

Affinities with Amphibia

The circulator)' system of Sphenodon has some resemblances with Caudata or Urodela. These resemblances are —

1. Aortic arches arise from a short common stalk which is comparable to conus arteriosus of the amphibians.
2. Presence of ductus arteriosus and ductus caroticus
3. The pattern of distribution of blood vessels almost similar but, these resemblances are mainly because of the reason that Sphenodon is an animal of primitive nature. Because of the presence of several reptilian features which have been discussed below, amphibians affinities can not be accepted.

Affinities with Dinosaurs

There are some resemblances with the dinosaurs. These are—

1. Diapsid condition of the skull i.e., the skull has two vacuities (fossae).
2. Fixed quadrate bone
3. Abdominal ribs.
4. Uncinate process of the ribs.

But, there are some **differences** also — e.g.

1. In Dinosaurs, double headed ribs are present – not single headed like Sphenodon.
2. Dentition is thecodont in dinosaurs
3. Absence of clavicle and interclavicle
4. Absence of Parietal foramen
5. Absence of third eye elements

So, *Sphenodon* can not be linked with the dinosaurs, however, its reptilian ancestry is evident.

Chelonian Affinities

The following are the important resemblances with the chelonians:

1. Fixed quadrate bone
2. Ribs of caudal region fused with the vertebrae
3. Pecten not present in the eye
4. Urinary bladder present

The differences with chelonians include the following —

1. Sphenodon is terrestrial but chelonians are aquatic.
2. Teeth absent in the chelonians, only horny beaks present
3. Parietal foramen in chelonians
4. Unpaired vomer
5. No sternum in chelonia
6. Cloacal opening longitudinal in chelonia but transverse in Sphenodon
7. Males have penis in chelonia
8. Opening of oviduct on ventral side

Affinities with Crocodilia

The following are the important resemblances.

Crocodilia

1. Diapsid condition of the skull
2. Fixed quadrate
3. Presence of Pro-atlas
4. Presence of abdominal ribs
5. Fusion of caudal ribs with vertebrae
6. Uncinate process of ribs
7. Presence of chevron bones
8. Cochlear process tubular

The differences between the two are:

1. Thecodont dention in crocodiles
2. Single nostril in Crocodile but double in Sphenodon
3. Procoelous vertebrae in Crocodile and amphicoelous inSphenodon
4. Clavides not present in Crocodilia.
5. Pecten present in Crocodiles
6. Penis present in male Crocodile

Affinities with Lacertilia

The important resemblances are the following —

1. The body plan is similar
2. Pro-atlas present
3. Amphicoelous vertebrae in certain geckos
4. Single headed ribs
5. Chevron bone present in both
6. Structure of the respiratory organs
7. Parietal organ common
8. Cloacal glands present

However, there are some **differences** also, which are —

1. Quadrate fixed in Sphenodon
2. Most lizards have procoelous vertebrae
3. Rami of the jaw united by symphysis in lizards
4. Erect ilium in Sphenodon
5. Clavicles and interclavicles present in Sphenodon
6. Absence of conus arteriosus in Lacertilia
7. Presence of Pecten
8. Uncinate process of true ribs absent in Lacertilia
9. Copulatory organs present in Lacertilia
10. Presence of lower temporal arch

It is thought that the arrival of Polynesians roughly 1000 years ago and of Europeans over the last several centuries is responsible for the tuatara's disappearance from the mainland. This is because settlers brought rats and dogs, which preyed on tuataras and cattle, trampled them. *Sphenodon* has been unable to tolerate the impact of humans because of its incredibly limited reproductive potential [Crook, 1973].

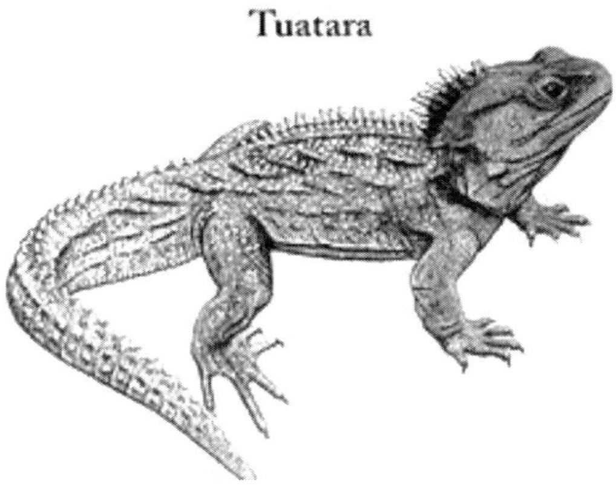

Figure 12. Sphenodon (Tuatara).

Figure 13. Male northern tuatara (Spenodon punctatus), still reproductively active at 111 years of age.

Figure 14. Tuatara juvenile (*Sphenodon punctatus*).

Figure 15. Tuatara at the Karori Sanctuary are given coloured markings on the head for identification.

Thus, we see that Rhynchocephalia retains many primitive characters and resembles chelonians, crocodilia and dinosaurs in many features. But it is convincing that Sphenodon is more closely allied to the lacertilia than any other group discussed above.

Because of the occurrence of certain peculiar features it seems justified that the Rhynchocephalia should be placed as a separate order of class-Reptilia as suggested by Romer.

3.13. Reptile Study Guide

Enclosing the Pond

A. Amphibians lack shelled eggs and often has gill-breathing larvae.

B. Lineage leading to reptiles, birds, and mammals developed an egg or other strategies that allowed them to be truly terrestrial.

C. Conquest of the land by vertebrates was ensured, and this monophyletic assemblage is called the Amniota, named after the innermost of the three extraembryonic membranes, the amnion.

D. Before the end of the Paleozoic era, the amniotes had diversified into multiple lineages that gave rise to all reptilian groups, birds, and mammals.

Diversity

A. Amniotes

1. Any animal with a shell-less egg remains tied to water.
2. The development of a shelled egg freed the reptilian groups to exploit land.
3. **Extraembryonic membranes** from previous evolutionary aquatic stages are maintained.
4. The **allantois** serves as a respiratory surface and a chamber to store nitrogenous wastes.
5. The chorion allows oxygen and carbon dioxide to freely pass.
6. Surrounding the whole young organism is a porous, parchment-like or leathery shell.
7. The **Amniota** is a single, **monophyletic lineage** of Paleozoic tetrapods with this developmental pattern.
8. Before the end of the Paleozoic Era, amniotes diverged into reptilian groups, birds and mammals.

B. Diversity

1. Members of the paraphyletic class Reptilia include the first truly terrestrial vertebrates.
2. Nearly 8000 species are described; about 340 occur in the United States and Canada.
3. The Age of Reptiles lasted over 165 million years and included the dinosaurs.
4. A mass extinction occurred at the end of the Mesozoic; modern reptiles represent surviving lineages.
5. The tuatara is the sole survivor of a group that otherwise disappeared 100 million years ago.
6. Lizards and snakes radiated into diverse and abundant groups.
7. 300-million-year-old history of reptile life on earth is complicated by widespread convergent and parallel evolution among the many lineages.

Origin and Adaptive Radiation of Reptilian Groups

A. History

1. Amniotes arose from amphibian-like tetrapods, the **anthracosaurs**, during the Carboniferous.
2. By the late Carboniferous, amniotes had separated into three lineages:
   a. **Anaspids** have a skull with no temporal opening behind the orbits; modern turtles are anaspids.
   b. **Diapsids** gave rise to all other reptilian groups and to the birds.
      1) The diapsid skull has two temporal openings; one pair below the cheeks and another above.

2) **Lepidosaurs** include ichthyosaurs and modern reptiles except for turtles and crocodilians.
3) The more derived archosaurs included dinosaurs, living crocodilians, and birds.
4) **Sauropterygians** included extinct aquatic groups including the long-necked plesiosaurs.

c. **Synapsids** are mammal-like reptiles with a single pair of temporal openings.

B. Changes in Traditional Classification of Reptilian Groups

1. **Cladistic methodology** insists on hierarchical arrangement of **monophyletic groups**.
2. This disqualifies the traditional **class Reptilia** as a valid taxon because it is **not monophyletic**.
3. **Class Reptilia** excludes birds, which descend from the most recent common ancestor of reptiles.
4. This makes the **class Reptilia a paraphyletic group** because it does not include all descendants and their most recent common ancestor.
5. Reptiles are therefore identified as amniotes that are not birds.
6. Based solely on shared derived characteristics, crocodilians and birds are sister groups belonging to a monophyletic group apart from other reptiles, the Archosauria.
7. Some taxonomists defined Reptilia as the Archosauria plus the lepidosaurs, thus including birds.
8. Evolutionary taxonomists argue that birds represent a novel adaptive zone and grade of organization; they **maintain the class Aves** based on the morphological and ecological novelty of birds.
9. "Reptilian group" refers to members of four monophyletic groups formerly considered class Reptilia.

Characteristics Distinguishing Reptiles from Amphibians

A. Skin

1. Reptiles have a tough, dry, scaly skin that offers protection against desiccation and injury.
2. The thin epidermis is shed periodically.
3. A thicker, well-developed dermis underneath has chromatophores that provide color.
4. The dermis is converted to snakeskin leather for shoes and pocketbooks.
5. Reptile scales are primarily of keratin, formed from epidermis, and not homologous with fish scales.
6. Scales may grow gradually to replace wear, as in alligators.
7. Snakes and lizards replace old scales with new and "shed their skins."
8. Turtles add new layers underneath old layers of platelike scutes.

B. Shelled Egg

1. A shelled egg **contains food and protective membranes** to support embryonic development on land.
2. The chorion and allantois exchange carbon dioxide and oxygen with the environment.
3. The amnion and shell support the growing embryo and reduce water loss.
4. The shelled egg widened the division between evolving amphibians and reptiles.

C. Reptilian Jaws

1. The jaws of fish allowed fast jaw closure to seize food but little force for chewing.
2. In reptiles, jaw muscles became larger and **arranged for the mechanics of chewing**.

D. Internal Fertilization

1. A shelled egg requires internal fertilization; the sperm must reach the egg before it is enclosed.
2. Reptiles have some form of **copulatory organ**, permitting internal fertilization.
3. Paired testes produce sperm that are carried by **vasa deferentia** to the copulatory organ.
4. The female has **paired ovaries** and oviducts.
5. Glandular walls of the **oviducts secrete albumin and shells** for the relatively large eggs.

E. Circulatory System Modifications

1. Reptiles have an efficient circulatory system and higher blood pressure than amphibians.
2. The right atrium that receives deoxygenated blood is completely partitioned from the left atrium.
3. Crocodilians have **separated ventricles** dividing **pulmonary and systemic circulation**
4. Other reptiles have an incompletely separated ventricle but little mixture of blood occurs and the effect is two functionally separate circulations.
5. Incomplete separation between heart permits blood to bypass the lungs during diving or aestivation

F. Lungs

1. Reptile lungs are better developed than those of amphibians.
2. Reptiles depend almost exclusively on lungs for respiration, supplemented by pharyngeal membranes.
3. Inspiration occurs by enlarging the **thoracic cavity**; some aquatic species use the pharynx, cloaca, and skin.
4. Snakes and lizards use expansion of the rib cage; turtles and crocodilians use movement of internal organs to create negative pressure to inhale (i.e., reptiles "suck" air into the lungs, unlike amphibians, which "force" air into the lungs)
5. Reptiles lack diaphragms.

G. Water Conservation

1. All amniotes have a **metanephric kidney** drained by the ureter.
2. Nephrons of the reptilian metanephros **lack the loop of Henle** that allows concentration of solutes.
3. Many reptiles have **salt glands** near the nose or eyes to secrete salty fluid hyperosmotic to body fluids.
4. Nitrogenous wastes are excreted as **uric acid** rather than urea or ammonia.
5. Uric acid has **low solubility** and precipitates readily; this allows water to be conserved.

H. Support of Limbs for Locomotion

1. Except for limbless members, all reptiles have better body support than amphibians.
3. Most dinosaurs and some modern lizards have more efficient legs directed beneath the body.
4. Bipedal locomotion is superior to quadrupedal locomotion if speed and better support of the body is required.

I. Nervous System

1. The reptilian nervous system is considerably more complex than the amphibian.
2. The reptile brain is still small but the cerebrum is relatively larger.
3. Sense organs are generally well developed, except for hearing.
4. The Jacobson's organ is highly developed in lizards and snakes to detect odors carried by the tongue.

Characteristics and Natural History of Reptilian Orders

A. Anapsid Reptiles: Subclass Anapsida

1. **Order Testudines (Chelonia)** includes the turtles.
2. Turtles descended from one of the earliest anapsid lineages, probably in the late Permian.
3. Fossils appear in the Upper Triassic, 200 million years ago, and they have occurred ever since.
4. The shells consist of the **dorsal carapace** and the **ventral plastron**.
5. The bony layer is a fusion of ribs, vertebrae, and **dermally-ossified** elements – the shell.
6. The shell offers protection for the head and appendages.
7. Turtles lack teeth and have tough, horny plates for gripping food
8. **Breathing**
   a. One consequence of having a rigid shell is the turtle **cannot expand its chest** to breathe.
   b. Turtles solve this problem by using abdominal and pectoral muscles as a "diaphragm."
   c. Air is drawn in by contraction of the limb flank muscles, increasing abdominal cavity volume.

d. Exhalation is accomplished by drawing back the shoulder girdle to compress the viscera.

9. **Nervous System and Senses**
   a. Turtles have a middle and an inner ear but sound perception is poor; they make few sounds aside from during mating
   b. Turtles have a good sense of smell, acute vision, and color perception about equal to humans.

10. **Reproduction and Development**
   a. Turtles are oviparous; fertilization is internal and all turtles bury their eggs in the ground.
   b. In some turtle families, as in crocodilians and some lizards, nest temperature determines the sex of hatchlings; low temperatures produce males and high temperatures produce females.

11. **Giant Turtles**
   a. Buoyed by the water, marine turtles may reach two meters long and 725 kilograms in weight.
   b. Giant land tortoises, such as those on the Galápagos Islands, weigh several hundred kilograms.
   c. Low metabolic activity may explain their longevity believed to exceed 150 years.

B. Diapsid Reptiles: Subclass Diapsida (skull with two pairs of temporal openings)

1. Consists of superorders:
   a. Lepidosauria – lizards, snakes, worm lizards, and *Sphenodon*.
   b. Archosauria – crocodilians (and birds in classic taxonomy).
2. **Order Squamata**: 3 lineages: Lizards, Snakes and Worm Lizards
   a. Squamates are the most recent and diverse of diapsids; they make up **95% of living reptiles**.
   b. Lizards appeared in the fossil record in the Permian but did not radiate until the Cretaceous.
   c. Snakes appeared in the Jurassic from a group whose descendants include monitor lizards.
   d. **Snakes** gained specializations for losing their legs and therefore for engulfing large prey.
   e. **Amphisbaenians** probably evolved from lizards and are specialized for burrowing.
   f. Diapsid skulls have lost dermal bone ventral and posterior to the lower temporal opening.
      1) This permitted evolution in lizards of a mobile skull with movable joints, a **kinetic skull**.
      2) The quadrate, fused to the skull in other reptiles, has a joint at the dorsal end and articulates with the lower jaw.
      3) Joints in the palate and across the roof of the skull permit the snout to be tilted up.
      4) Squamates seize and manipulate prey; they close the jaw with force.

5) Exceptional skull mobility of snakes is considered a major factor in their diversification.

l. **Viviparity**

1) Viviparity in reptiles is limited to squamates.
2) It has evolved at least 100 separate times.
3) It is associated with cold climates.
4) Viviparity involves increasing the length of time eggs are kept in the oviduct.
5) Developing young respire through extraembryonic membranes.
6) Young obtain nutrition from yolk sacs or via the mother, or a combination of both.

3. **Lizards: Suborder Sauria**

a. Lizards are a diverse group with terrestrial, burrowing, aquatic, arboreal, and some aerial members.
   1) **Geckos** are small, agile, nocturnal forms; adhesive toe pads with extremely fine filaments that allow them to walk on ceilings.
   2) **Iguanids** include many New World lizards as well as the marine iguana of the Galápagos.
   3) **Chameleons** are arboreal lizards of Africa and Madagascar; many have an extendible tongue.
   4) **Skinks** have elongate bodies and reduced limbs.
   5) Some lizards have degenerate limbs; the **glass lizards** are nearly limbless.

b. Lizards have movable eyelids whereas snakes have a transparent covering.

c. Nocturnal geckos have retinas with only rods; day-active lizards have both rods and cones.

d. Lizards have an external ear that snakes lack.

e. Geckos use vocal signals to announce territory and drive away males.

f. Some lizards survive well in hot and dry regions.
   1) They conserve water by producing crystalline uric acid.
   2) Water loss is also minimized with lipids in thick skin.
   2) Some lizards store fat in their tails to provide energy and metabolic water during drought.

g. The Gila monster and beaded lizard are the only lizards capable of a venomous bite.

h. Lizards are ectothermic, few live in cold climates.
   1) Ectotherms use less energy than endotherms; they survive in habitats with low productivity and warm climates such as tropical deserts and grasslands.
   3) Lizards keep their body temperature relatively constant by behavioral thermoregulation.

4. **Worm Lizards: Suborder Amphisbaenia**

a. These are highly specialized burrowing forms that are not true lizards.

b. They generally lack any trace of external limbs; eyes and ears are hidden under the skin.

c. The skin is divided into numerous rings resembling earthworms.

d. One species occurs in Florida but most live in South America and tropical Africa.

5. **The Mesozoic World of Dinosaurs**
   a. **History**
      1) Richard Owen coined term "dinosaur"—terrible lizard—in 1841.
      2) Two basic groups: "lizard-hipped" **Saurischia** and "bird-hipped" **Ornithischia**
      3) Dinosaurs and their living relatives, the birds, are **archosaurs**, which includes **thecodonts**, **crocodilians**, and **pterosaurs**.
      4) Walking upright is well illustrated in genus *Herrerasaurus*, from Argentina.
   b. **Saurischian Diversity**:
      1) Two groups of saurischian dinosaurs existed based on difference in feeding habits and locomotion: the carnivorous and bipedal theropods, and the herbivorous and quadrupedal sauropods.
      2) Some small sauropods reached 25 m (82 feet in length), but *Argentinosaurus* was adapted to feed on trees and reached sizes of 40 meters (132 ft).
      3) *Coelophysis* was a typical theropod with powerful hind legs, slender grasping forelimbs, a flexible neck, long counterbalancing tail, and large head with jaws and dagger like teeth.
   c. **Ornithischian Diversity**
      1) Quite varied, they were all herbivores.
      2) Many diverse forms are united by several derived skeletal features indicating common ancestry (e.g., *Triceratops*).
      3) The huge back-plated *Stegosaurus* of the Jurassic is an armored ornithischian.
      4) Hadrosaurs probably communicated with elaborate skull crests used as vocal resonators.
   d. **Parental Care**
      1) Recent fossil discoveries reveal dinosaurs exhibited complex parental care.
      2) An adult theropod *Oviraptor* was found probably with its own eggs.
      3) Wear on teeth of baby hadrosurs suggest they were fed by parents in the nest.
   e. **Extinction**
      1) Dinosaurs became extinct 65 million years ago leaving crocodiles and birds as survivors.
      2) A large asteroid impact or changing climate and landforms may have caused their demise.
      3) Paleontologists are still puzzled by dinosaur extinction yet persistence of other vertebrates.
6. **Snakes: Suborder Serpentes**
   a. Snakes are **limbless** and have lost the pectoral and pelvic girdles (except for vestigial structures in pythons).
      1) The many vertebrae are shorter and wider than in other tetrapods, allowing undulation.
      2) Ribs increase vertebral column rigidity thereby improving resistance to lateral stresses.
      3) Elevation of the neural spine gives the musculature more leverage.

b. Their feeding apparatus allows them to eat prey several times their own diameter.
    1) The two halves of the lower jaw are loosely joined, allowing them to spread apart.
    2) The skull boncs also are loosely articulated so the mouth can accommodate large prey.
c. Eyeballs have reduced mobility and a permanent corneal membrane for protection.
d. Most snakes have **poor vision** but arboreal snakes in tropical forests have highly developed vision.
e. Snakes lack external ears but respond to low frequency vibrations and ground vibrations.
f. Chemical senses rather than vision or hearing, are the main senses snakes use to hunt prey.
g. **Jacobson's organs** are a pair of pits in the roof of the mouth; they are lined with olfactory epithelium and the **forked tongue** picks up scent particles and conveys them past this organ.
h. Many snakes swallow their prey alive.
    1) Smaller prey may cause less injury due to struggles.
    2) Worms, insects, frogs, and small mammals are eaten.
    3) Some snakes locate prey by actively foraging.
    4) Constrictors often feed on larger mammals using ambush tactics.
    5) Muscle arrangements in constrictors reduce their travel speeds.

i. **Venomous Snakes**
    1) Less than 20% of all snakes are venomous except in Australia.
    2) Venomous snakes are divided into five families.
    3) Family Viperidae includes New World and Old World vipers with and without pits.
    4) Family Elapidae includes cobras, mambas, coral snakes, and kraits.
    5) Family Hydrophiidae includes the highly venomous sea snakes
    6) **Fossorial** mole vipers are found in the family Atractaspididae.
    7) Most members of the family Colubridae are non-venomous; several including the African boomslang and African twig snake are rear-fanged and their bite can be fatal to humans.

k. **Pit vipers**, such as rattlesnakes, have "**pits**" with nerve endings sensitive to heat emitted by warm-bodied birds and mammals.
    1) Viper fangs are hollow and hinged to inject venom when the snake strikes.
    2) Of an average 8,000 snake bites each year in the U.S., only 5-10 deaths result from pit vipers.
    3) A pair of modified teeth on the maxillary bones serves as fangs.
    4) Fangs become erect during a strike; venom is injected through their canals
    5) Prey are paralyzed or die after the bite then swallowed whole by the viper.

l. **Snakebite and Toxicity**
    1) Saliva of harmless snakes contains limited toxins; this provided a basis for natural selection of venom.
    2) Most snake **venoms** are a complex combination of venom types.

3) **Neurotoxins** act on the nervous system, causing blindness or stopping respiration.
4) **Hemorrhagin** type venoms breaks down blood vessels and red blood cells; much blood is leaked into tissue spaces.
5) Sea snakes and the Australian tiger snake have the most deadly venom per unit.
6) Large venomous snakes deliver more venom; a king cobra may be the most dangerous.
7) In India, Pakistan, and other nearby countries, dense human populations with poor footwear and delayed medical care contributes to most deaths from snakebites.
8) Worldwide, approximately 50,000-60,000 persons die each year from snakebites.

m. **Reproduction**
1) Most snakes are **oviparous** and lay shelled eggs under logs or rocks or in holes in the ground.
2) Others, including pit vipers, are **ovoviviparous** (the egg hatches internally to facilitate a live birth).
3) A few snakes are **viviparous**, having a primitive "placenta" to exchange nutrients with the young.

7. **The Tuatara: Order Sphenodonta**
a. Only two living species in New Zealand represent this ancient lineage.
b. Sphenodontids radiated modestly in the early Mesozoic but then declined.
c. Once widespread across New Zealand, the two species are now restricted to small islands.
d. They are lizard-like and live in burrows often shared with petrels.
e. The tuatara is slow growing and may live to 77 years of age.
f. Its skull is nearly identical to diapsid skull of 200 million years ago.
g. A well-developed median parietal eye is buried beneath its skin, function is unknown.
h. *Sphenodon* represents one of the slowest rates of evolution known among vertebrates.

8. **Order Crocodilia: Crocodiles and Alligators**
a. Modern crocodilians are the only surviving reptiles of the **archosaurian lineage**.
b. This lineage gave rise to the Mesozoic radiation of dinosaurs and to birds.
c. Modern crocodilians differ little from primitive crocodilians of the early Mesozoic.
d. Modern crocodilians are classified in three families.
1) **Alligators and caimans** are found primarily in the New World; they have a broader snout.
2) **Crocodiles** are widely distributed and include the huge saltwater crocodile.
3) **Gavials** have one species in India and Burma; it has a very narrow snout.

e. All have a long, well-reinforced skull and jaw musculature for a powerful bite.
f. Teeth are set in sockets typical of archosaurs and the earliest birds.
g. They have a complete secondary palate, a feature only shared with mammals.
h. They also share a **four-chambered heart** with birds and mammals.

i. The estuarine crocodile in southern Asia and the Nile crocodile are both very large.
j. Crocodiles may attack cattle, deer, and people; alligators are less aggressive.
k. Alligators and crocodiles are oviparous; usually 20-50 eggs are laid in a mass of vegetation.
   1) Alligators emit loud bellows during the mating season.
   2) Females of crocodiles and alligators guard their eggs then open nest sites when young hatch.
   3) Nests left unguarded are easily discovered and raided by predators.
   4) High nest temperatures produce males; low temperatures produce females.

C. Classification of Living Reptilian Groups

Subclass Anapsida
Order Testudines
Subclass Diapsida
Superorder Lepidosauria
Order Squamata
Suborder Lacertilia
Suborder Amphisbaenia
Suborder Serpentes
Order Sphenodonta
Superorder Archosauria
Order Crocodilia

REFERENCES

Aurélien, B., Christophe, L., Peggy, V., Romain, A., Nathalie, B., Eric, B., Gilles, C., François, F., François, M., Jean-Michel, M. & Abel, P. (2010). Regulation of body temperature by some Mesozoic marine reptiles. *Science*, *328*, 1379–1382.

Barlow, G. (2002). The cichlid fishes: nature's grand experiment in evolution.

Benton, M. J. (1993). *The fossil record 2*. Chapman & Hall, London.

Benton, M. J. (2000). *Vertebrate Palaeontology* (2nd ed.). London: Blackwell Science Ltd. ., 3rd ed. 2004.

Carroll, R. L. (1991). The origin of reptiles. In Schultze H.P., Trueb, L. *Origins of the higher groups of tetrapods — controversy and consensus*. Ithaca: Cornell University Press. 331–53.

Carroll, R. L. (1970). The ancestry of reptiles. Philosophical transactions of the Royal society London, *257*, 267-308.

Carroll, R. L. (1997). *Patterns and processes of vertebrate evolution*. Cambridge.

Cockrem, J. F. & Guillette, L. J. (1992). Reproductive cycles of male and female tuatara (*Sphenodon punctatus*) on Stephens Island, New Zealand. *J Zool*, *226*, 199–217.

Colenso, W. (1885). Notes on the Bones of a Species of *Sphenodon*, (*S. diversum*, Col.,) apparently distinct from the Species already known. *Tran Proc Roy Soc New Zealand*, *18*, 118–128.

Cree, A., Thompson, M. B. & Daugherty, C. H. (1995). Tuatara sex determination. *Nature*, *375*, 543.

Crook, I. G. (1973). The tuatara, *Sphenodon punctatus* Gray, on islands with and without populations of the Polynesian rat, *Rattus exulans* (Peale). *Proc New Zealand Ecol Soc*, *20*, 115–120.

Daugherty, C. & Keall, S. Tuatara: Tuatara islands In: *Te Ara: The Encyclopedia of New Zealand.*

Daugherty, C. H., Cree, A., Hay, J. M. & Thompson, M. B. (1990). Neglected taxonomy and continuing extinctions of tuatara *(Sphenodon). Nature*, *347*, 177–179.

Dean, D. R. (1999). *Gideon Mantell and the Discovery of Dinosaurs*. Cambridge University Press. 97–98.

Erickson, G. M., Gignac, P. M., Steppan, S. J., Lappin, A. K., Vliet, K. A., Brueggen, J. D., Inouye, B. D., Kledzik, D., Webb, G. J. W. & Claessens, L. (2012). Insights into the Ecology and Evolutionary Success of Crocodilians Revealed through Bite-Force and Tooth-Pressure Experimentation. *PLoS ONE*, *7* (3), e31781.

Falcon-Lang, H. J., Benton, M. J. & Stimson, M. (2007). Ecology of early reptiles inferred from Lower Pennsylvanian trackways. *J Geol Soc*, *164*, 1113–1118.

Gand, C., Gillingham, J. C. & Clark, D. L. (1984). Courtship, mating and male combat in Tuatara, *Sphenodon punctatus*. *J Herpetol*, *18*, 194–197.

Gauthier J. A. (1994). *The diversification of the amniotes*. In: Prothero, D. R. & Schoch, R. M. (ed.) Major Features of Vertebrate Evolution: 129-159. Knoxville, Tennessee: The Paleontological Society.

Gauthier, J. A., Kluge, A. G. & Rowe, T. (1988). The early evolution of the Amniota. In Benton, M.J. (ed.). *The Phylogeny and Classification of the Tetrapods* **1**. Oxford: Clarendon Press. 103–155.

Goodrich, E. S. (1916). On the classification of the Reptilia. *Proc Roy Soc London*, *89B* (615), 261–276.

Graciela, P., Jorge, F., Alejandro, R. & Michel, L. (2012). Cranial morphology of the Early Permian mesosaurid *Mesosaurus tenuidens* and the evolution of the lower temporal fenestration reassessed. *Comp Ren Palevol*, *11*, 379–391.

Günther, A. (1867). Contribution to the anatomy of Hatteria (Rhynchocephalus, Owen). *Sep Philosoph Transac*, *II*, 1-36.

Haines, T. (2000). *Walking with Dinosaurs: A Natural History*, New York: Dorling Kindersley Publishing, Inc., p. 65.

Hay, J. M., Sarre, S. D., Lambert, D. M., Allendorf, F. W. & Daugherty, C. H. (2010). Genetic diversity and taxonomy: a reassessment of species designation in tuatara (Sphenodon: Reptilia).*Conserv Genet*, *11*, 1063–1081.

Huey, R. B. (1982). Temperature, physiology, and the ecology of reptiles. Side 25-91. In Gans, C. & Pough, F.H. (red), *Biol Rep*, *12*, Physiology (C). Academic Press, London.

Jacobs, L. L. (1997). African Dinosaurs". *Encyclopedia of Dinosaurs*. Edited by Phillip J. Currie and Kevin Padian. Academic Press. p. 2-4.

Jaeckel, O. (1902): Über *Gephyrostegus bohemicus*, n.g. n.sp. *Zeitschrift der Deutschen Geologischen Gesellschaft* no 54, 127–132.

John, W. M. (1997). A phylogenetic analysis of the euryapsid reptiles. *J Vert Pal* 17 (Supplement to 3): 65A.

Jones, M. E. (2008). Skull shape and feeding strategy in Sphenodon and other Rhynchocephalia (Diapsida: Lepidosauria). *J Morphol*, *269*, 945–66.

Jones, M. E. (2009). Dentary tooth shape in Sphenodon and its fossil relatives (Diapsida: Lepidosauria: Rhynchocephalia). *Front Oral Biol*, *13*, 9–15.

Jones, M. E., Tennyson, A. J., Worthy, J. P., Evans, S. E. & Worthy, T. H. (2009). A sphenodontine (Rhynchocephalia) from the Miocene of New Zealand and palaeobiogeography of the tuatara (Sphenodon). *Proc Biol Sci*, *276*, 1385–90.

Juan, C. C., Ross, D., Cesar, S., Átila da, R., Cibele, S., Leopoldo, W. N. & Pedro, L. P. A. (2004). A procolophonoid reptile with temporal fenestration from the Middle Triassic of Brazil. *Proc Royal Soc B: Biol Sc*, *271*, 1541–1546.

Langer, M. C., Martin, D. E., Jonathas, S. B. & Fernando, E. N. (2010). The origin and early evolution of dinosaurs. *Biol Rev*, *85*, 65–66, 82.

Laurin, M. & Reisz, R. R. (1995). A re-evaluation of early amniote phylogeny". *Zool J Linn Soc*, *113*, 165–223.

Laurin, M. (1996). A reevaluation of *Ariekanerpeton*, a Lower Permian seymouriamorph (Vertebrata: Seymouriamorpha) from Tadzhikistan. *J Vert Paleontol*, *16*, 653–65.

Linda, A. T. & Johannes, M. (2009). Assembling the history of the Parareptilia: phylogeny, diversification, and a new definition of the clade. *Fossil Rec*, *12*, 71–81.

Mayer, E. (2001). *What evolution is*. Basic Books, New York, NY.

Modesto, S. P. (1999). Observations of the structure of the Early Permian reptile *Stereosternum tumidum* Cope. *Palaeontol Afr*, *35*, 7–19.

Monastersky, R. (1999). Out of the Swamps, How early vertebrates established a foothold—with all 10 toes—on land. *Science News*, *155*, 328.

Musico, B. (1999). *Sphenodon punctatus*. *Animal Diversity Web*. University of Michigan Museum of Zoology.

Olson, E. C. (1947). The family Diadectidae and its bearing on the classification of reptiles". *Field Geol*, *11*, 1–53.

Osborn, H. F. (1903). The Reptilian subclasses Diapsida and Synapsida and Early History of Diaptosauria. *Mem Am Mus Nat His*, *1*, 451–507.

Palmer, D. ed. (1999). *The Marshall Illustrated Encyclopedia of Dinosaurs and Prehistoric Animals*. London: Marshall Editions. p. 62.

Paton, R. L., Paton, T. R. & Clack, J. A. (1999). An amniote-like skeleton from the Early Carboniferous of Scotland. *Nature*, *398*, 508–513.

Pawley, K. (2006). Walking with early tetrapods: evolution of the postcranial skeleton and the phylogenetic affinities of the Temnospondyli (Vertebrata: Tetrapoda). *In*: "The postcranial skeleton of temnospondyls (Tetrapoda: temnospondyli)." PhD Thesis. La Trobe University, Melbourne.

Pope, K. O., Baines, K. H., Ocampo, A. C. & Ivanov, B. A. (1997). Energy, volatile production, and climatic effects of the Chicxulub Cretaceous/Tertiary impact. *J Geophys Res*, *102*, 21645–21664.

Ramstad, K. M., Nelson, N. J., Paine, G., Beech, D., Paul, A., Paul, P., Allendorf, F. W. & Daugherty, C. H. (2007). Species and Cultural Conservation in New Zealand: Maori Traditional Ecological Knowledge of Tuatara. *Conserv Biol*, *21*, 455–464.

Raup, D. M. (1995). The role of extinction in evolution. In *Tempo and mode in evolution*, eds Fitch & Ayala. Washington D.C. National Academy of Sciences.

Rieppel, O. & DeBraga, M. (1996). Turtles as diapsid reptiles. *Nature*, *384*, 453–5.

Romer, A. S. & Parsons, T. S. (1977). *The Vertebrate Body.* 5th ed. Saunders, Philadelphia. (6th ed. 1985)

Russell, M. (1998). Tuatara, Relics of a Lost Age. *Cold Blooded News*. Colorado Herpetological Society.

Ruta, M., Coates, M. I. & Quicke, D. L. J. (2003). Early tetrapod relationships revisited. *Biol Rev*, *78*, 251–345.

Schluter, D. (2000). *The ecology of adaptive radiation*. Oxford University Press.

Sean, M., Robert, R. & Diane, S. (2011). A neodiapsid reptile from the Lower Permian of Oklahoma. *Soc Vert Pal 71st Annual Meeting Program and Abstracts*: p. 160.

Simpson, G. G. (1944). *Tempo and mode in evolution*. Columbia, N.Y.

Simpson, G. G. (1953). *The major features of evolution*. Columbia, N.Y.

Stanley, S. M. (1979). *Macroevolution: pattern and process*. Freeman, S.F. p65

Thompson, M. B. & Daugherty, C. H. (1998). Metabolism of tuatara, *Sphenodon punctatus*. *Comp biochem physiol A*, *119*, 519–522.

Van, T. M. & Hadly, E. A. (2004). Error in Estimation of Rate and Time Inferred from the Early Amniote Fossil Record and Avian Molecular Clocks. *J Mol Biol*, *59*, 267–276.

Wang, S. C. & Dodson, P. (2006). Estimating the Diversity of Dinosaurs. *Proc Nat Acad Sci USA*, *103*, 13601–13605.

Watson, D. M. S. (1957). On Millerosaurus and the early history of the sauropsid reptiles. *Philos Trans Royal Soc London, Series B, Biol Sci*, *240*, 325–400.

Wu, X. C. (1994). *Late Triassic-Early Jurassic sphenodontians from China and the phylogeny of the Sphenodontia" in Nicholas Fraser & Hans-Dieter Sues (eds) In the Shadow of the Dinosaurs: Early Mesozoic Tetrapods*. Cambridge University Press.

Chapter 4

IS ARCHAEOPTERYX THE MISSING LINK BETWEEN DINOSAURS AND BIRDS?

4.1. ARCHAEOPTERYX

Archaeopteryx, sometimes referred to by its German name ***Urvogel*** ("original bird" or "first bird"), is a genus of early bird that is transitional between feathered dinosaurs and modern birds. The name derives from the ancient Greek meaning "feather" or "wing". Since the late nineteenth century, it had been generally accepted by palaeontologists, and celebrated in lay reference works, as being the oldest known bird (member of the group Avialae) [Xu et al., 2011]. However, older potential avialans have since been identified, including *Anchiornis*, *Xiaotingia*, and *Aurornis* [Godefroit, 2013].

Archaeopteryx lived in the Late Jurassic period around 150 million years ago, in what is now southern Germany during a time when Europe was an archipelago of islands in a shallow warm tropical sea, much closer to the equator than it is now. Similar in shape to a European Magpie, with the largest individuals possibly attaining the size of a raven [Erickson et al., 2009], *Archaeopteryx* could grow to about 0.5 m (1 ft 8 in) in length (Figure 1). Despite its small size, broad wings, and inferred ability to fly or glide, *Archaeopteryx* has more in common with other small Mesozoic dinosaurs than it does with modern birds. In particular, it shares the following features with the deinonychosaurs (dromaeosaurs and troodontids): jaws with sharp teeth, three fingers with claws, a long bony tail, hyperextensible second toes ("killing claw"), feathers (which also suggest homeothermy), and various skeletal features [Yalden, 1984; Chiappe and Witme, 2002].

These features make *Archaeopteryx* a clear candidate for a transitional fossil between dinosaurs and birds. Thus, *Archaeopteryx* plays an important role, not only in the study of the origin of birds, but in the study of dinosaurs. It was named from a feather in 1861. That same year, first complete specimen of *Archaeopteryx* was announced. Over the years, ten more fossils of *Archaeopteryx* have surfaced. Despite variation among these fossils, most experts regard all the remains that have been discovered as belonging to a single species, although this is still debated [Welnhofer, 2004].

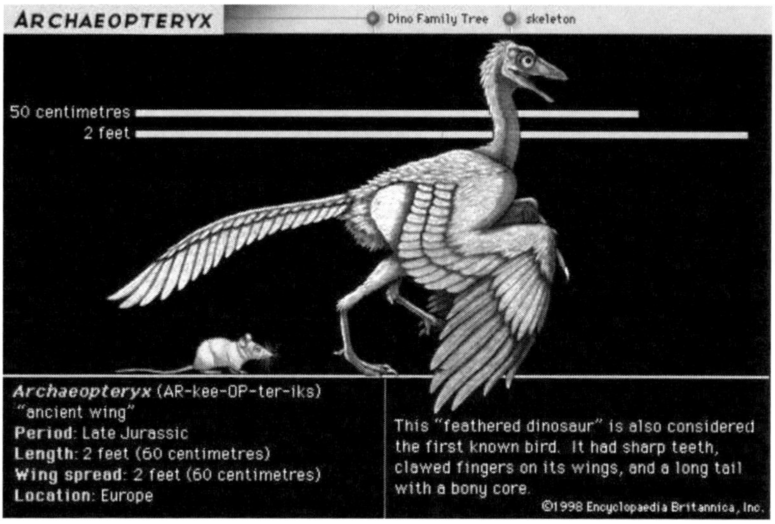

Figure 1. *Archaeopteryx* (adapted from University of California, Museum of Palaeontology).

4.2. Morphology

Archaeopteryx lived during the early Tithonian stage of the Jurassic period, approximately 150.8–148.5 million years ago [Schweigert, 2007]. Most of the specimens of *Archaeopteryx* that have been discovered come from the Solnhofen limestone in Bavaria, southern Germany, which is a *lagerstätte*, a rare and remarkable geological formation known for its superbly detailed fossils [Lambert, 1993].

Archaeopteryx was roughly the size of a raven, with broad wings that were rounded at the ends and a long tail compared to its body length. It could reach up to 500 millimetres (20 in) in body length, with an estimated weight of 0.8 to 1 kilogram (1.8 to 2.2 lb) [Erickson et al., 2009], *Archaeopteryx* feathers, although less documented than its other features, were very similar in structure and design to modern-day bird feathers (Lambert, 1993). Despite the presence of numerous avian features [Holtz, 1995], *Archaeopteryx* had many theropod dinosaur characteristics. Unlike modern birds, *Archaeopteryx* had small teeth [Lambert, 1993], as well as a long bony tail, features which *Archaeopteryx* shared with other dinosaurs of the time.

Because it displays a number of features common to both birds and dinosaurs, *Archaeopteryx* has often been considered a link between them [Lambert, 1993]. In the 1970s, John Ostrom, following T. H. Huxley's lead in 1868, argued that birds evolved within theropod dinosaurs and *Archaeopteryx* was a critical piece of evidence for this argument; it had a number of avian features, such as a wishbone, flight feathers, wings, and a partially reversed first toe along with a number of dinosaur and theropod features. For instance, it has a long ascending process of the ankle bone, interdental plates, an obturator process of the ischium, and long chevrons in the tail. In particular, Ostrom found that *Archaeopteryx* was remarkably similar to the theropod family Dromaeosauridae (Figure 2) [Owen, 1863; Huxley, 1868, 1870; Nedin, 1999; Ostrom, 1976, 1985; Olson and Feduccia, 1979; Bühler and Bock, 2002].

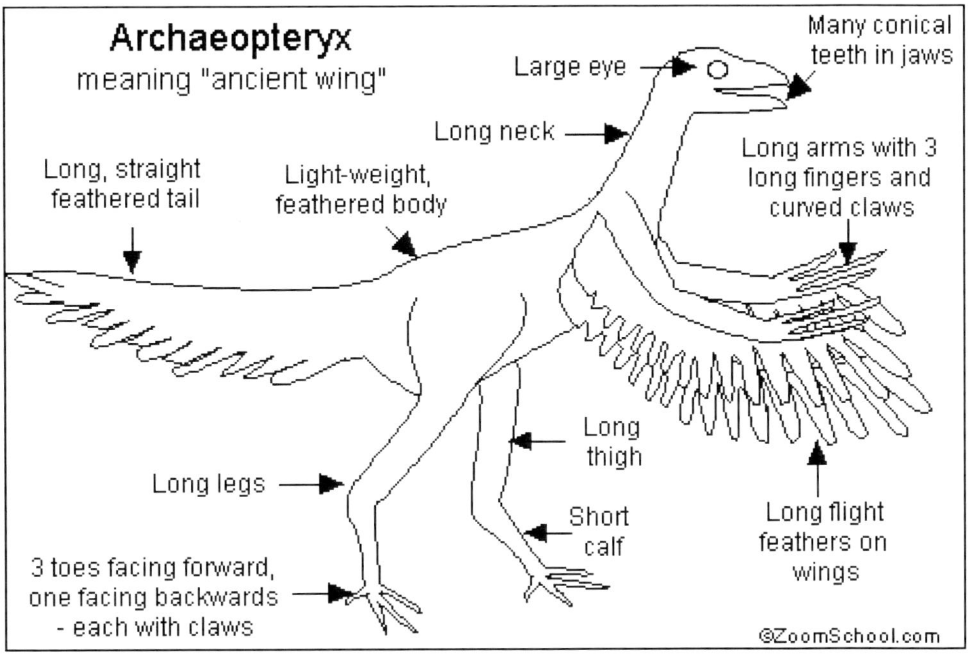

Figure 2. Different body parts of *Archaeoteryx* (Reconstructed; Adapted from Gauthier, J.A. 1986).

4.3. *ARCHAEOPTERYX* FEATURES

Much has been made in pseudoscientific circles about the position of Archae within the evolutionary scheme of things. The usual "argument" put forward is that Archae cannot be a transitional fossil between birds and dinosaurs because it is a bird. This simplistic line belies the fact that, whilst Archae is indeed classified as a bird, it has been done so on the strength of 4 main characters - 2 of which are not unique to birds. This classification ignores the fact that Archae has numerous characters which *are* unique, unique in that they are **not** possessed by birds. Archae's avian affinities are allowable on the strength of the following 4 main characters:

4.4. ARCHAEOPTERYX'S AVIAN FEATURES

Feathers

Feathers are *the* diagnostic feature of modern birds. This is one of the main criterions for classifying Archae as a bird, as no other modern animal has feathers. The possession of feathers is a characteristic of birds, so strike one up for the birds. However, in late 1996, a discovery in China has changed this view. A small theropod dinosaur *Sinosauropteryx* [Chen et al. 1998] was found with what appear to be feathers preserved along the back. The identification of the structures was equivocal however, [Unwin 1998], with some doubting that the structures are feathers.

Feathered Dinosaurs Found

Two species of dinosaur have recently been found in northeast China which possess feathers [Qiang et al. 1998]. *Protoarchaeopteryx robusta* and *Caudipteryx zoui* show regiges, rectrices and plumulaceous feather impressions. Further, they are not birds, lacking a reverted (backwards facing) big toe (see number 2 below) and a quadrratojugal squamosal contact, having a quadrojugal joined to the quatrate by a ligament and a reduced or absent process of the ishium. These and other characters group *Protoarchaeopteryx* and *Caudipteryx* with maniraptoran coelurosaurs rather than birds (Figure 3).

Systematists *define* the names of organisms by their ancestry, in this case birds (Aves) consist of *Archaeopteryx* plus living birds and all the descendants of their most recent common ancester. Birds are *diagnosed* by unique features that only they possess and which are inherited from that common ancestor. Even if feathers are shared by a wider group than just birds, birds are still defined as *Archaeopteryx* and later relatives. *Protoarchaeopteryx* and *Caudipteryx* are not birds even though they have feathers because the suites of morphological characters they possess mark them as belong to the maniraptoran coelurosaur dinosaurs [Padian 1998].

It appears that feathers can no longer be used as a unique feature of birds.

Opposable Hallux (Big Toe)

This is also a character of birds and not of dinosaurs. Although opposable big toes are found in other groups, they are notfound in dinosaurs. A reversed big toe is found in some dinosaurs however, and the condition is approached in some theropod dinosaurs.

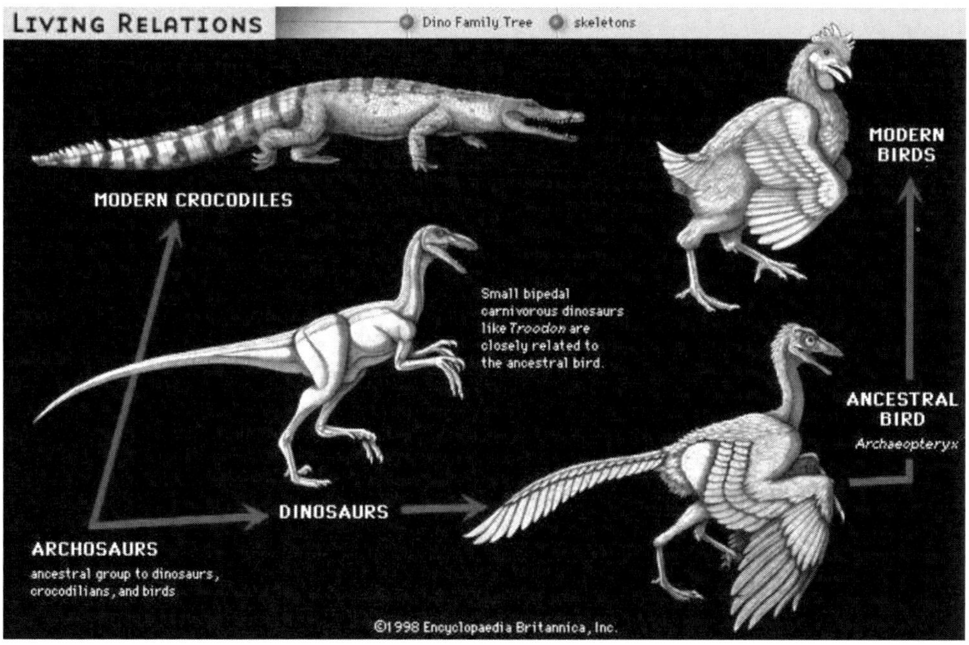

Figure 3. Arrow diagram depicting evolution of *Archaepteryx* (Source: Encyclopedia Britannica, Inc).

Furcula (Wishbone) Formed of Two Clavicles Fused Together in the Midline

It used to be thought that the possession of a furcula distinguished birds from dinosaurs. Indeed, up until recently even clavicles were few and far between in even theropod dinosaurs, the suggested closest group to the birds and from which the birds evolved [Ostrom 1976]. However, it has been found that theropod dinosaurs did indeed have clavicles [Bryant & Russell 1993] and they have been found in several species, e.g., *Segisaurus, Velociraptor, Euparkeria, Ornithosuchus, Saltoposuchus, Ticinosuchus.* Also, Chure & Madson (1996) reported furculae in a non-maniraptoran, allosaurid dinosaur.

It has been found that the clavicles are often small and poorly ossified. This is no surprise, since they are of little evolutionary advantage to average theropod dinosaur. However, birds too show this variation in ossification, especially amongst the carniates and some parrots, where clavicles are reduced or even missing. Therefore, the apparent absence of clavicles in some theropod dinosaurs may well be due to poor ossification rather than true absence. However, furculas have been found in some theropod dinosaurs, namely the Oviraptorosauria [Barsbold et al. 1990, Bryant & Russell 1993], for example *Oviraptor* and *Ingenia.* Thus furculas do not appear to be diagnostic to birds and certain members of the suggested closest group to the birds now appear to possess furculas so it is a neutral character. The presence of clavicles shows that this character is a feature of theropod dinosaurs and thus was probably present in early theropods.

Pubis Elongate and Directed Backward

This is a feature of birds, but it is also a feature of some theropod dinosaurs so is not diagnostic of birds - another neutral character. However, the pubic shafts of *Archaeopteryx* and dromaeosaurs (a group of theropod dinosaurs which are thought to be closely linked to birds) share a plate-like, slightly angled transverse cross-section not found in any other archosaurs.

4.5. ARCHAEOPTERYX'S REPTILE FEATURES

Premaxilla and Maxilla Are Not Horn-Covered

The premaxilla does not have a keratinized covering, so *Archaeopteryx* has no bill. The bill is produced via the process of 'cornification' which involves the mucus layer of the epidermis [Romanoff 1960] and thus its formation is independent of jaw bone formation.

Trunk Region Vertebra Are Free

In birds the trunk vertebrae are always fused.

Bones Are Pneumatic

Have air-sacs, as they do in birds and in some dinosaurs [Witmer 1990, Brooks 1993]. It should be pointed out that previous claims suggesting the bones of Archae were not pneumatic [Lambrecht 1933; de Beer 1954], was based on negative evidence, i.e., that the bones do not exhibit pneumatic pores (through which the air sacs enter the bones) and the bones show none of the plumpness and bulges which characterise the pneumatic bones of modern birds. Britt et al. (1998) found evidence for the presence of pneumatic bones in *Archaeopteryx*. These specimens show evidence of vertebral pneumaticity in the cervical and anterior thoracic vertebrae, thus confirming the phylogenetic continuity between the pneumatic systems of non-avialan theropods and living birds [Britt et al. 1998].

Pubic Shafts with a Plate-like, and Slightly Angled Transverse Cross-Section

A Character shared with dromaeosaurs but not with other dinosaurs or birds

Cerebral Hemispheres Elongate, Slender and Cerebellum Is Situated Behind the Mid-Brain and Doesn't Overlap it from Behind or Press Down on It

This again is a reptilian feature. In birds the cerebral hemispheres are stout, cerebellum is so much enlarged that it spreads forwards over the mid-brain and compresses it downwards. Thus, the shape of the brain is not like that of modern birds, but rather an intermediate stage between dinosaurs and birds [Alexander 1990].

Neck Attaches to Skull from the Rear as in Dinosaurs Not from below As in Modern Birds

The site of neck attachement (from below) is characteristic in birds, Archaeopteryx does not have this character. This coelurosaurian-like neck extended back from the rear of the skull in Archaeopteryx - as it does in coelurosaurs [theropod dinosaurs], rather than from beneath as in later birds [Ostrom, 1976]. Skull and brain of Archae is basically reptilian and is not "totally birdlike" (contrary to a certain creationist's claim).

Center of Cervical Vertebrae Have Simple Concave Articular Facets

This is the same as the archosaur pattern. In birds the vertebrae are different, they have a saddle-shaped surface. The most striking feature of the vertebrae is the simple disk-like facets of their centra, without any sign of the saddle-shaped articulations found in other birds [de Beer, 1954].

Long Bony Tail with Many Free Vertebrae up to Tip (No Pygostyle)

Birds have a short tail and the caudal vertebrae are fused to give the pygostyle.

Premaxilla and Maxilla Bones Bear Teeth

No modern bird possess teeth [Romanoff, 1960; Orr, 1966]. Bird embryos form tooth buds, but do not actually produce teeth. Some birds subsequently produce ridges in the bill, but there is no connection between them and the embryonic tooth buds, since the ridges also form in other areas of the bill where no tooth buds have previously formed. Some birds produce hook-like structures which are papillae, and appear to be related to the process of keratinization of the beak [Romanoff 1960], and have nothing to do with teeth. They do not possess blood vessel or nerve connections, nor do they produce dentine.

The expression of tooth buds in the bird embryo has a simple evolutionary explanation, since it suggests that the ancestors of modern birds possessed teeth and that this character has been suppressed in modern birds. The presence of tooth buds in the embryos of organisms which do not possess teeth in the adult is a difficulty for anti-evolutionists, as why should a character be expressed that is never used in the organism? Some fossil birds exhibit a reduction in the number of bones which have teeth. Both *Hesperornis* and *Baptornis* lack teeth on the premaxilla (*Archaeopteryx* and theropod dinosaurs have teeth on both the maxilla and premaxilla). Not only that, *Hesperornis* has a beak, but on the upper jaw only [Gingerich 1975]. It therefore has **half** beak **and** teeth, a good example of a morphologicaly intermediate structure between toothed birds which lack a beak, and beaked, toothless birds.

Ribs Slender, without Joints or Uncinate Processes and Do Not Articulate with the Sternum

Birds have stout ribs with uncinate processes (braces between them) and articulate with the sternum.

Pelvic Girdle and Femur Joint Is Archosaurian Rather than Avian (Except for the Backward Pointing Pubis as Mentioned above)

The pelvic girdle as a whole is basically free and similar to archosaur girdles, the pubis points backward - a character shared with birds and some other bird-like theropod dinosaurs.

What *is* interesting is that with the bird pelvis, the ischium lies beneath the posterior part of the ilium and beneath this again is the pubis, which is directed backwards. Embryological studies show that the peculiar position of these bones is the result of secondary rotation and that the pectineal process, in front of the acetabulum, is not the true pubis as some workers have maintained [Bellairs and Jenkin, 1960].

In other words, the embyonic pelvis of the bird, when first formed, looks, in shape and angle between the ilium and the pubis (45 degrees), very similar to the "A"-frame pelvis of

Archaeopteryx (i.e., like this: <) [Romanoff 1960]. The fully formed pelvis with all bones lying parallel is the result of secondary rotation of the pubis from "<" to "=". This supports the view that birds had an ancester with a saurischian pelvis such as the type possessed by *Archaeopteryx* and other theropod dinosaurs.

The Sacrum (the Vertebrae Developed for the Attachment of Pelvic Girdle) Occupies 6 Vertebra

This is the same as in reptiles and especially ornithipod dinosaurs. The bird sacrum covers between 11-23 vertebrae! So, while the variation seen in modern birds is large, it is nowhere near the number found in *Archaeopteryx.*

Metacarpals (Hand) Free (Except 3rd metacarpal), Wrist Hand Joint Flexible

In birds the metacarpals are fused together with the distal carpals in the carpo-metacarpus (wrist /hand fused). All modern birds have a carpo-metacarpus, as in fossil birds. However, the carpals of several coelurosaur dinosaur groups show a trend towards fusion, and in the Late Cretaceous form *Avimimus*, a true carpo- metacarpus is formed.

It has been suggested that the ostrich and/or other Ratites also possess unfused wrist/hand bones, which is not correct. The ostrich, emus, rheas, cassowaries and kiwis are often referred to together as the Ratites, though they may not be closely related to each other. They have tiny wings and cannot fly, but the bones of their hands are fused together in the same peculiar way as in flying birds, which suggest that they evolved from flying birds [Alexander, 1990].

Some similarity between the hand of the ostrich and some of the more derived theropod dinosaurs was once used to suggest that the Ratites were 'primitive' and evolved before the advent of flight in birds. However, Tucker (1938b) showed that such similarities are entirely superficial.

He has directed attention to the bird-like characters of the hand of the dinosaur *Ornitholestes* as evidence that a bird-like hand can be developed independently of flight, but the [Tucker 1938b] resemblance is utterly superficial and that the peculiar bowing and terminal fusion of metacarpals 2 and 3 which characterize both the Carnate and the Ratite hand are in no way reproduced in the dinosaur [Tucker 1938a].

Nasal Opening Far Forward, Separated from the Eye by a Large Preorbital Fenestra (Hole)

This is typical of reptiles, but not of birds. Where a fenestra is present in birds, it is always greatly reduced, and is involved in prokinesis (movement of the beak).

Deltoid Ridge of the Humerus Faces Anteriorly as Do the Radial and Ulnar Condyles

Typical of reptiles but not found in birds.

Claws on 3 Unfused Digit

No modern adult bird has 3 claws, nor do they have unfused digits. The juvenile hoatzin and Touracos do have 2 claws but loose them as they grow, the ostrich appears to retain its 2 claws into adulthood, due to the early termination of development. In case of the hoatzin it is thought that these claws allow the juvenile to climb. It had been claimed that since these birds do have claws, even in the juvenile stage, the presence of claws cannot be used as a reptilian character. In fact almost all birds exhibit claws, but in the embryonic stage and they are lost by the time the bird leaves the egg. In case of the few which do retain claws into the juvenile stage, this is merely the extension of the condition into the post-embryonic stage. As McGowan says:

> "*In retaining a primitive reptilian feature which other birds lose just before leaving the egg [the hoatzin] is showing its reptilian pedigree. Far from being evidence to the contrary, the hoatzin is additional evidence for the reptilian ancestry of birds.*"

The Fibula is Equal in Length to the Tibia in the Leg

This is again a typical character of reptiles. In birds the fibula is shortened and reduced.

Metatarsals (Foot Bones) Free

In birds these are fused to form tarso-metatarsus. However, in modern bird embryos, the foot bones are initially separate as in the adult *Archaeopteryx* and is another character supporting a reptilian ancestry for birds. After all, why bother producing separate bones in the embryo and then fuse them? Why not produce a fused mass to start with? No adult modern bird has separate metatarsals, but they are separated, initially, in the embryo. This can be explained in terms of evolution - birds evolved from a group which had unfused metatarsals.

Ceratosaurians, *Avimimus*, and Elmisauridae all show true tarso-metatarsi. Archae itself only shows the beginning of this structure.

Gastralia Present

Gastralia are "ventral ribs," elements of dermal bone in the ventral wall of the abdomen. Typical of reptiles, they are absent in birds. In addition to the true ribs the British Museum specimen shows a large number of so-called ventral ribs or gastralia, elements of dermal bone lying in the ventral wall of the abdomen [de Beer, 1954]. The gastralia of the Berlin specimen are identical with those of the British Museum specimen, but more have been preserved [de Beer, 1954].

"The "new" specimen consists of two small slabs (specimens 6928 & 6929), part and counterpart which contain impressions or parts of the left manus and forearm, pelvis, both legs and feet, and some gastralia. Also present were numerous fragments of gastralia, faint impressions of three or four dorsal vertebrae, gastralia, phalanges [Ostrom, 1970, 1972, 1976; Wellnhofer, 1974].

4.6. A Tale of Two Pelvises

Dinosaurs can be divided into two groups based on the shape of the pelvis. The **Saurischia**, or 'lizard hipped' dinosaurs, and the **Ornithischia**, or 'bird hipped' dinosaurs. Strange as it may seem, *Archaeopteryx* and all modern birds are thought to have evolved from the Saurischia, not from the bird hipped, Ornithischia! Odd? Well, not really. The pelvises in question are shown in Figure 4.

The saurischian pelvis (Figure 4a) is the more primitive structure, and has the pubis pointing forwards and the ischium backwards. In ornithischians (Figure 4b), the pubis runs back in parallel with the ischium. Additionally, a prepubic process is present, which is assumed to have been used as a point of muscle attachment in the absence of the pubis (probably to support the gut). This situation is superficially similar to that in birds (hence the name 'bird hipped' - the pelvis of birds was well known to anatomists by the time dinosaurs were being described). However, on closer examination, clear differences are evident. In *Archaeopteryx* (Figure 4c) and all birds, the prepubic process is absent and, unlike in saurischians, the pubis does not meet along the midline. Also, in *Archaeopteryx* the pubis still maintains the club-like termination, as in many saurischians. It should also be noted that several saurischian dinosaurs have the pubis facing backward, e.g., *Troodon*. The pelvis of *Archaeopterys* is morphologically intermediate between saurischian dinosaurs and birds [Romanoff ,1960].

The primary thrust generator in earlier theropods was the Caudofemoralis longus, which streched from the leg to some distance along the tail. Later forms such as the dromaeosaurids (the group closely related to *Archaeopteryx*) rearranged this so that the primary thrust was produced by muscles attaching from the leg to the pelvis and the base of the tail. This allowed the tail to be braced and used as a stabilizing device, such as in allosaurs and 'raptors, or to be lost as in modern birds [Romanoff ,1960]. In ornithischians, the backwards pointing pubis probably occurred to allow for the expansion of the stomach and gut in order to better digest plant material.

Study of the embryonic development of the modern bird pelvis is also instructive. The embryonic pelvis in modern birds starts out in a very similar configuration to *Archaeopteryx*, with the ilium and pubis bones set apart at an angle of approx. 45 degrees. The embryonic bones then secondarily rotate through angles seen in modern Ratites until reaching the parallel alignment typifing adult, modern birds [Romanoff ,1960]:

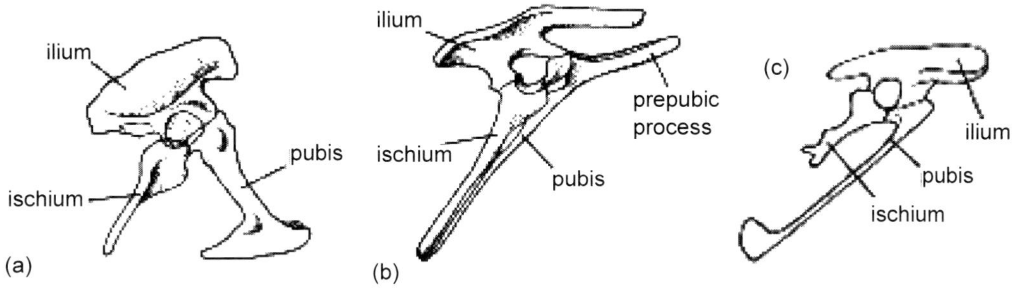

Figure 4. a, b, c showing shape of pelvis in dinosaurs (Source: Romanoff, A.L. 1960).

Chicken embryo	Angle ilium-pubis	Adults with same angle
7.0 days	45 degrees	*Archaeopteryx*
7.3 days	35 degrees	Kiwi
8.0 days	30 degrees	Ostrich
Hatching	<5 degrees	Modern chicken

The angle between the ilium and the pubis bones in the pelvis of modern Ratites is intermediate between *Archaeopteryx* and modern birds, as would be expected if:

A) Modern birds are descended from reptile-like ancestors similar to *Archaeopteryx*
B) Ratites are neotenous descendents of flying birds, in which development has been halted prematurely.

4.7. CRANIAL FEATURES OF ARCHAEOPTERYX

Using both specimens, Ostrom (1976) delineated 10 characters on the skull of Archae which it shares with other theropod dinosaurs such as *Ornitholestes*, *Compsognathus*, *Velociraptor* and perhaps *Saurornithoides*. These are:

1. A sharply tapered snout. In birds the maxilla makes up the "snout". Although the snout is shortened in Archae.
2. Long elliptical external nares (nostrils) bound almost exclusively above and below by the premaxilla and nasal (a bone at the front of the upper jaw). In birds, the nares are reduced and removed posteriorly (closer to the eyes) due to the increase in size of the premaxilla.
3. A large preorbital fossa and a large triangular postorbital fossa which contains two small openings (holes in the skull, one is a single opening in front of the orbit or eye socket, the other is behind the eye socket and has two openings). In other fossil birds, the preorbital is reduced, the postorbital has only a single opening. In modern birds, the preorbital is greatly reduced and the postorbital has a single opening.
4. A slender, nearly vertical preorbital bar separating the antorbital fossa and the orbit. Fossil and modern birds have a much more robust preorbital bar - due to the reduction of the preorbital fossa.

5. A large circular orbit which contains a large sclerotic ring (sclerotic ring is common to reptiles, birds and actinoptygian fish, and most fossil reptiles and all fossil birds have them. They are a series of ossified plates which circle the eye).
6. A thin, straight jugal bone makes up the zygomatic arch (this is the bone that runs under the eye, the cheek-bone as it extends back towards ear in humans). In modern birds the arch is composed of the quadratojugal, jugal and maxilla bones.
7. A stout quadrate of moderate length which is inclined forward. This is the bone in the upper jaw which forms part of the jaw articulation - with the articular - in reptiles. The stapes is the bone which vibrates in the ear so that we can all hear. Thus, the stapes and quadrate were attached in reptiles and it is not a great leap forward to have both the stapes and the quadrate in the ear as it is in mammals. Thus, the claim made that - in order for the bones to enter the ear in the transition from reptiles to mammals they had to go through a stage whereby the 'mammal' would have deafened itself every time it opened its mouth - is not accurate, since the condition of stapes+quadrate articulating in the jaw joint **is found in fossil and extant reptiles**.
8. A lower jaw which is unusually shallow and has a conspicuous bend behind the tooth row. Not in birds.
9. A long retroarticular process. Again a classic reptile jaw-joint feature, but shared with birds.
10. Premaxilla, lacrimal and jugal bones separate. In birds the premaxilla bone is connected to the, jugal bone; the jugal bone is connected to the, lacrimal bone.

Recently Elzanowski & Wellenhofer (1996) have highlighted some more similarities between the skull of *Archaeopteryx* and those of theropod dinosaurs:

1. A single vomer.
2. Strong curvature of the ectopterygoid hook.
3. A dorsal recess in the prootic.

Although *Archaeopteryx* does have certain skull characters which are avian, these are insufficiant to allow the characterization "totally birdlike".

Heilmann (1926) described an external mandibular fenestra (a hole in the lower jaw) bordered below by the dentary bone, which is the condition in birds rather than by the angular bone as it is in reptiles. This would indicate that Archae has a lower jaw which contained both avian (mandibular fenestra) and reptilian features (toothed lower jaw). Another important feature of the skull of Archae is the occipital condyle and the foramen magnum. In Archae these are well above the dorsal end of the quadrate. This primitive condition is characteristic of both pseudosuchians and theropods, in contrast to all later birds where the occipital condyle and foramen magnum are at the base of the skull, well below the level of the upper extremity of the quadrate. In this feature, *Archaeopteryx* was far from avian [Ostrom 1976]. Whetstone (1983) does describe the **braincase** (as opposed to the **skull**) of *Archaeopteryx* as, "typically avian", however, he also describes skull features found in *Archaeopyteryx* and not in birds.

More recently there has been some conjecture as to the structure of the quadrate bone in Archae. The quadrate from the Eichstatt specimen was described as being single headed i.e., the top of the quadrate has only one rounded knob which articulates with the skull in only one place. In all modern birds, the quadrate is double headed i.e., the top of the quadrate has two rounded knobs and therefore articulates with the skull in two places. Based on computer tomography a buried portion of the skull of the Eichstatt specimen, Haubiltz et al. (1988) suggested that Archae possessed a double headed quadrate. However, the image is poor and the presence of another bone underlying the quadrate cannot be ruled out. The only quadrate found in place in any specimen is single headed and a similar bone from the London specimen which appears to be the quadrate, is also single headed. Identification of the quadrate in the seventh specimen has confirmed that the bone was single headed [Elzanowski & Wellnhofer 1996]. Thus, the quadrate of Archae appears to be reptilian and not avian.

One is that the differences seen between Archaeopteryx, fossil birds and modern birds can be explained as just variation within birds. Allied to this is the claim that some modern birds (Ratites) share a number of morphological features with Archaeopteryx.

4.8. Colouration

In 2011, graduate student Ryan Carney and colleagues performed the first colour study on an *Archaeopteryx* specimen (Figure 5) [Carney et al., 2012]. Using scanning electron microscopy technology and energy-dispersive X-ray analysis, the team was able to detect the structure of melanosomes in the single-feather specimen described in 1861. The resultant structure was then compared to that of 87 modern bird species and was determined with a high percentage of likelihood to be black in colour. The feather studied was most probably a single covert, which would have partly covered the primary feathers on the wings. While the study does not mean that *Archaeopteryx* was entirely black, it does suggest that it had some black colouration which included the coverts. Carney pointed out that this is consistent with what we know of modern flight characteristics, in that black melanosomes have structural properties that strengthen feathers for flight [Switek, 2011]. A new analyses of *Archaeopteryx's* feathers published in the *Journal of Analytical Atomic Spectrometry* revealed that the animal may have had complex coloring in the form of light and dark colored plumage, with the tips of its flight feathers being primarily black as opposed to the entire feather being dark in color. This may have been integral for display and flight, but this remains unknown at present [Manning et al., 2013].

Figure 5. Artist's restoration illustrating one interpretation of Carney's study (Carney et al., 2012).

4.9. PLUMAGE

Specimens of *Archaeopteryx* were most notable for their well-developed flight feathers. They were markedly asymmetrical and showed the structure of flight feathers in modern birds, with vanes given stability by a barb-barbule-barbicel arrangement [Feduccia and Tordoff, 1979]. The tail feathers were less asymmetrical, again in line with the situation in modern birds and also had firm vanes. The thumb, however, did not yet bear a separately movable tuft of stiff feathers (Figure 6).

The body plumage of *Archaeopteryx* is less well documented and has only been properly researched in the well-preserved Berlin specimen. As more than one species seems to be involved, the research into the Berlin specimen's feathers does not necessarily hold true for the rest of the species of *Archaeopteryx*. In the Berlin specimen, there are "trousers" of well-developed feathers on the legs; some of these feathers seem to have a basic contour feather structure, but are somewhat decomposed (they lack barbicels as in ratites) [Christensen and Bonde, 2004]. However, in part they are firm and thus capable of supporting flight [Longrich, 2006].

A patch of pennaceous feathers is found running along its back, which was quite similar to the contour feathers of the body plumage of modern birds in being symmetrical and firm, although not so stiff as the flight-related feathers. Apart from that, the feather traces in the Berlin specimen are limited to a sort of "proto-down" not dissimilar to that found in the dinosaur *Sinosauropteryx*, being decomposed and fluffy, and possibly even appeared more like fur than like feathers (though not in their microscopic structure). These occur on the remainder of the body, as such structures are preserved and not obliterated by preparation, and the lower neck (Figure 7) [Christensen and Bonde, 2004].

Figure 6. 1880 photo of the Berlin *Archaeopteryx* specimen, showing leg feathers that were removed subsequently, during preparation (Source: Vogt, C. 1880. Archaeopteryx macrura, an Intermediate Form between Birds and Reptiles. Ibis 4:434-456).

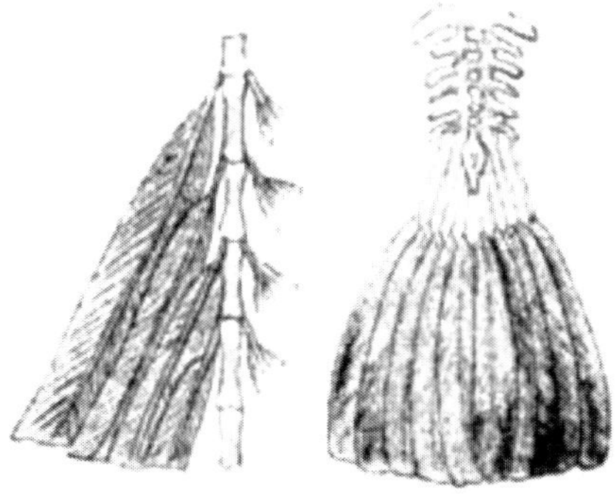

Figure 7. Anatomical illustration comparing the "frond-tail" of *Archaeopteryx* with the "fan-tail" of a modern bird.

There is no indication of feathering on the upper neck and head. While these conceivably may have been nude, this may still be an artifact of preservation. It appears that most *Archaeopteryx* specimens became embedded in anoxic sediment after drifting some time on their backs in the sea — the head and neck and the tail are generally bent downward, which suggests that the specimens had just started to rot when they were embedded, with tendons and muscle relaxing so that the characteristic shape of the fossil specimens was achieved [Reisdorf and Wuttke, 2012]. This would mean that the skin already was softened and loose, which is bolstered by the fact that in some specimens the flight feathers were starting to detach at the point of embedding in the sediment. So, it is hypothesized that the pertinent specimens moved along the sea bed in shallow water for some time before burial, the head and upper neck feathers sloughing off, while the more firmly attached tail feathers remained [Elżanowski, 2002].

4.10. Feathers

Feathers are composed of a long, tapering, central rachis, which bears closely spaced side branches called barbs. The barbs on either side of the rachis constitute a surface called the vane. Two vanes on the feather may be symmetrical (i.e., the same width) or asymmetrical, in which case the rachis appears closer to one edge of the feather than the other. The flight feathers of modern birds are typically asymmetrical, whilst body contour and semiplume feathers are symmetrical.

In modern birds, remiges, or wing feathers, are highly modified for power flight [McFarland et al. 1985], primarily in that the rachis is shifted towards the leading edge of the feather (i.e., the leading vane is thinner than the trailing vane), resulting in an asymmetric feather. The thinner or leading vane of the feather overlaps the wider or trailing vane of the feather in front of it. The trailing vane contains zones of friction barbules, which grip the overlapping feather and stop the feathers from slipping apart.

Speakman & Thomas (1994) compared the asymmetry of some of the flight feathers of Archae (flight feathers 4, 5 and 6) with those of modern flying and flightless birds and the isolated feather from Solnhofen. They found that the average asymmetry for Archae feathers was 1.25, which was lower than that of modern flying birds (lowest around 2.2), but which overlapped that of modern flightless birds. The isolated feather exhibited an asymmetry of 2.2, just within the range of modern flying birds. However, it is not known whether this feather *is* from Archae, or where on Archae the feather was situated if it is an Archae feather. Norberg (1995) correctly pointed out, that the curvature of the feather is also important in flight ability, and Archie's feathers exhibit significant curvature. It appears as though Speakmean and Thomas may have measured the wrong feathers. Speakman and Thomas appear to have numbered the primary feathers starting at the wing tip and moving towards the body, whereas current practice is to number the feathers starting from the body and moving outwards along the wing [Rietschel, 1985]. This is important, since the asymmetry of feathers decreases towards the wing tip, thus Speakman and Thomas appear to have measured the feathers with lower asymmetry. Measurements of the higher asymmetry feathers close to the body yield value of around 2.32 - which overlap the lower range for extant flying birds. The first flight feathers of birds are always highly asymmetric however, and so this result is equivocal. The bulk of the flight feathers on Archae appear to have a slightly lower asymmetry than that found in flying birds today. In reality however, feather asymmetry is a bit of a red herring when it comes to flight capability. Feathers are asymmetrical to maintain aerodynamic integrity as they deform under aerodynamic loads in flight, and thus the greatest asymmetry occurs in birds in which such loads are greatest i.e., birds with high manouverability and climbing flight. Asymmetry is a derived condition for a derived flight capability, one *Archaeopteryx* never possessed. Thus, it is not surprising that *Archaeopteryx* possessed low asymmetry feathers, but this does not preclude *Archaeopteryx* from flying.

4.11. Wing Flexibility

Consider the carniates (birds with a keeled sternum) wing. A marvel of adaptation, will able to sustain quite large animals in power flight for long distances. But what about the anatomy of the wing that allows the possibility of flight? Well simply put, 'it's all in the wrist action.'

In modern birds the shoulder joint acts as a universal joint, with a high degree of mobility for adduction (wing movement down), abduction (wing movement up) and extension (extending wing away from body) and retraction (pulling wing into body), and also significant rotation along the long axis of the humerus.

The elbow, by contrast, is a very restrictive joint. It only allows planar extension and flexure (unfolding and folding of the wing), there is no rotational component.

However, it is the wrist joint which is important. The wrist is composed of two bony carpals which control the movement of the joint. Distally beyond this, is the fused carpometacarpus - produced from a fusion of the remaining carpal bones and the metacapal bones, sometimes followed by incompletely fused manus composed of three digits. This complex, distal of the wrist joint, comprises a near-rigid platform for the attachment of the primary remiges (flight feathers), the secondary remiges attaching to the forearm.

The configuration of the wrist joint, articulating as it does on two small, round carpels, allows an amazing range of movement. This rotation is carried out by the wrist joint in birds. Note that in humans, this action is carried out by rotation of the forearm and not the wrist. The carniate wrist is much more flexible than the human wrist. This flexibility is vital to powered flight, since it allows the wrist/hand to describe a lazy figure-of-eight during a complete flap. Remember that the elbow joint allows no rotation, but the wing must be able to present a solid open wing on the down (power) stoke, and be able to rotate the wing during the recovery (up) stroke to minimise resistance. The birds accomplish this by rotating the wrist and shoulder.

At the top of the power stoke, the wrist is orientated to present the maximum surface area to the air (palm facing downward). During the power stroke, the wrist and hand describes an "S", or two arcs representing the left hand side of the upper loop and the right hand side of the lower loop of an "8". At the bottom of the power stroke the wrist and hand is at the bottom of the "8". During the start of the recovery stroke the wrist reorients at 90 degrees, thus presenting the minimum surface area to the air. The wing then completes the "8" by describing the left hand side of the lower loop and the right hand side of the upper loop.

Folding of the wing occurs by contraction of the M. biceps brachii, which is attached to the ulna. This movement automatically folds the metacarpus and hand downward and inward, so that folding at the wrist joint involves the M. biceps brachii.

Now, the point of all this is: Does this arrangement occur in Archae? Does this arrangement occur in maniraptoran theropods? If so, what possible use could this particular wrist joint be to maniraptoran theropods? Could this be an example of preadaptation?

Firstly, the wrist joint of Archae does indeed exhibit *some* (but not all) of the features seen in the modern carniate wrist joint. Most notably a carpal, hemispherical in shape, the flat portion of which articulated with the carpometacarpus, leaving the rounded end to articulate with the wrist joint. This provides a flexible wrist joint, although not as flexible as in modern carniates.

Secondly, this configuration is found in several theropod taxa, such as *Deinonychus*, *Velociraptor* and *Stenonychosaurus*. There are two problems here, firstly all these forms postdate Archae, they are all Cretaceous forms; secondly, what would be the use of a nifty flexible wrist to a bunch of theropods which are all teeth and toenails? It was a problem that this feature occurred in Cretaceous theropods and in the Late Jurassic Archae only, since to be an important feature and to show a close link between theropods and Archae, theropods coeval with Archae must have had this particular arrangement. Introducing *Coelurus fragilis* a late Jurassic theropod from America. This form does indeed possess the a very similar wrist joint (Ostrom 1980) strengthening not only the theropod-bird link, but also the theropod-Archae-bird link.

The second problem is slightly harder. Why would such a wrist joint be retained in maniraptoran theropods which are dentally and toenail minded when it comes to killing? Ostrom (1980, 1995) suggests that the flexibility would come in useful where a grasping action of the hands could augment the killing action of the teeth and/or toenails. In modern birds the biceps are involved in the folding of the wing. If this were extended to the theropods, the action of pulling the prey into the body would tend to *strengthen* the grip of the theropod by bringing into play the largest muscle on the arm - the biceps, rather than just the wrist muscles. Such a scenario would provide a distinct advantage and thus the retention of the wrist joint in the theropods is not surprising. Thus, it appears that the flexible wrist joint may have first appeared in theropod dinosaurs as an aid in grasping and retaining prey and was secondarily co-opted by the birds as a mechanism which allowed flight. It may well be that not only were the theropod dinosaurs the ancestors of birds, but that they are the only group which *could* be the ancestor to modern birds.

However, the wrist of Archae does not appear to have the flexibility to allow power flight.

4.12. Shoulder Joint

The shoulder joint is also important to modern birds because, not only does it allow full adduction (wing movement down), but also full abduction (wing movement up) i.e., the wing can be drawn up over the back (try it and see how difficult it is). This full range of movement is important in allowing modern birds to power fly. It turns out that the shoulder joint of Archae appears to be intermediate in orientation between some coelurosaurs (theropod dinosaurs e.g., *Deinonychus*) and birds [Jenkins 1993], just as one would expect from a transitional form. However, whilst the shoulder joint provided a substantial degree of arm (wing) elevation, it is not sufficient to allow full abduction as in modern birds. Thus, the shoulder joint of Archae does not appear to allow the full range of movement needed for power flying.

4.13. Muscle Mass

In modern birds the flight muscles (the pectoralis) comprise up to 35% of the weight of the animal. It is this massive muscle mass which provides the power needed for flapping powered flight and the ability to take off from a standing start. Ruben (1991) used a pectoralis muscle mass estimate for Archae of approximately 9% of the body weight. This is much lower than in modern birds, but assumed that the muscle attached to the cartilaginous chest wall and not a keel since a keel is unknown in Archae. The keel is a bony extension of the sternum for the attachment of excess flight muscle in modern birds. If the muscles did attach to the chest wall in Archae, the length of the pectoralis muscles would have been too short to allow full abduction. However, the presence of a small, cartilagenous (and thus not preserved) keel is not out of the question. (Specimens of Archie from the Solnhofen Limestone show no evidence of a sternum. However, recent finds of Archie-like forms from China suggest that Archae did in fact possess a sternum. Therefore, the 9% figure is a minimum value. This value is well below that for modern birds. Rubin suggested that the maximum power output from the locomotor muscles of extant active reptiles is twice that of birds and mammals, indicating that, if Archae possessed a reptilian physiology, powered flight was a possibility, even with such a small muscle mass. Speakman (1993) questioned this, indicating that the power output of reptiles suggested by Rubin was over-exaggerated, and that Archae would not have had the power output to sustain power flight. Also Speakman pointed out that the theoretical lower limit of pectoralis muscle mass required for take-off from a standing start was approximately 16% of body weight. Thus, Archae was not capable of power flight or take-off from a standing start.

The supracoracoideus is an important muscle in modern birds because it is this muscle which elevates or abducts the wing during power flight. Archae did not possess the supracoracoideus arrangement found in modern birds [Ostrom 1974]. Thus, Archae does not appear to have had the muscle mass nor the muscle arrangement to allow power flight, nor be able to take off from a standing start.

Virtually all modern flying birds are capable of power flight. Power flight is controlled, flapping flight at low speeds with the ability to take off from a standing start. This type of flying requires a large amount of muscle mass and is energetically very expensive. Hence, modern birds have a pectoralis muscle mass which comprises approximately 20-35% of body mass. In order to accommodate this extra mass, a keel has developed on the sternum. In other evolutionary developments, flight feathers have become more asymmetric to assist in power flight, and the supracoracoideus muscle system took on the function of assisting in the lifting or abduction of the wing, since this is difficult at low flying speeds.

The structure of the wing of Archae suggests that it was incapable of power, flapping flight at low speeds. However, the amount of adduction and abduction, along with the increased flexibility of the wrist compared with theropod dinosaurs, suggests that some sort of flapping was possible, to keep Archae airborn at normal cruising speeds (approximately 7-9 m/s). Whilst the low pectoralis muscle mass appears to preclude power flight and take off from a standing start, it should be pointed out that some flying birds do possess a muscle mass lower than the 16% by lower weight limit. These include some grebes, which compensate for the lack of muscle mass by having a long running take-off [Speakman 1993]. Archae is well equipped with a terrestrial locomotion apparatus. Thus, it is possible that Archae could circumvent this problem in the same manner as the grebes.

Flying at normal cruising speeds requires much less energy than power, flapping flight and thus needs less muscle mass to sustain it. As an added bonus, at normal cruising speeds the wing is elevated (abducted) in the upstroke by aerodynamic forces [Rayner 1988], thus circumventing the need for the supracoracoideus arrangement necessary for power, flapping flight.

4.14. Fossil Record

Most of these eleven fossils include impressions of feathers. Because these feathers are of an advanced form (flight feathers), these fossils are evidence that the evolution of feathers began before the Late Jurassic [Wellnhofer, 2004]. The type specimen of *Archaeopteryx* was discovered just two years after Charles Darwin published '*On the Origin of Species*'. *Archaeopteryx* seemed to confirm Darwin's theories and has since become a key piece of evidence for the origin of birds, the transitional fossils debate, and confirmation of evolution. Indeed, further research on dinosaurs from the Gobi Desert and China has since provided more evidence of a link between *Archaeopteryx* and the dinosaurs, such as the Chinese feathered dinosaurs. *Archaeopteryx*is close to the ancestry of modern birds, and it shows most of the features one would expect in an ancestral bird. It may not be the direct ancestor of living birds, however, and it is uncertain how much evolutionary divergence already was present among other birds at the time.

4.15. *ARCHAEOPTERYX'S* ANCESTRY

Apart from the theropod ancestry, there have been three other suggested ancestral groups for Archae, the Ornithopoda, the Pseudosuchia and the Sphenosuchidae.

The Ornithopod ancestry theory was based on the tarsus and pelvis in various dinosaurs, which were supposed to approach the condition found in birds. However, most of the taxa used in the original theory in 1883-4 are now known to be saurischian rather than ornithischian. Whilst there was a tendency for the reduction of the hallux (big toe) in some ornithischians, it was never apparently opposable as it is in Archae. It was suggested that Archae had a pelvis which was very ornithischian, however, recent work has shown that this is an artifact of post mortem movement of elements of the pelvic region and there is very little similarity to the ornithischian pelvis. With this, the link between ornithischians and Archae disappears and so it is now thought that there is no link between them.

It was suggested that birds arose from an almost unknown Middle or Late Triassic reptile group, the Sphenosuchidae [Walker 1972, 1974], which also gave rise to the crocodiles. The theory is based on a single specimen of *Sphenosuchus*. However, in their general configurations, neither the skull and jaws, nor the scapulo-coracoid of *Sphenosuchus* is remotely suggestive of affinities with Archae [Ostrom, 1976]. More recently, the main proponent of the crocodile-bird hypothesis has stated that the hypothesis "has become so tenuous that it is very difficult to sustain." [Walker, 1985]

The pseudosuchian link was very much favoured earlier this century. However this was, by and large, based on Heilmann's reconstruction of the Berlin specimen, which has since proven to be inaccurate. According to Ostrom (1976):

> There are very few anatomical resemblances between Archaeopteryx and any pseudosuchian. In fact, only in one feature does any pseudosuchian resemble Archaeopteryx more closely than does any theropod, this being the tibia to femur ratio in Scleromochlus, Lagosuchus and Lagerpeton, where the tibia is from 20% to 30% longer than the femur. Amongst theropods, only in struthiomimids, Compsognathus, Microvenator and Deinonychus are the tibia longer than the femur but by only 10% to 15%. In all other features, the closest resemblance to the morphology preserved in Archaeopteryx is found in coelurosaurian theropods.

4.16. PROTOAVIS

Some people like to claim that the finding of a fossil bird from the Triassic of Texas (*Protavis*) proves that Archae cannot be transitional between dinosaurs and birds because *Protoavis* predates Archae by 75 million years. There are major problems with *Protoavis* (Figure 8). On the) has this to say: The only published material from the fossil is a monograph in the Philosophical Transactions of the Royal Society of London. However, this only describes the head. This is badly crushed and all the pieces have been extracted from the matrix, rendering precise placement of the pieces open to question. The description is done from an avian viewpoint, with no counterview used. The skull is so badly crushed that diagnostic features are not preserved. Therefore, the published material does not support the view that this is a bird. Indeed a viewing of the fossil showed that the diagnostic features which could identify the fossil either way are badly crushed and it is doubtful whether any definitive statement could be supported by the fossil. It may be a bird, it may not [Chatterjee 1991; Ostrom (1991].

4.17. PALEOECOLOGY

The richness and diversity of the Solnhofen limestones in which all specimens of *Archaeopteryx* have been found have shed light on an ancient Jurassic Bavaria strikingly different from the present day. The latitude was similar to Florida, though the climate was likely to have been drier, as evidenced by fossils of plants with adaptations for arid conditions and a lack of terrestrial sediments characteristic of rivers. Evidence of plants, although scarce, include cycads and conifers while animals found include a large number of insects, small lizards, pterosaurs, and *Compsognathus* [Chiappe, 2007].

The excellent preservation of *Archaeopteryx* fossils and other terrestrial fossils found at Solnhofen indicates that they did not travel far before becoming preserved [Davis and Briggs, 1998]. The *Archaeopteryx* specimens found are likely therefore, to have lived on the low islands surrounding the Solnhofen lagoon rather than to have been corpses that drifted in from farther away. *Archaeopteryx* skeletons are considerably less numerous in the deposits of Solnhofen than those of pterosaurs, of which seven genera have been found [Bartell et al., 1994]. The pterosaurs included species such as *Rhamphorhynchus* belonging to the Rhamphorhynchidae, the group which dominated the niche currently occupied by seabirds, and which became extinct at the end of the Jurassic. The pterosaurs, which also included *Pterodactylus*, were common enough that it is unlikely that the specimens found are vagrants from the larger islands 50 km (31 mi) to the north [Paul, 2002].

The islands that surrounded the Solnhofen lagoon were low lying, semi-arid, and sub-tropical with a long dry season and little rain [Buisonjé, 1985]. The closest modern analogue for the Solnhofen conditions is said to be Orca Basin in the northern Gulf of Mexico, although that is much deeper than the Solnhofen lagoons [Bartell et al., 1994]. The flora of these islands was adapted to these dry conditions and consisted mostly of low (3 m [10 ft]) shrubs [Paul, 2002]. Contrary to reconstructions of *Archaeopteryx* climbing large trees, these seem to have been mostly absent from the islands; few trunks have been found in the sediments and fossilized tree pollen is also absent.

The lifestyle of *Archaeopteryx* is difficult to reconstruct and there are several theories regarding it. Some researchers suggest that it was primarily adapted to life on the ground [Ostrom, 1976], while other researchers suggest that it was principally arboreal [Feduccia, 1993]. The absence of trees does not preclude *Archaeopteryx* from an arboreal lifestyle as several species of extant bird live exclusively in low shrubs. Various aspects of the morphology of *Archaeopteryx* point to either an arboreal or ground existence, including the length of its legs and the elongation in its feet. Some authorities consider it likely to have been a generalist capable of feeding in both shrubs and open ground, as well as alongside the shores of the lagoon [Paul, 2002]. It most likely hunted small prey, seizing it with its jaws if it was small enough or with its claws if it was larger.

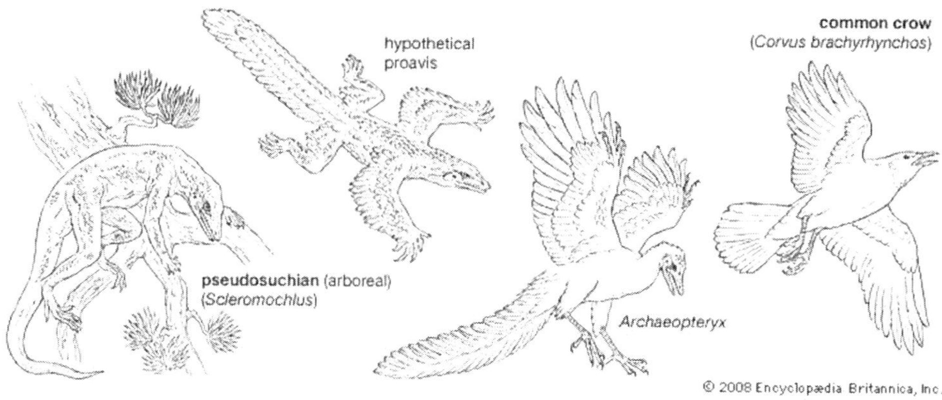

Figure 8. Evolution of *Archaeopteryx* from reptilian ancestor via Protoavis ancestor (Source: Encyclopedia Britannica, Inc.).

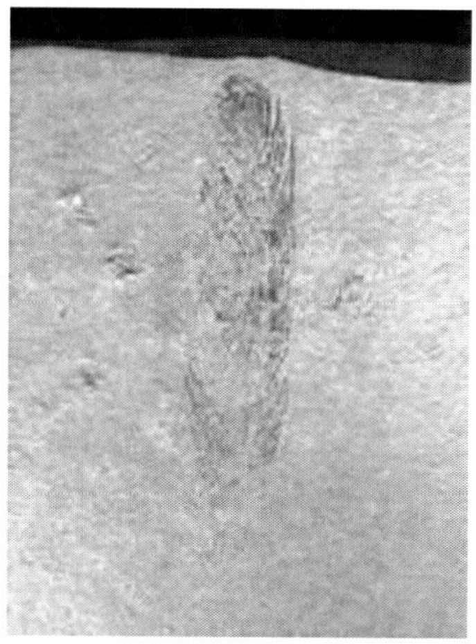

Figure 9. The single feather (Raab, H.).

Over the years, twelve body fossil specimens of *Archaeopteryx* and a feather that may belong to it have been found. All of the fossils come from the limestone deposits, quarried for centuries, near Solnhofen, Germany (Figure 9) [Chiappe, 2007].

The initial discovery, a single feather, was unearthed in 1860 or 1861 and described in 1861 by Christian Erich Hermann von Meyer. It is currently located at the Humboldt Museum für Naturkunde in Berlin. This is generally assigned to *Archaeopteryx* and was the initial holotype, but whether it is a feather of this species, or another, as yet undiscovered, proto-bird is unknown. There are some indications it is indeed not from the same animal as most of the skeletons (the "typical" *A. lithographica*) [Griffiths, 1996].

The first skeleton, known as the **London Specimen** (BMNH 37001), was unearthed in 1861 near Langenaltheim, Germany, and perhaps given to a local physician Karl Häberlein in return for medical services. He then sold it for £700 to the Natural History Museum in London [Chiappe, 2007]. Missing most of its head and neck, it was described in 1863 by Richard Owen as *Archaeopteryx macrura*, allowing for the possibility it did not belong to the same species as the feather. In the subsequent fourth edition of his *On the Origin of Species*,

Charles Darwin described how some authors had maintained "that the whole class of birds came suddenly into existence during the eocene period; but now we know, on the authority of professor Owen, that a bird certainly lived during the deposition of the upper greensand; and still more recently, that strange bird, the Archeopteryx, with a long lizard-like tail, bearing a pair of feathers on each joint, and with its wings furnished with two free claws, has been discovered in the oolitic slates of Solnhofen. Hardly any recent discovery shows more forcibly than this how little we as yet know of the former inhabitants of the world [Darwin, 1859].

The Greek term "pteryx" primarily means "wing", but can also designate merely "feather". Von Meyer suggested this in his description. At first he referred to a single feather which appeared to resemble a modern bird's remex (wing feather), but he had heard of and been shown a rough sketch of the London specimen, to which he referred as a "*Skelett eines mit ähnlichen Federn bedeckten Tieres*" ("skeleton of an animal covered in similar feathers"). In German, this ambiguity is resolved by the term *Schwinge* which does not necessarily mean a wing used for flying. *Urschwinge* was the favored translation of *Archaeopteryx* among German scholars in the late nineteenth century. In English, "ancient pinion" offers a rough approximation.

Since then twelve specimens have been recovered:

The **Berlin Specimen** (HMN 1880) was discovered in 1874 or 1875 on the Blumenberg near Eichstätt, Germany, by farmer Jakob Niemeyer (Figure 11). He sold this precious fossil for the money to buy a cow in 1876, to inn-keeper Johann Dörr, who again sold it to Ernst Otto Häberlein, the son of K. Häberlein. Placed on sale between 1877 and 1881, with potential buyers including O. C. Marsh of Yale University's Peabody Museum, it eventually was bought for 20,000 Goldmark by the Humboldt Museum für Naturkunde, where it now is displayed. The transaction was financed by Ernst Werner von Siemens, founder of the famous company that bears his name [Chiappe, 2007]. Described in 1884 by Wilhelm Dames, it is the most complete specimen, and the first with a complete head. In 1897 it was named by Dames as a new species, *A. siemensii*; though often considered a synonym of *A. lithographica,* A. siemensii *several 21st century studies have concluded that it is a distinct species including the Berlin, Munich, and Thermopolis specimens* [Mayr et al., 2007].

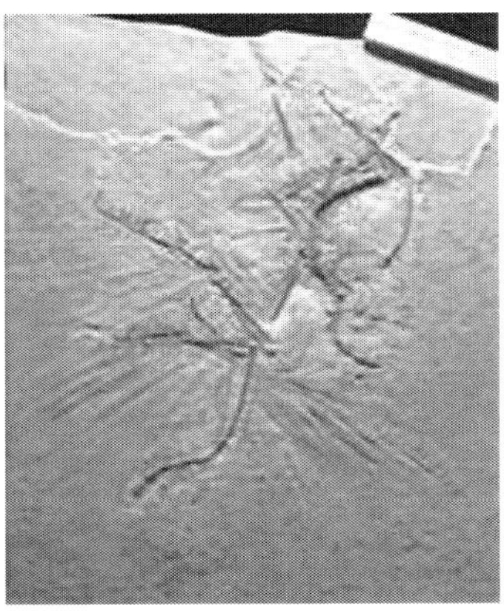

(Source: Callaway, E., NPG).

Figure 10. Cast of the Maxberg Specimen .

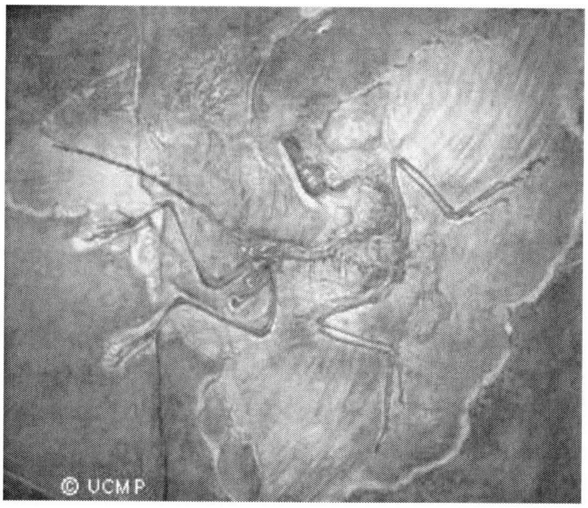

Figure 11. Cast of the Berlin specimen of *Archaeopteryx lithographica*, from the collections of UCMP. Original at Humboldt University, Berlin.

(Source: von Meyer, 1857).

Figure 12. Slab of the Haarlem Specimen.

Composed of a torso, the **Maxberg Specimen** (S5) was discovered in 1956 near Langenaltheim. It was brought to the attention of professor Florian Heller in 1958 and described by him in 1959 (Figure 10). The specimen is missing its head and tail, although the rest of the skeleton is mostly intact. Although it was once exhibited at the Maxberg Museum in Solnhofen, it is currently missing. It belonged to Eduard Opitsch, who loaned it to the museum until 1974. After his death in 1991, it was discovered that the specimen was missing and may have been stolen or sold.

The **Haarlem Specimen** (TM 6428, also known as the *Teyler Specimen*) was discovered in 1855 near Riedenburg, Germany, and described as a *Pterodactylus crassipes* in 1857 by von Meyer. It was reclassified in 1970 by John Ostrom and is currently located at the Teylers Museum in Haarlem, Netherlands. It was the very first specimen, despite the classification error. It is also one of the least complete specimens, consisting mostly of limb bones, isolated cervical vertebrae, and ribs (Figure 12).

The **Eichstätt Specimen** (JM 2257) was discovered in 1951 near Workerszell, Germany and described by Peter Wellnhofer in 1974. Currently located at the Jura Museum in Eichstätt, Germany, it is the smallest known specimen and has the second best head. It is possibly a separate genus (*Jurapteryx recurva*) or species (*A. recurva*).

The **Solnhofen Specimen** (BSP 1999) was discovered in the 1970s near Eichstätt, Germany, and described in 1988 by Wellnhofer. Currently located at the Bürgermeister-Müller-Museum in Solnhofen, it originally was classified as *Compsognathus* by an amateur collector, the same mayor Friedrich Müller after which the museum is named. It is the largest specimen known and may belong to a separate genus and species, *Wellnhoferia grandis*. It is missing only portions of the neck, tail, backbone, and head.

The **Munich Specimen** (S6, formerly known as the *Solenhofer-Aktien-Verein Specimen*) was discovered on 3 August 1992 near Langenaltheim and described in 1993 by Wellnhofer (Figure 13). It currently is located at the Paläontologisches Museum München in Munich, to which it was sold in 1999 for 1.9 million Deutschmark. What was initially believed to be a bony sternum turned out to be part of the coracoid [Wellnhofer and Tischlinger, 2004], but a cartilaginous sternum may have been present. Only the front of its face is missing. It has been used as the basis for a distinct species, *A. bavarica* [Wellnhofer, 2008], but more recent studies suggest it belongs to *A. siemensii* [Mayr et al., 2007].

(Source: Wikipedia) .

Figure 13. Munich Specimen.

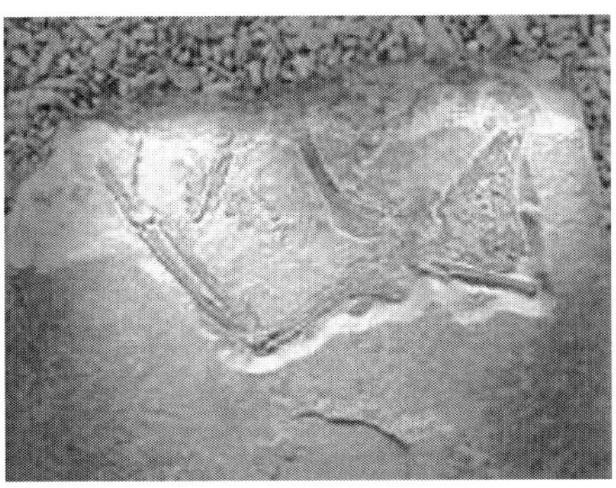

(Source: Wikipedia).

Figure 14. Daiting Specimen.

An eighth, fragmentary specimen was discovered in 1990, not in Solnhofen limestone, but in somewhat younger sediments at Daiting, Suevia. Therefore it is known as the **Daiting Specimen**, and had been known since 1996 only from a cast, briefly shown at the *Naturkundemuseum* in Bamberg (Figure 14). The original was purchased by palaeontologist Raimund Albertsdörfer in 2009. It was on display for the first time with six other original fossils of *Archaeopteryx* at the *Munich Mineral Show* in October 2009. A first, quick look by scientists indicates that this specimen might represent a new species of *Archaeopteryx*. It was found in a limestone bed that was a few hundred thousand years younger than the other finds.

Another fragmentary fossil was found in 2000. It is in private possession and, since 2004, on loan to the Bürgermeister-Müller Museum in Solenhofen, so it is called the **Bürgermeister-Müller Specimen**; the institute itself officially refers to it as the "Exemplar of the families Ottman and Steil, Solnhofen". As the fragment represents the remains of a single wing of *Archaeopteryx*, the popular name of this fossil is "chicken wing" (Figure 15).

Long in a private collection in Switzerland, the **Thermopolis Specimen** (WDC CSG 100) was discovered in Bavaria and described in 2005 by Mayr, Pohl, and Peters. Donated to the Wyoming Dinosaur Center in Thermopolis, Wyoming, it has the best-preserved head and feet; most of the neck and the lower jaw have not been preserved. The "Thermopolis" specimen was described in the December 2, 2005 *Science* journal article as "A well-preserved *Archaeopteryx* specimen with theropod features"; it shows that the *Archaeopteryx* lacked a reversed toe—a universal feature of birds—limiting its ability to perch on branches and implying a terrestrial or trunk-climbing lifestyle [Mayr et al., 2005]. This has been interpreted as evidence of theropod ancestry. In 1988, Gregory S. Paul claimed to have found evidence of a hyperextensible second toe, but this was not verified and accepted by other scientists until the Thermopolis specimen was described [Paul, 1988]. Until now, the feature was thought to belong only to the species' close relatives, the deinonychosaurs [Switek, 2011]. The Thermopolis Specimen was assigned to *Archaeopteryx siemensii* in 2007. The specimen is considered the most complete and well preserved *Archaeopteryx* remains [Mayr et al., 2007].

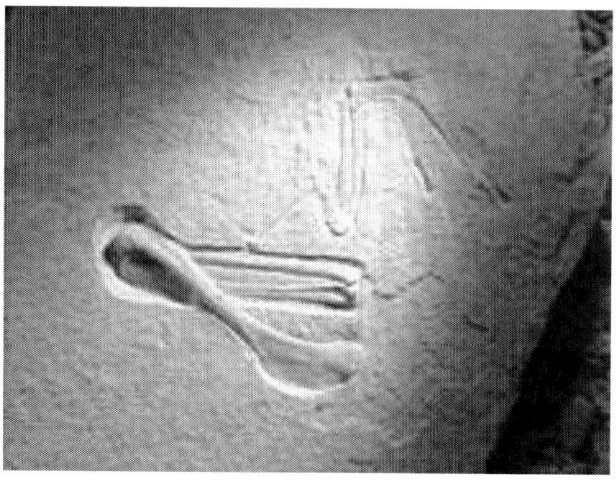

(Source: Wikipedia).

Figure 15. Bürgermeister-Müller ("chicken wing") Specimen.

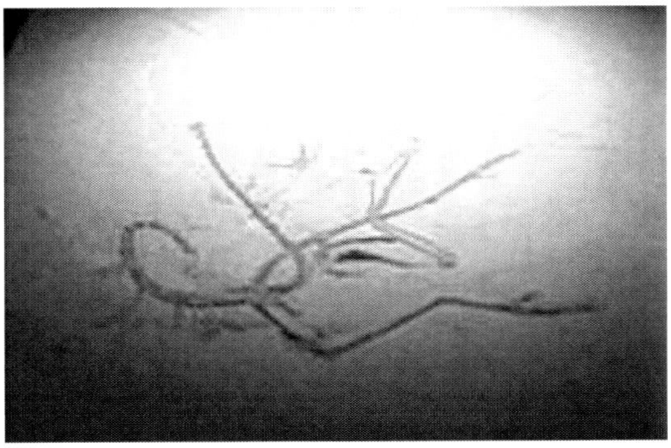

(Source: Wikipedia).

Figure 16. The eleventh specimen.

The discovery of an eleventh specimen was announced in 2011. It is said to be one of the more complete specimens, but is missing the skull and one forelimb. It is privately owned and has yet to be given the name or described scientifically (Figure 16).

A twelfth specimen was discovered by amateur collectors already in 2010 at the Schamhaupten quarry, but the finding was only announced in February 2014. It is not yet scientifically described [Swinton, 1970].

4.18. Replica of the London Specimen

In 2004, scientists analyzing a detailed CT scan of the braincase of the London *Archaeopteryx* concluded that its brain was significantly larger than that of most dinosaurs, indicating that it possessed the brain size necessary for flying. The overall brain anatomy was reconstructed using the scan. The reconstruction showed that the regions associated with vision took up nearly one-third of the brain. Other well-developed areas involved hearing and muscle coordination [Witmer, 2004]. The skull scan also revealed the structure of its inner ear. The structure more closely resembles that of modern birds than the inner ear of non-avian reptiles. These characteristics taken together, suggest that Archaeopteryx had the keen sense of hearing, balance, spatial perception, and coordination needed to fly [Alonso et al., 2004]. *Archaeopteryx* had a cerebrum-to-brain-volume ratio 78% of the way to modern birds from the condition of non-coelurosaurian dinosaurs such as *Carcharodontosaurus* or *Allosaurus*, which had a crocodile-like anatomy of the brain and inner ear [Larsson, 2001]. Newer research show that while the species' brain was more complex than that of more primitive theropods, it had a more generalized brain volume among maniraptoran dinosaurs, even smaller than that of other non-avian dinosaurs in several instances, which indicates the neurological capabilities required for flight, already a common trait in the maniraptoran clade.

Archaeopteryx continues to play an important part in scientific debates about the origin and evolution of birds. Some scientists see it as a semi-arboreal climbing animal, following the idea that birds evolved from tree-dwelling gliders (the "trees down" hypothesis for the evolution of flight proposed by O. C. Marsh). Other scientists see *Archaeopteryx* as running quickly along the ground, supporting the idea that birds evolved flight by running (the "ground up" hypothesis proposed by Samuel Wendell Williston). Still others suggest that *Archaeopteryx* might have been at home both in the trees and on the ground, like modern crows, and this latter view is what currently, is considered best-supported by morphological characters. Altogether, it appears that the species was not particularly specialized for running on the ground or for perching. A scenario outlined by Elżanowski in 2002 suggested that *Archaeopteryx* used its wings mainly to escape predators by glides punctuated with shallow downstrokes to reach successively higher perches, and alternatively, to cover longer distances (mainly) by gliding down from cliffs or treetops [Elżanowski, 2002].

4.19. Palentology

Flight

As in the wings of modern birds, the flight feathers of *Archaeopteryx* were somewhat asymmetrical and the tail feathers were rather broad. This implies that the wings and tail were used for lift generation, however, it is unclear whether *Archaeopteryx* was simply a glider, or capable of flapping flight. The lack of a bony breastbone suggests that *Archaeopteryx* was not a very strong flier, but flight muscles might have attached to the thick, boomerang-shaped wishbone, the plate like coracoids, or perhaps, to a cartilaginous sternum. The sideways orientation of the glenoid (shoulder) joint between scapula, coracoid, and humerus — instead of the dorsally angled arrangement found in modern birds—might indicate that *Archaeopteryx* was unable to lift its wings above its back—a requirement for the upstroke found in modern flapping flight. According to a study by Philip Senter in 2006, *Archaeopteryx* was indeed unable to use flapping flight as modern birds do, but it may well have used a downstroke-only flap-assisted gliding technique [Senter, 2006].

Archaeopteryx wings were relatively large, which would have resulted in a low stall speed and reduced turning radius. The short and rounded shape of the wings would have increased drag, but also could have improved the *Archaeopteryx'* ability to fly through cluttered environments such as trees and brush (similar wing shapes are seen in birds which fly through trees and brush, such as crows and pheasants). The presence of "hind wings", asymmetrical flight feathers stemming from the legs similar to those seen in dromaeosaurids such as *Microraptor*, also have added to the aerial mobility of *Archaeopteryx*. The first detailed study of the hind wings by Longrich in 2006, suggested that the structures formed up to 12% of the total airfoil. This would have reduced stall speed by up to 6% and turning radius by up to 12% [Longrich, 2006].

The feathers of *Archaeopteryx* were asymmetrical. This has been interpreted as evidence that it was a flyer, as flightless birds tend to have symmetrical feathers, however some scientists, including Thomson and Speakman, have questioned this. They studied more than 70 families of living birds, and found that some flightless types do have a range of asymmetry in their feathers, and that the feathers of *Archaeopteryx* fall into this range [Speakman and Thomson, 1994]. However, the degree of asymmetry with *Archaeopteryx* is more typical for slow flyers than for flightless birds [Norberg, 1995].

In 2010, Robert L. Nudds and Gareth J. Dyke in the journal *Science* published a paper in which they analysed the rachises of the primary feathers of *Confuciusornis* and

Archaeopteryx. The analysis suggested that the rachises on these two genera were thinner and weaker than those compared to modern birds relative to body weight. The authors determined that *Archaeopteryx*, along with *Confuciusornis*, were unable to use flapping flight [Nudds, 2010]. This study was criticized by Philip J. Currie and Luis Chiappe, they suggested that it is difficult to measure the rachises of fossilized feathers, and Currie speculated that *Archaeopteryx* and *Confuciusornis* must have been able to fly to some degree, as their fossils are preserved in what is believed to have been in marine or lake sediments, suggesting that they must have been able to fly over deep water [Balter, 2010]. Gregory Paul also disagreed with the study, arguing in a 2010 response that Nudds and Dyke had overestimated the weights of these early birds, and that more accurate weight estimates allowed powered flight even with relatively narrow rachises. Nudds and Dyke had assumed a weight of 250 grams for the Munich specimen *Archaeopteryx*, a young juvenile, based on published weight estimates of larger specimens. Paul argued that a more reasonable body weight estimate for the Munich specimen is about 140 grams. Paul also criticized the measurements of the rachises themselves, noting that the feathers in the Munich specimen are poorly preserved. Nudds and Dyke reported a diameter of 0.75 millimetre (0.030 in) for the longest primary feather, which Paul could not confirm using photographs. Paul measured some of the inner primary feathers, finding rachises 1.25–1.4 mm across [Paul, 2010]. Despite these criticisms, Nudds and Dyke (2010) stood by their original conclusions. They claimed that Paul's statement, that an adult *Archaeopteryx* would have been a better flyer than the juvenile Munich specimen, was dubious. This, they reasoned, would require even thicker rachis, evidence for which has not yet been presented [Dyke and Nudds, 2010]. Another possibility is that they hadn't achieved true flight, but instead used their wings as aid for extra lift while running over water after the fashion of the basilisk lizard, which could explain their presence in lake and marine deposits [Videler, 2005].

4.20. Growth

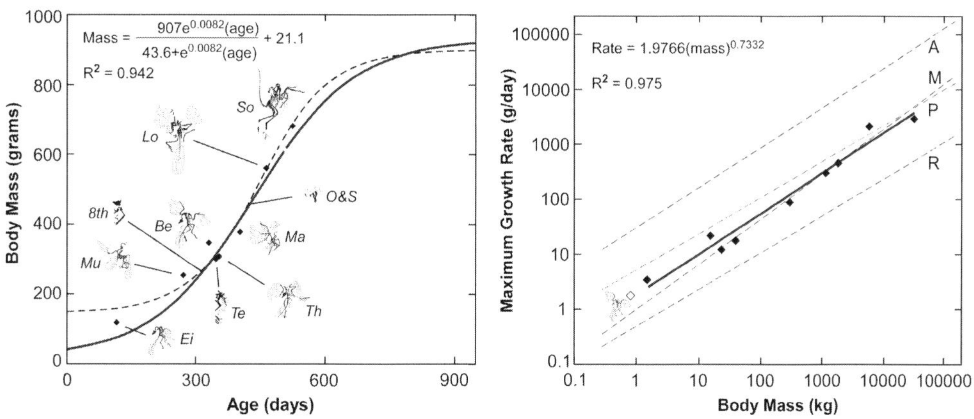

Figure 17. Growth trends compared with other dinosaurs and birds (Erickon et al., 2009).

A histological study by Erickson, Norell, Zhongue, and others in 2009 showed that *Archaeopteryx* grew relatively slowly compared to modern birds, based on growth lines found in slices of *Archaeopteryx* bones [Erickson et al., 2009]. According to their study, all known skeletons of *Archaeopteryx* came from juvenile specimens. They estimated that *Archaeopteryx* reached adult size in about 970 days (there were 375 days in a Late Jurassic year), and weighed between 0.8 – 1 kilogram. This study also found that the birds *Jeholornis* and *Sapeornis* grew slowly, as did the dromaeosaurid *Mahakala*. The birds *Confuciusornis* and *Ichthyornis*, however, grew very quickly, following a growth trend similar to that of modern birds. One of the few modern birds that exhibit slow growth is the flightless kiwi, and the authors speculated that *Archaeopteryx* and the kiwi had similar metabolisms (Figure 17) [Erickson et al., 2009].

4.21. Daily Activity Patterns

Comparisons between the scleral rings of *Archaeopteryx* and modern birds and reptiles indicate that it may have been diurnal, similar to most modern birds [Scmitz and Motani, 2011].

4.22. Classification

Today, fossils of the genus *Archaeopteryx* are usually assigned to one or two species, A. *lithographic* and *A. siemensii*, but their taxonomic history is complicated. Dozens of names have been published for the handful of specimens, most of which are simply spelling errors (lapsus). As interpreted today, the name *A. lithographica* only referred to the single feather described by von Meyer. In 1954, however, Gavin de Beer concluded that the London specimen was the holotype. In 1960, Swinton accordingly proposed that the name *Archaeopteryx lithographica* be placed on the official genera list making the alternative names *Griphosaurus* and *Griphornis* invalid [Swinton, 1960]. The ICZN, implicitly accepting de Beer's standpoint, did indeed suppress the plethora of alternative names initially proposed for the first skeleton specimens (1961), which mainly resulted from the acrimonious dispute between von Meyer and his opponent Johann Andreas Wagner (whose *Griphosaurus problematicus* — "problematic riddle-lizard", was a vitriolic sneer at von Meyer's *Archaeopteryx*) [Wagner, 1861]. In addition, in 1977 the first specific name of the Haarlem specimen, *crassipes*, described by von Meyer as a pterosaur before its true nature was realized, was also suppressed [ICZN, 1977].

It has been noted that the feather, the first specimen of *Archaeopteryx* described, does not correspond well with the flight-related feathers of *Archaeopteryx*. It certainly is a flight feather of a contemporary species, but its size and proportions indicate that it may belong to another, smaller species of feathered theropod, of which only this feather is known so far. As in the early twenty-first century, the feather was seen as the type specimen, this would have created significant nomenclatorial confusion because the name *Archaeopteryx* should then no longer be applied to the skeletons. In 2007, two sets of scientists therefore petitioned the ICZN requesting that the London specimen explicitly be made the type by designating it as the new holotype specimen, or neotype [Bock and Bühler, 2007]. This suggestion was upheld by the ICZN (2011) after four years of debate, and the London specimen was designated the neotype on October 3, 2011.

Classification

Kingdom: Animalia (animals)

Phylum: Chordata (having a hollow nerve chord ending in a brain)

Class: Archosauria (diapsids with socket-set teeth, etc.)

Order: Saurischia (lizard-hipped dinosaurs) [or Order Archaeopterygiformes]

Suborder: Theropoda - bipedal carnivores

Tetanura - advanced theropods with three fingers

Infraorder: Coelurosauria, lightly-built fast-running predators with hollow bones and large brains

Family: Archaeopteridae

Genus: Archaeopteryx

Species *A. lithographica* (type species named by Hermann von Meyer, 1861), A. bavarica [Wellnhofer, 1993].

(Source: Wikipedia).

Figure 18. The Thermopolis Specimen.

4.23. Species

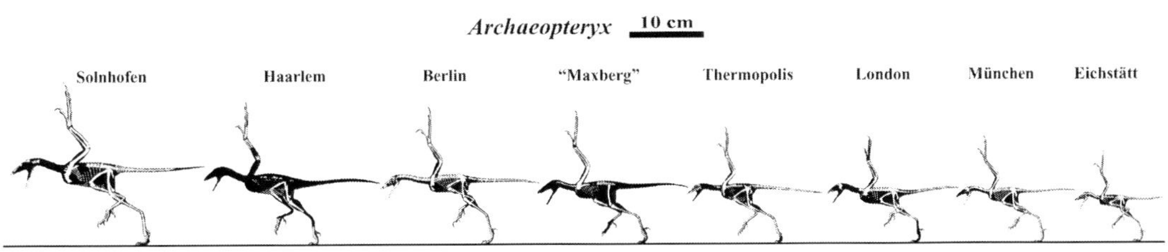

(Source: Wikipedia).

Figure 19. Skeletal restorations of 8 specimens.

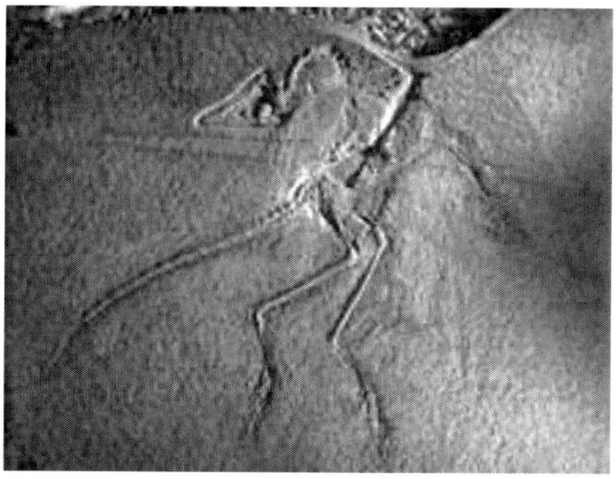

(Source: Wikipedia).

Figure 20. Eichstätt Specimen, once considered a distinct genus, *Jurapteryx*.

(Source: Wikipedia).

Figure 21. The Solnhofen Specimen, by some considered as belonging to the genus *Wellnhoferia*.

It has been argued that all the specimens belong to the same species, *A. lithographica* [Godefroit et al., 2013]. However, differences do exist among the specimens, and while some researchers regard these as due to the different ages of the specimens, some may be related to actual species diversity. In particular, the Munich, Eichstätt, Solnhofen, and Thermopolis specimens differ from the London, Berlin, and Haarlem specimens in being smaller or much larger, having different finger proportions, having more slender snouts lined with forward-pointing teeth, and possible presence of a sternum (Figure 18,19). Due to these differences, most individual specimens have been given their own species name at one point or another. The Berlin specimen has been designated as *Archaeornis siemensii*, the Eichstätt specimen as *Jurapteryx recurva* (Figure 20), the Munich specimen as *Archaeopteryx bavarica*, and the Solnhofen specimen as *Wellnhoferia grandis* (Figure 21) [Elżanowski, 2002].

In 2007, a review of all well-preserved specimens including the then newly discovered Thermopolis specimen concluded that two distinct species of *Archaeopteryx* could be *A. lithographica* (consisting of at least the London and Solnhofen specimens), and *A. siemensii* (consisting of at least the Berlin, Munich, and Thermopolis specimens). The two species are distinguished primarily by large flexor tubercles on the foot claws in *A. lithographica* (the claws of *A. siemensii* specimens being relatively simple and straight). *A. lithographica* also had a constricted portion of the crown in some teeth and a stouter metatarsus. A supposed additional species, *Wellnhoferia grandis* (based on the Solnhofen specimen), seems to be indistinguishable from *A. lithographica* except in its larger size [Mayr et al., 2007].

Synonyms

If two names are given, the first denotes the original describer of the "species", the second the author on whom the given name combination is based. As always in zoological nomenclature, putting an author's name in parentheses denotes that the taxon was originally described in a different genus.

- ***Archaeopteryx lithographica*** Meyer, 1861 [conserved name]
    - *Pterodactylus crassipes* [Meyer, 1857] (suppressed in favor of *A. lithographica* 1977 per ICZN Opinion 1070)
    - *Rhamphorhynchus crassipes* [Meyer, 1857] (as *Pterodactylus (Rhamphorhynchus) crassipes*) [suppressed in favor of *A. lithographica* 1977 per ICZN Opinion 1070]
    - *Scaphognathus crassipes* [Meyer, 1857; Wagner, 1861] (rejected in favor of *A. lithographica* 1977 per ICZN Opinion 1070)
    - *Archaeopterix lithographica* [Anon., 1861] (*lapsus*)
    - *Griphosaurus problematicus* [Wagner, 1862] (rejected name 1961 per ICZN Opinion 607)
    - *Griphornis longicaudatus* [Woodward, 1862] (rejected name 1961 per ICZN Opinion 607)
    - *Griphosaurus longicaudatum* [Woodward, 1862] (*lapsus*)
    - *Griphosaurus longicaudatus* [Owen, 1862] (rejected name 1961 per ICZN Opinion 607)
    - *Archaeopteryx macrura* [Owen, 1862] (rejected name 1961 per ICZN Opinion 607)
    - *Archaeopterix macrura* [Owen, 1862] (*lapsus*)
    - *Archaeopterix macrurus* [Egerton, 1862] (*lapsus*)
    - *Archeopteryx macrurus* [Owen, 1863] (unjustified emendation)
    - *Archaeopteryx macroura* [Vogt, 1879] (*lapsus*)
    - *Archaeopteryx oweni* [Petronievics, 1917] (rejected name 1961 per ICZN Opinion 607)
    - *Archeopteryx macrura* [Ostrom, 1970] (*lapsus*)
    - *Archaeopteryx crassipes* [Meyer, 1857; Ostrom, 1972] (rejected in favor of *A. lithographica* 1977 per ICZN Opinion 1070)

    - *Gryphornis longicaudatus* [Lambrecht, 1933] (*lapsus*)
    - *Gryphosaurus problematicus* [Lambrecht, 1933] (*lapsus*)
    - *Archaeopteryx macrourus* [Owen, 1862 *fide* Lambrecht, 1933] (*lapsus*)
    - *Archaeopterix lithographica* [di Gregorio, 1984] (*lapsus*)
    - *Archaeopteryx recurva* [Howgate, 1984]
    - *Jurapteryx recurva* [Howgate, 1984, 1985]
- ***Archaeopteryx siemensii*** [Dames, 1897]
    - *Archaeopteryx siemensi* [Dames, 1897] (*lapsus)*
    - *Archaeornis siemensii* [Dames, 1897 Petronievics, 1917]
    - *Archaeornis siemensi* [Dames, 1897 *fide* Lambrecht, 1933] (*lapsus)*
    - *Archaeopteryx bavarica* [Wellnhofer, 1993]
    - *Wellnhoferia grandis* [Elżanowski, 2001]

"Archaeopteryx" vicensensis [Anon. *fide* Lambrecht, 1933] is a *nomen nudum* for what appears to be an undescribed pterosaur.

4.24. Controversy

Beginning in 1985, a group including astronomer Fred Hoyle and physicist Lee Spetner published a series of papers claiming that the feathers on the Berlin and London specimens of *Archaeopteryx* were forged [Hoyle et al., 1985; Watlkins et al., 1985 a, b, c]. Their claims were repudiated by Alan J. Charig and others at the British Museum (Natural History) [Charig et al., 1986]. Most of their evidence for a forgery was based on unfamiliarity with the processes of lithification; for example, they proposed that, based on the difference in texture associated with the feathers, feather impressions were applied to a thin layer of cement [Watkins et al., 1985a], without realizing that feathers themselves would have caused a textural difference [Charig et al., 1986]. They also expressed disbelief that slabs would split so smoothly, or that one half of a slab containing fossils would have good preservation, but not the counters lab [Hoyle et al., 1985; Watkins et al., 1985b]. These, however, are common properties of Solnhofen fossils, because the dead animals would fall onto hardened surfaces, which would form a natural plane for the future slabs to split along, leaving the bulk of the fossil on one side and little on the other [Charig et al., 1986]. They also misinterpreted the fossils, claiming that the tail was forged as one large feather [Watkins et al., 1985b], when visibly this is not the case. In addition, they claimed that the other specimens of *Archaeopteryx* known at the time did not have feathers [Watkins et al., 1985 a,b], which is incorrect. The Maxberg and Eichstätt specimens have obvious feathers [Charig et al., 1986]. Finally, the motives they suggested for a forgery are not strong, and are contradictory; one is that Richard Owen wanted to forge evidence in support of Charles Darwin's theory of evolution, which is unlikely given Owen's views toward Darwin and his theory. The other is that Owen wanted to set a trap for Darwin, hoping the latter would support the fossils so Owen could discredit him with the forgery; this is unlikely because Owen wrote a detailed paper on the London specimen, so such an action would certainly backfire [Spetner et al., 1988].

Charig et al. (1986) pointed to the presence of hairline cracks in the slabs running through both rock and fossil impressions, and mineral growth over the slabs that had occurred before discovery and preparation, as evidence that the feathers were original. Spetner et al. (1988) then attempted to show that the cracks would have propagated naturally through their postulated cement layer, but neglected to account for the fact that the cracks were old and had been filled with calcite, and thus were not able to propagate. They also attempted to show the presence of cement on the London specimen through X-ray spectroscopy, and did find something that was not rock [Spetner et al., 1988], however, it was not cement either, and is most probably from a fragment of silicone rubber left behind when molds were made of the specimen. Their suggestions have not been taken seriously by palaeontologists, as their evidence was largely based on misunderstandings of geology, and they never discussed the other feather-bearing specimens, which have increased in number since then. Charig et al. (1986) reported a discolouration: a dark band between two layers of limestone – however, they say it is the product of sedimentation. It is natural for limestone to take on the colour of its surroundings and most limestones are coloured (if not colour banded) to some degree, the darkness was attributed to such impurities. They also mention that a complete absence of air bubbles in the rock slabs is further proof that the specimen is authentic [Spetner et al., 1988].

4.25. Archaeopteryx and Protoavis

In 1984, Sankar Chatterjee discovered fossils which, in 1991, he claimed belonged to a fossil bird far older than *Archaeopteryx*. These fossils, believed to be approximately 210 to 225 million years old, have been assigned the name *Protoavis* [Chatterjee, 1991]. The fossils are too badly preserved to allow an estimate of flying ability; although Chatterjee's reconstructions usually show feathers, many paleontologists, including Paul (2002) and Witmer (2002) have rejected the claims that *Protoavis* was an earlier bird (or, alternatively, that it existed at all). The fossils were found disarticulated, and were collected from different locations. Because the fossils are in poor condition, *Archaeopteryx* remains the earliest universally recognized bird [Ostrom, 1996].

4.26. Phylogenetic Position

Modern paleontology has consistently placed *Archaeopteryx* as the most primitive bird. It is not thought to be a true ancestor of modern birds, but rather, a close relative of that ancestor [Clarke et al., 2002]. Nonetheless, *Archaeopteryx* was often used as a model of the true ancestral bird. Several authors have done so [Witmer, 2002]; Lowe (1935) and Thulborn (1984) questioned whether *Archaeopteryx* truly was the first bird. They suggested that *Archaeopteryx* was a dinosaur that was no more closely related to birds than were other dinosaur groups. Kurzanov (1987) suggested that *Avimimus* was more likely to be the ancestor of all birds than *Archaeopteryx*. Barsbold (1983) and Zweers and Van den Berge (1997) noted that many maniraptoran lineages are extremely birdlike, and they suggested that different groups of birds may have descended from different dinosaur ancestors. The discovery of the closely related *Xiaotingia* in 2011 led to new phylogenetic analyses that suggested that *Archaeopteryx* is a deinonychosaur rather than an avialan, and therefore, not a "bird" under most common uses of that term. A more thorough analysis was published soon after to test this hypothesis, and failed to arrive at the same result. It found *Archaeopteryx* to be a bird in its traditional position at the base of Avialae, while *Xiaotingia* was recovered as a basal deinonychosaur, or troodontid. The authors of the follow-up study noted that uncertainties still exist, however, and that it may not be possible to state confidently whether or not *Archaeopteryx* is a member of Avialae or not, barring new and better specimens of relevant species [Michael, 2012]. Phylogenetic studies conducted by Senter, et al. (2012) and Turner, Makovicky, and Norell (2012) confirmed that *Archaeopteryx* was more closely related to living birds than to dromaeosaurids and troodontids. On the other hand, Godefroit, et al. (2013) recovered *Archaeopteryx* as more closely related to dromaeosaurids and troodontids in the analysis included in their description of *Eosinopteryx brevipenna*. The authors used a modified version of the matrix from the study describing *Xiaotingia*, adding *Jinfengopteryx elegans* and *Eosinopteryx brevipenna* to it, as well as adding four additional characters related to the development of the plumage. Unlike the analysis from the description of *Xiaotingia*, the analysis conducted by Godefroit, et al. did not find *Archaeopteryx* to be related particularly closely to *Anchiornis* and *Xiaotingia*, which were recovered as basal troodontids instead [Godefroit et al., 2013], Agnolín and Novas (2013) found *Archaeopteryx* and (possibly synonymous) *Wellnhoferia* to be the basal most avialans (Avialae being defined by the authors as including *Archaeopteryx lithographica* and *Passer*, their most recent common ancestor and all of its descendants), with Microraptoria, Unenlagiinae, and the clade containing *Anchiornis* and *Xiaotingia* being successively closer outgroups to the Avialae. A later phylogenetic study of Godefroit, et al., (2013) using a more inclusive matrix than the one from the analysis in the description of *Eosinopteryx brevipenna*, also found *Archaeopteryx* to be a member of Avialae (defined by the authors as the most inclusive clade containing *Passer domesticus*, but not *Dromaeosaurus albertensis* or *Troodon formosus*). It was found to form a grade at the base of Avialae with *Xiaotingia*, *Anchiornis*, and *Aurornis*, with *Xiaotingia* and the latter two taxa being respectively, more closely and more distantly related to the living birds than *Archaeopteryx* was [Godefroit et al., 2013].

4.27. Dethroned as First Bird

In 2011, scientists uncovered a fossil in China whose combination of features unexpectedly suggests Archaeopteryx was actually just a relative of the lineage that ultimately gave rise to birds. The new specimen, Xiaotingia zhengi, was found in Liaoning in China, where many other extraordinary specimens of feathered dinosaurs and early birds have been unearthed.

When the researchers analyzed features of Xiaotingia and Archaeopteryx, the resulting family tree clustered them together. Unexpectedly, it also yanked them out of the avialan category — dinosaur like birds and placed the duo with the deinonychosaurs, birdlike dinosaurs.

The analysis also suggested the earliest known avialan is currently a pigeon-size feathered creature known as Epidexipteryx hui recently discovered in Inner Mongolia, China.

REFERENCES

Alan, H. T., Peter, J. M. & Mark, N. (2012). A review of dromaeosaurid systematics and paravian phylogeny. *Bull Am Mus Nat Hist*, *371*, 1–206.

Alexander, R. M. (1990). Animals. Cambridge University Press, Cambridge, pp. 509.

Alonso, P. D., Milner, A. C., Ketcham, R. A., Cookson, M. J. & Rowe, T. B. (2004). The avian nature of the brain and inner ear of *Archaeopteryx*. *Nature*, *430*, 666–669.

Balter, M. (2010). Did First Feathers Prevent Early Flight? *Science Now*

Barsbold, R. (1983). Carnivorous dinosaurs from the Cretaceous of Mongolia. *Trans jnt Sov-Mong Paleontol Exped*, *19*, 5–119.

Barsbold, R. et al. (1990). Oviraptorosauria. In The Dinosauria, Weishampel, Dodson &p; Osmolska (eds) 249-258.

Bartell, K.W., Swinburne, N. H. M. & Conway-Morris, S. (1994). *Solnhofen: a study in Mesozoic palaeontology*. Cambridge University Press.

Bock, W. J. & Bühler, P. (2007). *Archaeopteryx lithographica* von Meyer, 1861 (Aves): proposed conservation of usage by designation of a neotype. *Bull Zool Nomencl*, *64*, 261–262.

Britt, B. B., Makovicky, P. J., Gauthier, J. & Bonde, N. (1998). Postcranial pneumatization in *Archaeopteryx*. *Nature*, *395*, 374-376.

Brooks, B. (1993). Pneumatic postcranial bones in dinosaurs and other archosaurs. Ph.D. Dissertation, Department of Geology and Geophysics University of Calgary, Calgary, Alberta, Canada.

Bryant, H. N. & Russell, A. P. (1993). The occurrence of clavicles within dinosauria: implications for the homology of the avian furcula and the utility of negative evidence. *J Vert Paleontol*, *13*,171-184.

Bühler, P. & Bock, W. J. (2002). Zur Archaeopteryx-Nomenklatur: Missverständnisse und Lösung. *J Ornithol*, *143*, 269–286.

Buisonjé, P. H. de (1985). Climatological conditions during deposition of the Solnhofen limestones. In Hecht, M.K., Ostrom, J.H., Viohl, G. & Wellnhofer, P. *The beginnings of Birds: Proc Int* Archaeopteryx *Conf, Eichstatt, 1984*. Eichstätt: Freunde des Jura-Museums Eichstätt. 45–65.

Carney, R., Vinther, J., Shawkey, M. D., d'Alba, L. & Ackermann, J. (2012). New evidence on the colour and nature of the isolated Archaeopteryx feather. *Nat Comm*, *3*, 637.

Charig, A. J., Greenaway, F., Milner, A. N., Walker, C. A. & Whybrow, P. J. (1986). *Archaeopteryx* is not a forgery. *Science*, *232*, 622–626.

Chatterjee, S. (1991). Cranial anatomy and relationships of a new Triassic bird from Texas. *Phil Trans R Soc Lond B*, *332*, 277-342.

Chatterjee, S. (1991). Cranial anatomy and relationships of a new Triassic bird from Texas. *Philos Trans R Soc B*, *332*, 277–342.

Chen, P. J., Dong, Z. M. & Zhen, S. (1998). An exceptionally well-preserved theropod dinosaur from the Yixian Formation of China. *Nature*, *391*, 147-152.

Chiappe, L. M. & Witme, L. M. (2002). University od California Press Berkeley and Los Angeles, California, ed. *Mesozoic birds: above the heads of dinosaurs*. University of California Press. p. 151.

Chiappe, L. M. (1995). The first 85 million years of avian evolution. *Nature*, *378*, 349-355.

Chiappe, L. M. (2007). *Glorified Dinosaurs*. Sydney: UNSW Press. 118–146.

Christensen, P. & Bonde, N. (2004). Body plumage in Archaeopteryx: a eview, and new evidence from the Berlin specimen. *Comptes Rendus Palevol*, *3*, 99–118.

Chure, D. J. & Madsen, J. H. (1996). On the presence of furculae in some non-maniraptoran theropods. *J Vert Paleontol*, *16*, 573-577.

Chure, D. J. & Madsen, J. H. (1996). On the presence of furculae in some non-maniraptoran theropods. *Journal of Vertebrate Paleontology*, *16*, 573-577.

Clarke, J. A. & Norell, M. A. (2002). The Morphology and Phylogenetic Position of Apsaravis ukhaana from the Late Cretaceous of Mongolia. *Am Mus Nov*, *3387*, 1–46.

Colbert, E. H. & Morales, M. (1991). Evolution of the vertebrates: a history of the backboned animals through time. 4th ed. Wiley-Liss, New York. 470 p.

Davis, P. & Briggs, D. (1998). The impact of decay and disarticulation on the preservation of fossil birds. *PALAIOS*, *13*, 3–13.

de Beer, G. A. (1954). *Archaeopteryx lithographica*. British Museum of Natural History, London.

de Beer, G. A. (1956). Evolution of the Ratites. Bulletin of the British Museum (Natural History). *Zoology*, *4*, 57-70.

Dyke, G. J. & Nudds, R. L. (2010). Response to Comments on "Narrow Primary Feather Rachises in Confuciusornis and Archaeopteryx Suggest Poor Flight Ability. *Science*, *330*, 320.

Elzanowski, A. & Wellnhofer, P. (1996). Cranial morphology of *Archaeopteryx*: Evidence from the seventh skeleton. *J Vert Paleontol*, *16*, 81-94.

Elżanowski, A. (2002). Archaeopterygidae (Upper Jurassic of Germany). *In:* Chiappe, L.M. & Witmer, L.M (eds.), *Mesozoic Birds: Above the Heads of Dinosaurs*: 129–159. University of California Press, Berkeley.

Erickson, G. M., Rauhut, O. W. M., Zhou, Z., Turner, A. H., Inouye, B. D. Hu, D. & Norell, M. A. (2009). Was Dinosaurian Physiology Inherited by Birds? Reconciling Slow Growth in *Archaeopteryx*". In Desalle, Robert. *PLoS ONE*, *4*, e7390.

Federico, L. A. & Fernando, E. N. (2013). Avian ancestors. A review of the phylogenetic relationships of the theropods Unenlagiidae, Microraptoria, *Anchiornis*, and Scansoriopterygidae. *SpringerBriefs in Earth System Sciences*. SpringerBriefs in Earth System Sciences: 1–96.

Feduccia, A. & Tordoff, H. B. (1979). Feathers of Archaeopteryx: Asymmetric vanes indicate aerodynamic function. *Science*, *203*, 1021–1022.

Feduccia, A. (1993). Evidence from claw geometry indicating arboreal habits of*Archaeopteryx*. *Science*, *259*, 790–793.

Gingerich, P. D. (1975). Evolutionary significance of the mesozoic toothed birds. *Smithsonian Contr Paleobiol*, *27*, 23-34.

Godefroit, P., Cau, A., Hu, D.Y., Escuillié, F., Wu, W. & Dyke, G. (2013). A Jurassic avialan dinosaur from China resolves the early phylogenetic history of birds. *Nature*, *498*, 352-362.

Griffiths, P. J. (1996). The Isolated Archaeopteryx Feather. *Archaeopteryx*, *14*, 1–26.

Haubiltz, B., Prokop, M., Dohring, W., Ostrom, J. H. & Wellnhofer, P. (1988). Computed tomography of *Archaeopteryx*. *Paleobiol*, *14*, 206- 213.

Heilmann, G. (1926). The Origin of Birds. Witherby, London. 208pp.

Holtz, T. Jr. (1995). *Archaeopteryxs* Relationship With Modern Birds. *J Dino Paleontol*

Hoyle, F., Wickramasinghe, N. C. & Watkins, R. S. (1985). *Archaeopteryx. British J Photo, 132*, 693–694.

Huxley, T. H. (1868) Remarks upon Archaeopteryx lithographica. *Proc Roy Soc, 16*, 243–48.

Huxley, T. H. (1868). On the animals which are most nearly intermediate between birds and reptiles. *Geol. Mag., 5*, 357–65.

Huxley, T. H. (1870). Further evidence of the affinity between the dinosaurian reptiles and birds. *Quart J Geol Soc, 26*, 32–50.

ICZN. (1961). Opinion 607, *Archaeopteryx* VON MEYER, 1861 (Aves); Addition to the Official list". *Bull Zool Nomencl, 18*, 260–261.

ICZN. (1977). "Opinion 1070. Conservation of *Archaeopteryx lithographica* VON MEYER 1861 (Aves)". *Bull Zool Nomencl, 33*, 165–166.

ICZN. (2011). OPINION 2283 (Case 3390) Archaeopteryx lithographica von Meyer, 1861 (Aves): conservation of usage by designation of a neotype. *Bull Zool Nomencl, 68*, 230–233.

Jenkins, F. A. (1993). The evolution of the avain shoulder joint. *Am J Sc, 293A*, 253-267.

Kurzanov, S. M. (1987). Avimimidae and the problem of the origin of birds. *Trans jnt Sov-Mong Paleontol Exped, 31*, 31–94.

Lambert, D. (1993). *The Ultimate Dinosaur Book*. New York: Dorling Kindersley. 38–81.

Lambrecht, K. (1933). Handbuch der Palaeornithologie. Berlin, pp.1024.

Larsson, H. C. E. (2001). Endocranial anatomy of Carcharodontosaurus saharicus (Theropoda: Allosauroidea) and its implications for theropod brain evolution. 19–33. In: Mesozioc Vertebrate Life. Ed.s Tanke, D.H., Carpenter, K. & Skrepnick, M.W. Indiana University Press.

Longrich, N. (2006). Structure and function of hindlimb feathers in Archaeopteryx lithographica. *Paleobiol, 32*, 417–431.

Lowe, P. R. (1935). On the relationship of the Struthiones to the dinosaurs and to the rest of the avian class, with special reference to the position of *Archaeopteryx. Ibis, 5*, 398–432.

Manning, P. L., Edwards, N. P., Wogelius, R. A., Bergmann, U., Barden, H. E., Larson, P. L., Schwarz-Wings, D., Egerton, V. M. & Sokaras, D. (2013). Synchrotron-based chemical imaging reveals plumage patterns in a 150 million year old early bird. *J Anal Atom Spec, 28*, 1024.

Mayr, G., Phol, B., Hartman, S. & Peters, D. S. (2007). The tenthArchaeopteryx. *Zool J Linn Soc, 149*, 97–116.

Mayr, G., Pohl, B. & Peters, D. S. (2005). A well-preserved Archaeopteryx specimen with theropod features. *Science, 310*, 1483–1486.

McFarland, W. N., Pough, H. F., Cade, H. F & Heiser, J. B. (1985). Vertebrate Life, 2nd ed. Macmillan Publishing, New York. 636 pp.

Michael, S. Y. L. & Trevor, H. W. (2012). Likelihood reinstates *Archaeopteryx*as a primitive bird. *Biol Lett, 8*, 299–303.

Nedin, C. (1999). All About Archaeopteryx. talk.origins archive. *Version of June, 10, 2002.*

Norberg, R. A. (1995). Feather asymmetry in *Archaeopteryx. Nature, 374*, 221.

Norberg, R. A. (1995). Feather asymmetry in *Archaeopteryx. Nature, 374*, 211.

Nudds, R. L. & Dyke, G. J. (2010). Narrow Primary Feather Rachises in *Confuciusornis* and *Archaeopteryx* Suggest Poor Flight Ability. *Science, 328*, 887–889.

Olson, S. L. & Feduccia, A. (1979). Flight capability and the pectoral girdle of Archaeopteryx. *Nature, 278*, 247–248.

Orr, R. T. (1966). Vertebrate Biology, 2nd ed. W.B. Saunders, Philadelphia. 483pp.

Ostrom, J. H. (1970). *Archaeopteryx*: Notice of a "new" specimen. *Science, 170*, 537-538.

Ostrom, J. H. (1972). Description of the *Archaeopteryx* specimen in the Teyler Museum, Haarlen. *Proc Koninklijke Nederlandse Akademie Van Wetenschappen, B, 75*, 289-305.

Ostrom, J. H. (1974). *Archaeopteryx* and the Origin of Flight. *Quart Rev Biol, 49*, 27-47.

Ostrom, J. H. (1976). Archaeopteryx and the origin of birds. *Biol J Linn Soc, 8*, 91–182.

Ostrom, J. H. (1980). *Coelurus* and *Ornitholestes*: Are they the same? In Jacobs, L. (ed) *Aspects of Vertebrate History*. 245-256.

Ostrom, J. H. (1985). *Introduction to Archaeopteryx*. In: Hecht, M.K.O., Ostrom, J.H., Viohl, G. & Wellnhofer, P. (eds.) The Beginnings of Birds: *Proc Int Archaeop Conf,* 9–20. Eichstätt, Freunde des Jura-Museums Eichstätt.

Ostrom, J. H. (1991). Bird in the Bush. *Nature, 353*, 212.

Ostrom, J. H. (1995). Wing biometrics and the origin of bird flight. *Neues Jahrbuch fur Geologie und Palaeontologie, 195*, 253-266.

Ostrom, J. H. (1996). The questionable validity of *Protoavis*. *Archaeopteryx, 14*, 39–42.

Ostrom, J. H. (1976). *Archaeopteryx* and the origin of birds. *Biol Linn Soc, 8*, 91–182.

Owen, R. (1863). On the Archaeopteryx of Von Meyer, with a description of the fossil remains of a long-tailed species from the lithographic stone of Solnhofen. *Philos Trans R Soc, 153*, 33–47.

Pascal, G., Helena, D., Gareth, D., Dongyu, H., François, E. & Philippe, C. (2013). Reduced plumage and flight ability of a new Jurassic paravian theropod from China. *Nature Comm, 4*,

Paul, G. S. (1988). *Predatory Dinosaurs of the World, a Complete Illustrated Guide*. Simon and Schuster, New York. 464 p.

Paul, G. S. (2010). Comment on 'Narrow Primary Feather Rachises in *Confuciusornis* and *Archaeopteryx* Suggest Poor Flight Ability. *Science, 330*, 320.

Paul, G. S. (2002). *Dinosaurs of the Air: the Evolution and Loss of Flight in Dinosaurs and Birds*. Baltimore: Johns Hopkins University Press.

Rayner, J. M. V. (1988). The evolution of vertebrate flight. *Biol J Linn Soc, 34*, 269-287.

Reisdorf, A. G. & Wuttke, M. (2012). Re-evaluating Moodie's Opisthotonic-Posture Hypothesis in fossil vertebrates. Part I: Reptiles - The taphonomy of the bipedal dinosaurs *Compsognathus longipes* and *Juravenator starki* from the Solnhofen Archipelago (Jurassic, Germany). *Palaeobiodiv Palaeoenv*.

Rietschel, S. (1985). In: The Beginnings of Birds. Hecht, M.K., Ostrom, J.H., Viohl, G. & Wellnhofer, P. (eds.). 303-318. Freunde Des Jura-Museums, Eichstatt.

Romanoff, A. L. (1960). The Avian Embryo. Macmillan, New York, pp. 1304.

Romer, A. S. (1950). Vertebrate Paleontology. p 261. University of Chicago Press, Chicago.

Ruben, J. (1991). Reptilian physiology and the flight capacity of *Archaeopteryx*. *Evolution, 45*, 1-17.

Schmitz, L. & Motani, R. (2011). Nocturnality in Dinosaurs Inferred from Scleral Ring and Orbit Morphology. *Science, 332*, 705–8.

Schweigert, G. (2007). Ammonite biostratigraphy as a tool for dating Upper Jurassic lithographic limestones from South Germany – first results and open questions. *Neues Jahrbuch für Geologie und Paläontologie - Abhandlungen, 245*, 117–125.

Senter, P. (2006). *Scapular orientation in theropods and basal birds and the origin of flapping flight. Acta Palaeontol Pol, 51*, 305–313.

Senter, P., James, I. K., Donald, D. D., Scott, M. & Natalie, T. (2012). New Dromaeosaurids (Dinosauria: Theropoda) from the Lower Cretaceous of Utah, and the Evolution of the Dromaeosaurid Tail. In Dodson, P. *PLoS ONE, 7*, e36790.

Speakman, J. R. & Thomson, S. C. (1994). Flight capabilities of *Achaeopteryx. Nature, 370*, 514.

Speakman, J. R. & Thomson, S. C. (1994). Flight capabilities of *Archaeopteryx. Nature, 370*, 514.

Speakman, J. R. (1993). Flight capabilities in *Archaeopteryx. Evolution, 47*, 336-340.

Spetner, L. M., Hoyle, F., Wickramasinghe, N. C. & Magaritz, M. (1988). *Archaeopteryx* - more evidence for a forgery. *British J Photo, 135*, 14–17.

Swinton, W. E. (1960). Opinion 1084, Proposed addition of the generic name *Archaeopteryx* VON MEYER, 1861 and the specific name *Lithographica*, VON MEYER, 1861, as published in the binomen *Archaeopteryx Lithographica* to the official lists (Class Aves). *Bull Zool Nomencl, 17*, 224–226.

Switek, B. (2011). Archaeopteryx was robed in black. *New Scientist* (Las Vegas)

Switek, B. (2011). *Paleontologists Unveil the 11th Archaeopteryx*. Dinosaur Tracking Blog • the Smithsonian Institution

Thulborn, R. A. (1984). The avian relationships of *Archaeopteryx*, and the origin of birds. *Zool J Linn Soc, 82*, 119–158.

Tucker, B. W. (1938a). Functional evolutionary morphology: the origin of birds. p. 321-336. In: Evolution: Essays on Aspects of Evolutionary Biology, de Beer, G.A. (ed.). Clarandon Press, Oxford.

Tucker, B. W. (1938b). Some observations on Dr. Lowe's theory of the relationship of the Struthoines to the dinosaurs and to other birds. *Proc 8th Int Ornithol Cong Oxford*, 1934, 222-224.

Unwin, D.M. (1998). Feathers, filaments, and theropod dinosaurs. *Nature, 391*, 119-120.

Videler, J. J. (2005). Avian Flight. Oxford University Press. Pp. 98–117

Wagner, A. (1861). Über ein neues, angeblich mit Vogelfedern versehenes Reptil aus dem Solnhofener lithographischen Schiefer. *Sitzungberichte der Bayerischen Akademie der Wissenschaften, mathematisch-physikalische Classe*, 146–154.

Walker, A. D. (1972). New light on the origins of birds and crocodiles. *Nature, 237*, 257-263.

Walker, A. D. (1974). Evolution, organic. In Yearbook of Science and Technology, Lapedes, D.N. (ed). 177-179. McGraw-Hill, New York.

Walker, A. D. (1985). The braincase of *Archaeopteryx*. In The Beginnings of Birds, Hecht, M.K., Ostrom, J.H., Viohl, G. & Wellnhofer, P. (eds.). 123-134. Freunde Des Jura-Museums, Eichstatt.

Watkins, R. S., Hoyle, F., Wickrmasinghe, N. C., Watkins, J., Rabilizirov, R. & Spetner, L. M. (1985a). *Archaeopteryx* - a photographic study. *British J Photo, 132*, 264–266.

Watkins, R. S., Hoyle, F., Wickrmasinghe, N. C., Watkins, J., Rabilizirov, R. & Spetner, L. M. (1985b). *Archaeopteryx* - a further comment. *British J Photo, 132*, 358–359, 367.

Watkins, R. S., Hoyle, F., Wickrmasinghe, N. C., Watkins, J., Rabilizirov, R. & Spetner, L. M. (1985c). *Archaeopteryx* - more evidence. *British J Photo*, 132, 468–470.

Wellnhofer, P. & Tischlinger, H. (2004). *Das "Brustbein" von Archaeopteryx Wellnhofer 1993 — eine Revision. Archaeopteryx. 22*, 3–15.

Wellnhofer, P. (2004). The Plumage of *Archaeopteryx*. In Currie, P.J., Koppelhus, E.B., Shugar, M.A. & Wright, J.L. *Feathered Dragons*. Indiana University Press. 282–300.

Wellnhofer, P. (2008). *Archaeopteryx. Der Urvogel Von Solnhofen*. München: Friedrich Pfeil, p. 54.

Whetstone, K. N. (1983). Braincase of Mesozoic birds: I. New preparation of the "London" *Archaeopteryx*. *J Vert Paleontol*, *2*, 439-452.

Witmer, L. M. (1990). The craniofacial air sac system of Mesozoic birds (Aves). *Zool J Linn Soc*, *100*, 327-378.

Witmer, L. M. (2002). The debate on avian ancestry. In Witmer, L. & Chiappe, L. *Mesozoic Birds: Above the Heads of Dinosaurs*. Berkeley: University of California Press. 3–30.

Witmer, L. M. (2004). Palaeontology: Inside the oldest bird brain. *Nature*, *430*, 619–620.

Xing, X., Hailu, Y., Kai, D. & Fenglu, H. (2011). An *Archaeopteryx*-like theropod from China and the origin of Avialae. *Nature*, *475*, 465–470.

Yalden, D. W. (1984). What size was *Archaeopteryx*? *Zool J Linn Soc*, *82*, 177–188.

Zweers, G. A. & Van den Berge, J. C. (1997). Evolutionary patterns of avian trophic diversification. *Zool: Anal Comp Syst*, *100*, 25–57.

Chapter 5

Origin and Evolution of Birds

5.1. Birds: Delayed Evolution of Dinosaurs

The **origin of birds** refers to the initial stages in the evolution of birds. The scientific consensus is that birds are a group of theropod dinosaurs that evolved during the Mesozoic Era.

A close relationship between birds and dinosaurs was first proposed in the nineteenth century after the discovery of the primitive bird *Archaeopteryx* in Germany. Birds share many unique skeletal features with dinosaurs [Chiappe, 2009]. Moreover, fossils of more than twenty species of dinosaur have been collected with preserved feathers. There are even very small dinosaurs, such as *Microraptor* and *Anchiornis*, which have long, vaned, arm and leg feathers forming wings. The Jurassic basal avialan *Pedopenna* also shows these long foot feathers. Witmer (2009) has concluded that this evidence is sufficient to demonstrate that avian evolution went through a four-winged stage [Witmer and Lawerence, 2009].

Fossil evidence also demonstrates that birds and dinosaurs shared features such as hollow, pneumatized bones, gastroliths in the digestive system, nest-building and brooding behaviors. The ground-breaking discovery of fossilized *Tyrannosaurus rex* soft tissue allowed a molecular comparison of cellular anatomy and protein sequencing of collagen tissue, both of which demonstrated that *T. rex*and birds are more closely related to each other than either is to *Alligator* [Asara et al., 2007]. A second molecular study robustly supported the relationship of birds to dinosaurs, though it did not place birds within Theropoda, as expected. This study utilized eight additional collagen sequences extracted from a femur of *Brachylophosaurus canadensis*, a hadrosaur [Schweitzer, 2009]. A study comparing juvenile and adult archosaur skulls concluded that birds derived from dinosaurs by neoteny [Bhullar et al., 2012].

The origin of birds has historically been a contentious topic within evolutionary biology. However, only a few scientists still debate the dinosaurian origin of birds, suggesting descent from other types of archosaurian reptiles. Among the consensus that supports dinosaurian ancestry, the exact sequence of evolutionary events that gave rise to the early birds within maniraptoran theropods is a hot topic. The origin of bird flight is a separate but related question for which there are also several proposed answers.

5.2. The Origin of Birds

Birds are generally thought to have evolved from small theropod dinosaurs in the mid-late Jurassic, over 150 million years ago. However, it has also been suggested that they evolved from thecodonts, which are primitive archosaurs, or crocodylomorphs. All three theories will be discussed here [Padian et al., 1998].

A cladogram showing the systematic position of birds according to all three hypotheses is illustrated in figure 1.

The Thecodont Hypothesis

The thecodont hypothesis for the origin of birds is characterised by being a default option. This is because it is not due to positive correlation of characters and taxa but due to the negative association with other taxa. It was originally thought that theropods shared more features with birds than any other group. However, at the time there was no evidence for clavicles in theropods, which are the equivalent of the furcula or wishbone in birds. It was thought that clavicles could not have been lost and then re-evolved into furcula and so a more ancient ancestor for birds was sought for. The candidate suggested was the thecodonts from which all other archosaurs are thought to have evolved. A problem is that the archosaurs are a "wastebasket" group containing all archosaurian reptiles that do not fit into dinosaurs, crocodiles or pterosaurs. Hence, they have no diagnostic characters of their own and are not a "good" phylogenetic group. This makes it difficult to compare them with other taxonomic groups [Padian et al., 1998].

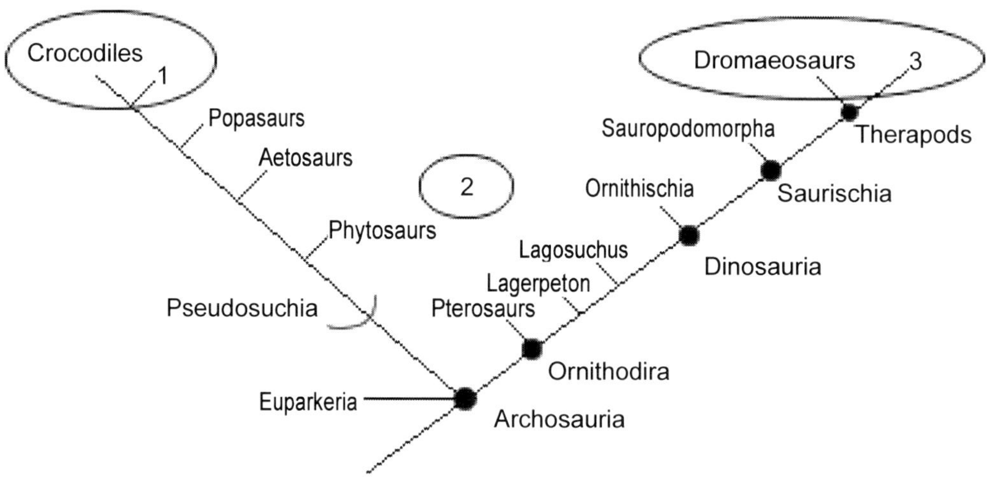

Figure 1. Possible systematic positions of birds. Redrawn from Padian et al. (1998).

Figure 2. Heilmann's hypothetical illustration of a pair of fighting 'Proaves' from 1916 (Source: Origin of Birds, Heilmann, G. 1926).

A turning point came in the early twentieth century with the writings of Gerhard Heilmann of Denmark. An artist by trade, Heilmann had a scholarly interest in birds and from 1913 to 1916 published the results of his research in several parts, dealing with the anatomy, embryology, behavior, paleontology, and evolution of birds (Figure 2) [Nieuwland, 2004]. His work, originally written in Danish as *Vor Nuvaerende Viden om Fuglenes Afstamning*, was compiled, translated into English, and published in 1926 as *The Origin of Birds*.

Like Huxley, Heilmann compared *Archaeopteryx* and other birds to an exhaustive list of prehistoric reptiles, and also came to the conclusion that theropod dinosaurs like *Compsognathus* were the most similar. However, Heilmann noted that birds had clavicles (collar bones) fused to form a bone called the furcula ("wishbone"), and while clavicles were known in more primitive reptiles, they had not yet been recognized in dinosaurs (Figure 3). Since he was a firm believer in Dollo's law, which states that evolution is not reversible, Heilmann could not accept that clavicles were lost in dinosaurs and re-evolved in birds. He was therefore forced to rule out dinosaurs as bird ancestors and ascribe all of their similarities to convergent evolution. Heilmann (1926) stated that bird ancestors would instead be found among the more primitive "thecodont" grade of reptiles [Heilmann, 1926]. Heilmann's extremely thorough approach ensured that his book became a classic in the field, and its conclusions on bird origins, as with most other topics, were accepted by nearly all evolutionary biologists for the next four decades [Padian, 2004].

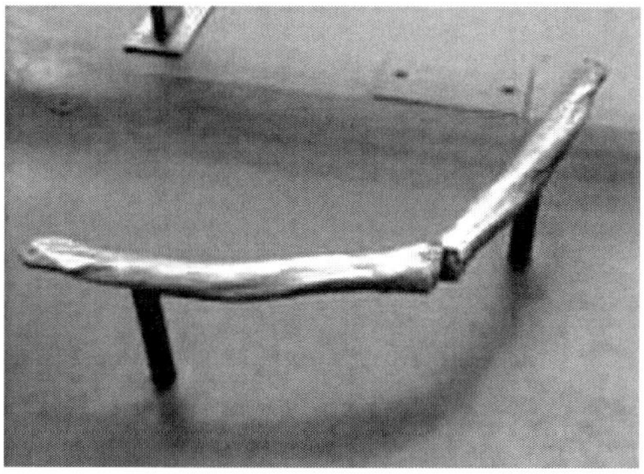

Figure 3. Bronze cast of the furcula of "Sue" the *Tyrannosaurus*, Field Museum.

Clavicles are relatively delicate bones and therefore in danger of being destroyed or at least damaged beyond recognition. Nevertheless, some fossil theropod clavicles had actually been excavated before Heilmann wrote his book but these had been misidentified [Paul, 2002]. The absence of clavicles in dinosaurs became the orthodox view despite the discovery of clavicles in the primitive theropod *Segisaurus* in 1936 [Camp, 1936]. The next report of clavicles in a dinosaur was in a Russian article in 1983 [Barsbold, 1983].

Contrary to what Heilmann believed, paleontologists now accept that clavicles and in most cases furculae are a standard feature not just of theropods but of saurischian dinosaurs. Up to late 2007, ossified furculae (i.e., made of bone rather than cartilage) have been found in all types of theropods except the most basal ones, *Eoraptor* and *Herrerasaurus* [Lipkin et al., 2007]. The original report of a furcula in the primitive theropod *Segisaurus* (1936) was confirmed by a re-examination in 2005 [Carrano et al., 2005]. Joined, furcula-like clavicles have also been found in *Massospondylus*, an Early Jurassic sauropodomorph [Yates et al., 2005].

The Crocodylomorph Hypothesis

Crocodylomorphs include crocodiles and some Triassic-Jurassic forms that are closely related but not true crocodiles. This hypothesis has fewer problems than the thecodont hypothesis, as crocodiles are a monophyletic group, which can be compared to other taxa. The synapomorphies of birds and crocodiles have been tested, but it has been found that most are general to the archosaurs. Even if these are accepted, there are only 15-20 synapomorphies compared to over 70 with the theropods [Padian et al., 1998].

The Theropod Hypothesis, Dinosaur Renaissance

Cladistic analysis supports this hypothesis and shows that the most closely related group (sister group) of theropods to the birds includes the dromaesaurids such as *Deinonychus*. There are numerous synapomorphies of the skeleton and skull that link birds and theropods. There have been found a series of skeletal changes in theropods that are considered to be avian. For example, basal theropod dinosaurs have lightly built bones and a foot reduced to three main toes, with the first held off the ground and the fifth lost. Moving through the theropod sequence towards birds, there is a reduction and loss of manual digits four and five, increasing lightness of the skeleton, and a reduction in the number, and partial interlocking, of the tail vertebrae. In coelurosaurs, which include birds, dromaeosaurids and other groups, the arms become longer and the first toe begins to rotate backwards behind the metatarsals. Fused clavicles (furcula) are apparently basal to Tetanurae (carnosaurs and coelurosaurs) and sternal plates are known in a variety of tetanurans. In the pelvis, the pubis and ischium begin to show a greater disparity in length. Finally, in the dromaeosaurids and *Archaeopteryx*, the pubis begins to point backwards instead of forwards, the anterior projection on the foot of the pubis is lost, the tail becomes even shorter and the hyperflexing wrist joint is present, which allows the action that is crucial to the flight stroke in birds. These features were passed to birds from their dinosaurian ancestors and not specifically evolved for an avian lifestyle [Padian et al., 1998].

The tide began to turn against the 'thecodont' hypothesis after the 1964 discovery of a new theropod dinosaur in Montana. In 1969, this dinosaur was described and named *Deinonychus* by John Ostrom of Yale University [Ostrom, 1969]. The next year, Ostrom redescribed a specimen of *Pterodactylus* in the Dutch Teyler Museum as another skeleton of *Archaeopteryx* [Ostrom, 1970]. The specimen consisted mainly of a single wing and its description made Ostrom aware of the similarities between the wrists of *Archaeopteryx* and *Deinonychus* [Chambers, 2002].

In 1972, British paleontologist Alick Walker hypothesized that birds arose not from 'thecodonts' but from crocodile ancestors like*Sphenosuchus* [Walker, 1972]. Ostrom's work with both theropods and early birds led him to respond with a series of publications in the mid-1970s in which he laid out the many similarities between birds and theropod dinosaurs, resurrecting the ideas first put forth by Huxley over a century before [Ostrom, 1973, 1975, 1976]. Ostrom's recognition of the dinosaurian ancestry of birds, along with other new ideas about dinosaur metabolism [Baker, 1972], activity levels, and parental care began what is known as the Dinosaur renaissance, which began in the 1970s and continues to this day [Horner, 1979].

Ostrom's revelations also coincided with the increasing adoption of phylogenetic systematics (cladistics), which began in the 1960s with the work of Willi Hennig [Hennig, 1966]. Cladistics is a method of arranging species based strictly on their evolutionary relationships, using a statistical analysis of their anatomical characteristics. In the 1980s, cladistic methodology was applied to dinosaur phylogeny for the first time by Jacques Gauthier and others, showing unequivocally that birds were a derived group of theropod dinosaurs [Gauthier, 1986]. Early analyses suggested that dromaeosaurid theropods like *Deinonychus* were particularly closely related to birds, a result that has been corroborated many times since [Senter, 2007; Turner et al., 2007].

5.3. HUXLEY, *ARCHAEOPTERYX* AND EARLY RESEARCH

Scientific investigation into the origin of birds began shortly after the 1859 publication of Charles Darwin's *On the Origin of Species* [Darwin, 1859]. In 1860, a fossilized feather was discovered in Germany's Late Jurassic Solnhofen limestone. Christian Erich Hermann von

Meyer described this feather as *Archaeopteryx lithographica* the next year [von Meyer and Hermann, 1861]. Richard Owen described a nearly complete skeleton in 1863, recognizing it as a bird despite many features reminiscent of reptiles, including clawed forelimbs and a long, bony tail [Owen, 1863].

Biologist Thomas Henry Huxley, known as "Darwin's Bulldog" for his ferocious support of the new theory of evolution, almost immediately seized upon *Archaeopteryx* as a transitional fossil between birds and reptiles. Starting in 1868, Huxley made detailed comparisons of *Archaeopteryx* with various prehistoric reptiles and found that it was most similar to dinosaurs like *Hypsilophodon* and *Compsognathus* [Huxley, 1868, 1870]. The discovery in the late 1870s of the iconic "Berlin specimen" of *Archaeopteryx*, complete with a set of reptilian teeth, provided further evidence. Huxley was the first to propose an evolutionary relationship between birds and dinosaurs. Although Huxley was opposed by the very influential Owen, his conclusions were accepted by many biologists, including Baron Franz Nopcsa [Nopcsa, 1901], while others, notably Harry Seeley [Seeley, 1907], argued that the similarities were due to convergent evolution.

5.4. Modern Research and Feathered Dinosaurs in China

The early 1990s saw the discovery of spectacularly preserved bird fossils in several Early Cretaceous geological formations in the northeastern Chinese province of Liaoning [Sereno and Chenggang, 1992; Hou et al., 1995]. In 1996, Chinese paleontologists described *Sinosauropteryx* (Figure 4) as a new genus of bird from the Yixian Formation [Quiang and Shu-an, 1995], but this animal was quickly recognized as a theropod dinosaur closely related to *Compsognathus*. Surprisingly, its body was covered by long filamentous structures. These were dubbed 'protofeathers' and considered homologous with the more advanced feathers of birds [Chen et al., 1998], although some scientists disagree with this assessment [Lingham et al., 2007]. Chinese and North American scientists described *Caudipteryx* and *Protarchaeopteryx* soon after. Based on skeletal features, these animals were non-avian dinosaurs, but their remains bore fully formed feathers closely resembling those of birds [Quiang et al., 1998]. "Archaeoraptor", described without peer review in a 1999 issue of *National Geographic* [Sloan, 1999], turned out to be a smuggled forgery [Monastersky, 2000] but legitimate remains continue to pour out of the Yixian, both legally and illegally. Feathers or "protofeathers" have been found on a wide variety of theropods in the Yixian [Xing et al., 1999, 2004] and the discoveries of extremely bird-like dinosaurs [Xing et al., 2003], as well as dinosaur-like primitive birds [Zhou and Zhang, 2002] have almost entirely closed the morphological gap between theropods and birds.

A small minority, including ornithologists Alan Feduccia and Larry Martin, continues to assert that birds are instead the descendants of earlier archosaurs, such as *Longisquama* or *Euparkeria* [Martin, 2004; Feduccia et al., 2005]. Embryological studies of bird developmental biology have raised questions about digit homology in bird and dinosaur forelimbs [Burke and Feduccia, 1997]. However, due to the cogent evidence provided by comparative anatomy and phylogenetics, as well as the dramatic feathered dinosaur fossils from China, the idea that birds are derived dinosaurs, first championed by Huxley and later by Nopcsa and Ostrom, enjoys near-unanimous support among today's paleontologists.

5.5. Thermogenic Muscle Hypothesis

A new theory of bird origins suggests that selection for the expansion of skeletal muscle, rather than the evolution of flight, was the driving force for the emergence of this clade [Newman, 2011; Newman et al., 2013]. Muscles became larger in prospectively endothermic saurians, according to this hypothesis, as a response to the loss of the vertebrate mitochondrial uncoupling protein, UCP1 [Mezentseva, 2008], which is thermogenic. In mammals, UCP1 functions within brown adipose tissue to protect newborns against hypothermia. In modern birds, skeletal muscle serves a similar function and is presumed to have done so in their ancestors. In this view, bipedality and other avian skeletal alterations were side effects of muscle hyperplasia, with further evolutionary modifications of the forelimbs, including adaptations for flight or swimming, and vestigiality, being secondary consequences of two-leggedness.

5.6. Phylogeny

Archaeopteryx has historically been considered the first bird, or *Urvogel*. Although newer fossil discoveries filled the gap between theropods and *Archaeopteryx*, as well as the gap between *Archaeopteryx* and modern birds, phylogenetic taxonomists, in keeping with tradition, almost always use *Archaeopteryx* as a specifier to help define Aves [Chiappe, 1997; Sereno, 1997]. Aves have more rarely been defined as a crown group consisting only of modern birds. Nearly all palaeontologists regard birds as coelurosaurian theropod dinosaurs. Within Coelurosauria, multiple cladistic analyses have found support for a clade named Maniraptora, consisting of therizinosauroids, oviraptorosaurs, troodontids, dromaeosaurids, and birds. Of these, dromaeosaurids and troodontids are usually united in the clade Deinonychosauria, which is a sister group to birds (together forming the node-cladeEumaniraptora) within the stem-clade Paraves [Holtz et al., 2004; Turner et al., 2007].

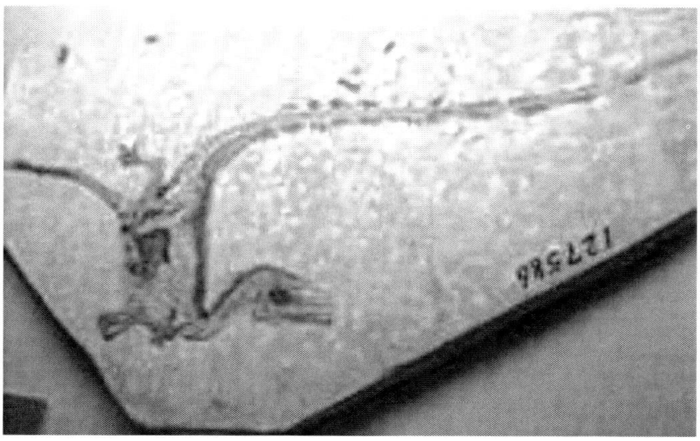

(Source: Wikipedia).

Figure 4. Fossil of *Sinosauropteryx prima*.

Other studies have proposed alternative phylogenies, in which certain groups of dinosaurs usually considered non-avian may have evolved from avian ancestors. For example, a 2002 analysis found that oviraptorosaurs were basal avians [Osmolska et al., 2007]. Alvarezsaurids, known from Asia and the Americas, have been variously classified as basal maniraptorans [Martinelli et al, 2007; Novas and Paul, 2002] paravians, the sister taxon of ornithomimosaurs [Sereno, 1999] as well as specialized early birds [Perle et al., 1993; Chiappe et al., 2002]. The genus *Rahonavis*, originally described as an early birds [Forster et al., 1998], has been identified as a non-avian dromaeosaurid in several studies. Dromaeosaurids and troodontids themselves have also been suggested to lie within Aves rather than just outside it [Paul, 2002; Makovicky, 2005; Mayr, 2005].

5.7. Anatomical Similarity between Birds and Theropod Dinosaur

Parts of a Feather

Archaeopteryx, the first good example of a "feathered dinosaur", was discovered in 1861. The first specimen was found in the Solnhofen limestonein southern Germany, which is a *lagerstätte*, a rare and remarkable geological formation known for its superbly detailed fossils. *Archaeopteryx* is a transitional fossil, with features clearly intermediate between those of modern reptiles and birds. Discovered just two years after Darwin's seminal *Origin of Species*, its discovery spurred the nascent debate between proponents of evolutionary biology and creationism. This early bird is so dinosaur-like that, without a clear impression of feathers in the surrounding rock, at least one specimen was mistaken for *Compsognathus* [Wellnhofer, 1988].

Since the 1990s, a number of additional feathered dinosaurs have been found, providing even stronger evidence of the close relationship between dinosaurs and modern birds. The first of these were initially described as simple filamentous *protofeathers*, which were reported in dinosaur lineages as primitive as compsognathids and tyrannosauroids [Xu et al., 2004]. However, feathers indistinguishable from those of modern birds were soon after found in non-avialan dinosaurs as well.

A small minority of researchers have claimed that the simple filamentous "protofeather" structures are simply the result of the decomposition of collagen fiber under the dinosaurs' skin or in fins along their backs, and that species with unquestionable feathers, such as oviraptorosaurs and dromaeosaurs are not dinosaurs, but true birds unrelated to dinosaurs [Feduccia, 2012]. However, a majority of studies have concluded that feathered dinosaurs are in fact dinosaurs, and that the simpler filaments of unquestionable theropods represent simple feathers. Some researchers have demonstrated the presence of color-bearing melanin in the structures—which would be expected in feathers but not collagen fibers (Figure 5) [Zhang et al., 2010]. Others have demonstrated, using studies of modern bird decomposition, that even advanced feathers appear filamentous when subjected to the crushing forces experienced during fossilization, and that the supposed "protofeathers" may have been more complex than previously thought [Foth, 2012]. Detailed examination of the "protofeathers" of *Sinosauropteryx prima* showed that individual feathers consisted of a central quill (*rachis*) with thinner *barbs* branching off from it, similar to but more primitive in structure than modern bird feathers (Figure 6) [Curie and Chen, 2001].

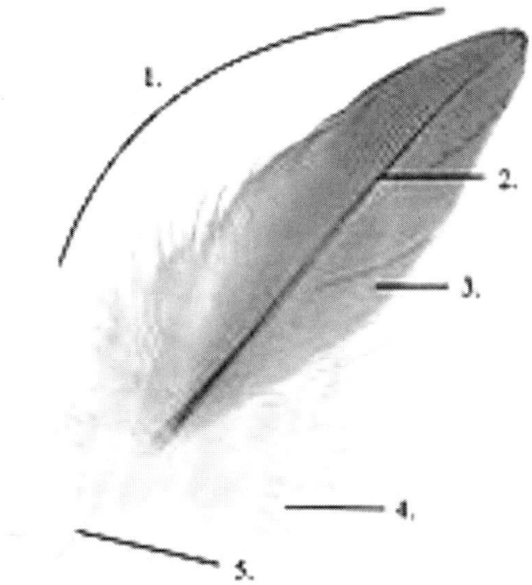

Figure 5. Parts of a Feather. 1. Vane, 2. Rachis, 3. Barb, 4. Aftershaft, 5. Calamus.

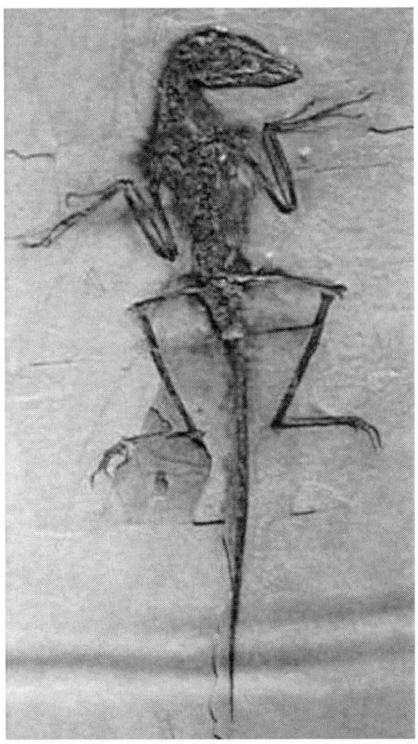

(Source: Wikipedia).

Figure 6. Fossil cast of NGMC 91, a probable specimen of *Sinornithosaurus*.

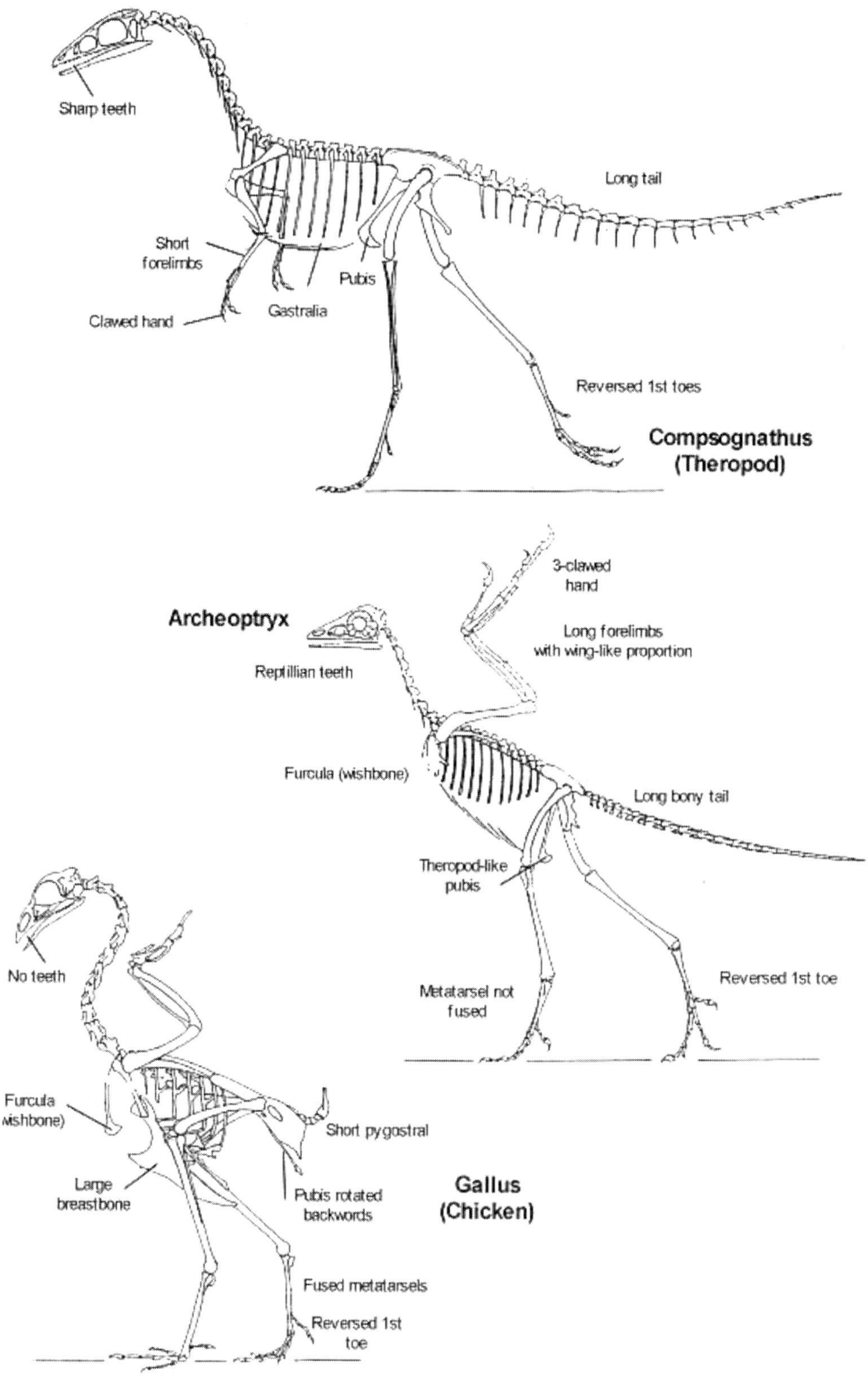

Figure 7. Comparison of skeletons of a theropod dinosaur, Archaeopteryx and a modern bird.

5.8. Skeleton

Because feathers are often associated with birds, feathered dinosaurs are often touted as the missing link between birds and dinosaurs. However, the multiple skeletal features also shared by the two groups represent the more important link for paleontologists. Furthermore, it is increasingly clear that the relationship between birds and dinosaurs, and the evolution of flight, are more complex topics than previously realized. For example, while it was once believed that birds evolved from dinosaurs in one linear progression, some scientists, most notably Gregory S. Paul, conclude that dinosaurs such as the dromaeosaurs may have evolved from birds, losing the power of flight while keeping their feathers in a manner similar to the modern ostrich and other ratites.

Comparisons of bird and dinosaur skeletons, as well as cladistic analysis, strengthens the case for the link, particularly for a branch of theropods called maniraptors. Skeletal similarities include the neck, pubis, wrist (semi-lunate carpal), arm and pectoral girdle, shoulder blade, clavicle, and breast bone (Figure 7).

5.9. Lungs

Large meat-eating dinosaurs had a complex system of air sacs similar to those found in modern birds, according to an investigation led by Patrick M. O'Connor of Ohio University (Figure 8). The lungs of theropod dinosaurs (carnivores that walked on two legs and had birdlike feet) likely pumped air into hollow sacs in their skeletons, as is the case in birds. "*What was once formally considered unique to birds was present in some form in the ancestors of birds*", O'Connor said [O'Connor et al., 2005; Sereno et al., 2008].

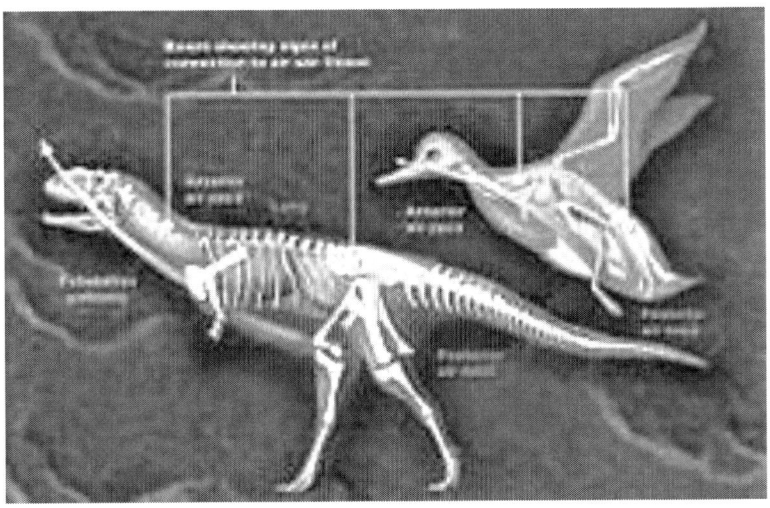

(Source: Wikipedia).

Figure 8. Comparison between the air sacs of *Majungasaurus* and a bird.

5.10. Heart

Computed tomography (CT) scans conducted in 2000 of the chest cavity of a specimen of the ornithopod *Thescelosaurus* found the apparent remnants of complex four-chambered hearts, much like those found in today's mammals and birds [Fisher et al., 2000]. The idea is controversial within the scientific community, coming under fire for bad anatomical science [Hillenius, 2004], or simply wishful thinking [Rowe et al., 2001].

A study published in 2011 applied multiple lines of inquiry to the question of the object's identity, including more advanced CT scanning, histology, X-ray diffraction, X-ray photoelectron spectroscopy, and scanning electron microscopy. From these methods, the authors found that: the object's internal structure does not include chambers but is made up of three unconnected areas of lower density material, and is not comparable to the structure of an ostrich's heart. The "walls" are composed of sedimentary minerals not known to be produced in biological systems—such as goethite, feldspar minerals, quartz, and gypsum, as well as some plant fragments. Carbon, nitrogen, and phosphorus, chemical elements important to life, were lacking in their samples and cardiac cellular structures were absent. There was one possible patch with animal cellular structures. The authors found their data supported identification as a concretion of sand from the burial environment, not the heart, with the possibility that isolated areas of tissues were preserved [Cleland et al., 2011].

The question of how this find reflects metabolic rate and dinosaur internal anatomy is moot, though, regardless of the object's identity [Cleland et al., 2011]. Both modern crocodilians and birds, the closest living relatives of dinosaurs, have four-chambered hearts (albeit modified in crocodilians), so dinosaurs probably had them as well; the structure is not necessarily tied to metabolic rate [Chinsamy et al., 2004].

5.11. Sleeping Posture

Fossils of the troodonts *Mei* and *Sinornithoides* demonstrate that the dinosaurs slept like certain modern birds, with their heads tucked under their arms [Xu and Norell, 2004]. This behavior, which may have helped to keep the head warm, is also characteristic of modern birds.

5.12. Reproductive Biology

When laying eggs, female birds grow a special type of bone in their limbs. This medullary bone forms as a calcium-rich layer inside the hard outer bone, and is used as a calcium source to make eggshells. The presence of endosteally derived bone tissues lining the interior marrow cavities of portions of a *Tyrannosaurus* rex specimen's hind limb suggested that T. rex used similar reproductive strategies, and revealed that the specimen is female [Schweitzer, 2005]. Further research has found medullary bone in the theropod *Allosaurus* and ornithopod *Tenontosaurus*. Because the line of dinosaurs that includes *Allosaurus* and *Tyrannosaurus* diverged from the line that led to *Tenontosaurus* very early in the evolution of dinosaurs, this suggests that dinosaurs in general produced medullary tissue [Li and Werning, 2008].

5.13. Brooding and Care of Young

Several *Citipati* specimens have been found resting over the eggs in its nest in a position most reminiscent of brooding (Figure 9) [Norell et al., 1994].

Numerous dinosaur species, for example *Maiasaura*, have been found in herds mixing both very young and adult individuals, suggesting rich interactions between them.

A dinosaur embryo was found without teeth, which suggests some parental care was required to feed the young dinosaur, possibly the adult dinosaur regurgitating food into the young dinosaur's mouth (altricial). This behaviour is seen in numerous bird species; parent birds regurgitate food into the hatchling's mouth.

5.14. Gizzard Stones

Both birds and dinosaurs use gizzard stones. These stones are swallowed by animals to aid digestion and break down food and hard fibres once they enter the stomach. When found in association with fossils, gizzard stones are called gastroliths [Wings, 2007]. Gizzard stones are also found in some fish (mullets, mud shad, and the gilaroo, a type of trout) and in crocodiles.

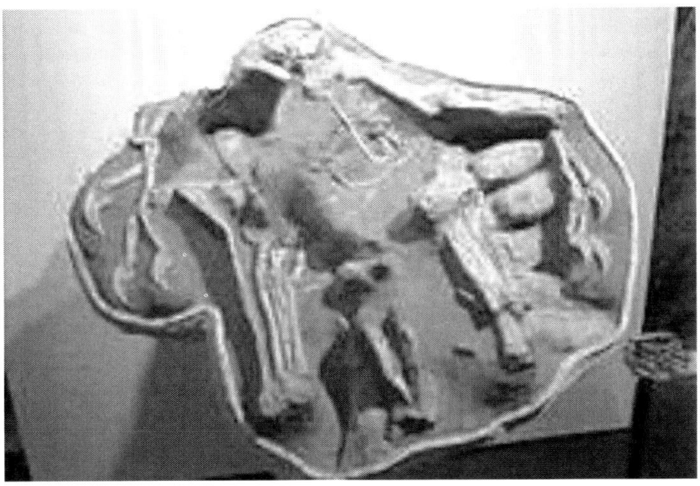

(Source: Wikipedia).

Figure 9. A nesting *Citipati osmolskae* specimen, at the American Museum of Natural History in New York.

5.15. Molecular Evidence and Soft Tissue

One of the best examples of soft tissue impressions in a fossil dinosaur was discovered in Petraroia, Italy. The discovery was reported in 1998, and described the specimen of a small, very young coelurosaur, *Scipionyx samniticus* (Figure 10). The fossil includes portions of the intestines, colon, liver, muscles, and windpipe of this immature dinosaur [Dal and Signore, 1998].

In the March 2005 issue of *Science*, Dr. Mary Higby Schweitzer and her team announced the discovery of flexible material resembling actual soft tissue inside a 68-million-year-old *Tyrannosaurus rex* leg bone from the Hell Creek Formation in Montana. After recovery, the tissue was rehydrated by the science team. The seven collagen types obtained from the bone fragments, compared to collagen data from living birds (specifically, a chicken), suggest that older theropods and birds are closely related.

When the fossilized bone was treated over several weeks to remove mineral content from the fossilized bone marrow cavity (a process called demineralization), Schweitzer found evidence of intact structures such as blood vessels, bone matrix, and connective tissue (bone fibers). Scrutiny under the microscope further revealed that the putative dinosaur soft tissue had retained fine structures (microstructures) even at the cellular level. The exact nature and composition of this material, and the implications of Dr. Schweitzer's discovery, are not yet clear; study and interpretation of the specimens is ongoing [Schweitzer et al., 2005].

The successful extraction of ancient DNA from dinosaur fossils has been reported on two separate occasions, but upon further inspection and peer review, neither of these reports could be confirmed [Wang et al., 1997]. However, a functional visual peptide of a theoretical dinosaur has been inferred using analytical phylogenetic reconstruction methods on gene sequences of related modern species such as reptiles and birds [Chang et al., 2002]. In addition, several proteins have putatively been detected in dinosaur fossils [Embery et al., 2003], including hemoglobin [Schweitzer, 1997].

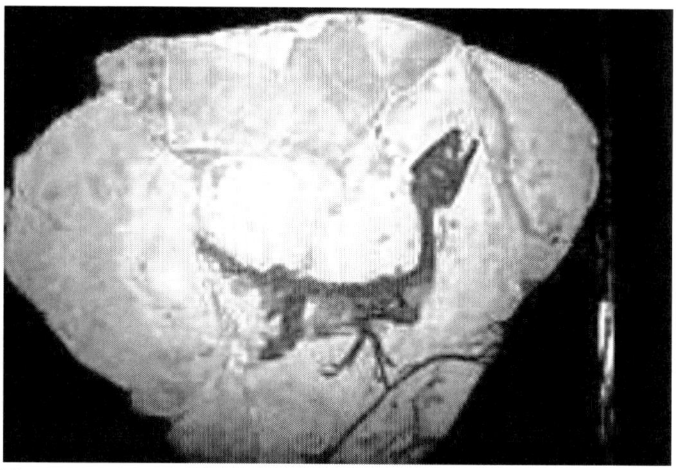

(Source: Wikipedia).

Figure 10. Fossil of a juvenile individual of *Scipionyx samniticus*. The fossil preserves clear traces of soft tissues.

Molecular Evidence of the Link between Birds and Dinosaurs

Scientists have confirmed the existence of protein in soft tissue recovered from the fossil bones of a 68 million-year-old Tyrannosaurus rex (T. rex) and a half-million-year-old mastodon. Schweitzer et al.. (2007) discovered soft tissue in the leg bone of a *T. rex* and other fossils recovered from the Hell Creek sediment formation in Montana. After her chemical and molecular analyses of the tissue indicated that original protein fragments might be preserved, she turned to colleagues John Asara and Lewis Cantley of Harvard Medical School, to see if they could confirm her suspicions by finding the amino acid used to make collagen, a fibrous protein found in bone. Bone is a composite material, consisting of both protein and mineral. In modern bones, when minerals are removed, a collagen matrix--fibrous, resilient material that gives the bones structure and flexibility--is left behind. When Schweitzer demineralized the *T. rex* bone, she was surprised to find such a matrix because current theories of fossilization held that no original organic material could survive that long. To see if the material had characteristics indicating the presence of collagen, which is plentiful, durable and has been recovered from other fossil materials, the scientists examined the resulting soft tissue with electron microscopy and atomic force microscopy. They then tested it against various antibodies that are known to react with collagen. Identifying collagen would indicate that it is original to *T. rex* -- that the tissue contains remnants of the molecules produced by the dinosaur. "*This is the breakthrough that says it's possible to get sequences beyond 1 million years*," said Cantley. "*At 68 million years, it's still possible*." Asara et al. (2007) successfully sequenced portions of the dinosaur and mastodon proteins, identifying the amino acids and confirming that the material was collagen. When they compared the collagen sequences to a database that contains existing sequences from modern species, they found that the *T. rex* sequence had similarities to those of chickens, and that the mastodon was more closely related to mammals, including the African elephant. The protein fragments in the *T. rex* fossil appear to most closely match amino acid sequences found in collagen of present-day chickens, lending support to the idea that birds and dinosaurs are evolutionarily related. "*Most people believe that birds evolved from dinosaurs, but that's based on the 'architecture' of the bones. This finding allows us the ability to say that they really are related because their sequences are related*" Asara said.

Origin of Avian Genome Size and Structure in Non-avian Dinosaurs

Avian genomes are small and streamlined compared with those of other amniotes, with fewer repetitive elements and less non-coding DNA (a typical bird genome consists of about 1.45 billion base pairs; human genomes are another billion base pairs longer). This condition has been suggested to represent a key adaptation for flight in birds, by reducing the metabolic costs associated with having large genome and cell sizes. However, the evolution of genome architecture in birds, or any other lineage, is difficult to study because genomic information is often absent for long-extinct relatives. Organ et al. (2007) found that bone-cell size correlates well with genome size in extant vertebrates, and used that relationship to estimate the genome sizes of 31 species of extinct dinosaur, including several species of extinct birds. Their results indicate that the small genomes typically associated with avian flight evolved in the saurischian dinosaur lineage between 230 and 250 million years ago, long before this lineage gave rise to the first birds. By comparison, ornithischian dinosaurs were inferred to have had much larger genomes, probably typical of ancestral Dinosauria. Using comparative genomic data, Organ et al. (2007) estimated that genome-wide interspersed mobile elements, a class of repetitive DNA, comprised 5–12% of the total genome size in the saurischian dinosaur lineage, but was 7–19% of total genome size in ornithischian dinosaurs, suggesting that repetitive elements became less active in the saurischian lineage. These genomic characteristics should be added to the list of attributes previously considered avian, but now thought to have arisen in non-avian dinosaurs, such as feathers, pulmonary innovations, and parental care and nesting.

5.16. Origin of Bird Flight

Debates about the origin of bird flight are almost as old as the idea that birds evolved from dinosaurs, which arose soon after the discovery of *Archaeopteryx* in 1862. Two theories have dominated most of the discussion since then: the cursorial ("from the ground up") theory proposes that birds evolved from small, fast predators that ran on the ground; the arboreal ("from the trees down") theory proposes that powered flight evolved from unpowered gliding by arboreal (tree-climbing) animals. A more recent theory, "wing-assisted incline running" (WAIR), is a variant of the cursorial theory and proposes that wings developed their aerodynamic functions as a result of the need to run quickly up very steep slopes such as trees, which would help small feathered dinosaurs escape from predators.

The Evolution of Feathers

The first known feathers are from a small coelurosaurian dinosaur called *Sinosauropteryx*, which has a row of small fringed structures along its vertebral column. Therefore, feathers are not a synapomorphy of birds, they are shared by a broader group within the theropod dinosaurs. This shows that feathers did not evolve specifically for flight [Padian et al., 1998].

Hypotheses for the original function of feathers include insulation, display and camouflage. These hypotheses are difficult to test, and it seems likely that feathers were used for several different purposes as they are now [Padian et al., 1998].

The Origin of Flight

The origin of flight argument is split into two theories; arboreal, "from the trees down", and cursorial, "from the ground up". These are summarised, and possible landing precision shown, in figure 11.

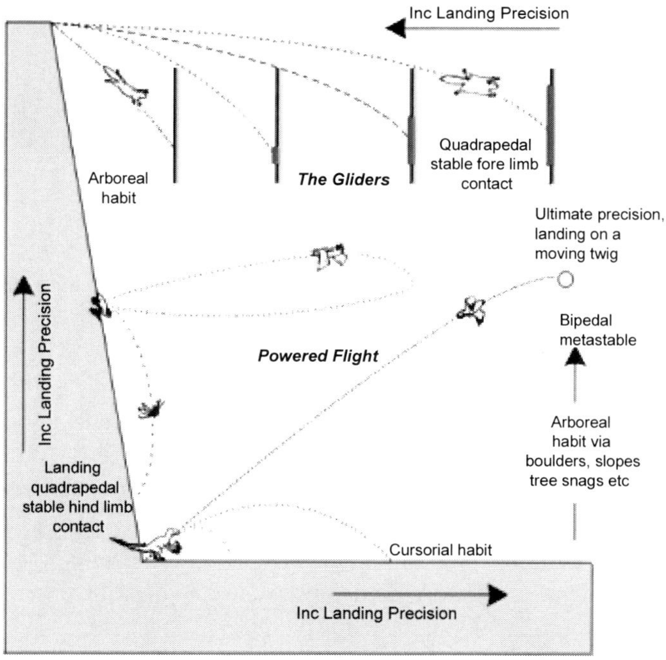

Figure 11. Flying modes showing how, mode and type of habit affects landing precision. Redrawn from Padian et al. (1998).

The Arboreal ("from the trees down") Theory

The arboreal theory is more intuitive in that flight evolving from an arboreal gliding stage would seem to be relatively easy. However, this theory has little support from comparative biology as it requires the ability to climb trees and to glide. Neither capacity seems to be present in *Archaeopteryx* or in theropod dinosaurs. *Archaeopteryx* has none of the features of typical vertebrate gliders, nor is it aerodynamically designed to fly. It has been argued that the lateral grooves and curvature of the claws are an arboreal specialisation, but these have also been compared to those of ground-dwelling birds. The evidence for an arboreal lifestyle overall is weak. It should also be considered that large theropods such as *Allosaurus* and *Tyrannosaurus* have curved claws with deep lateral grooves, and they were obviously not arboreal. Also, palaeobotanical evidence shows an absence of large trees anywhere near the Solnhofen lagoons in which *Archaeopteryx* is preserved [Padian et al., 1998].

Most versions of the arboreal hypothesis state that the ancestors of birds were very small dinosaurs that lived in trees, springing from branch to branch. These small dinosaur already had feathers, which were co-opted by evolution to produce longer, stiffer forms that were useful in aerodynamics, eventually producing wings. Wings would have then evolved and become increasingly refined as devices to give the leaper more control, to parachute, to glide, and to fly in stepwise fashion. The arboreal hypothesis also notes that, for arboreal animals, aerodynamics are far more energy efficient, since such animals simply fall to achieve minimum gliding speeds [Chatterjee et al., 2004; Samuel et al., 2000].

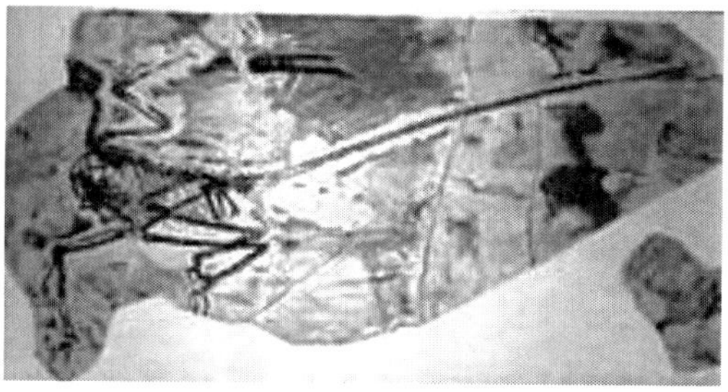

Figure 12. The four-winged *Microraptor*, a member of the Dromaeosauridae, a group of dinosaurs closely related to birds.

Several small dinosaurs from the Jurassic or Early Cretaceous, all with feathers, have been interpreted as possibly having arboreal and/or aerodynamic adaptations. These include *Scansoriopteryx*, *Epidexipteryx*, *Microraptor*, *Pedopenna*, and *Anchiornis*. *Anchiornis* is particularly important to this subject, as it lived at the beginning of the Late Jurassic, long before Archaeopteryx (Figure 12) [Hu et al., 2009].

Analysis of the proportions of the toe bones of the most primitive birds *Archaeopteryx* and *Confuciusornis*, compared to those of living species, suggest that the early species may have lived both on the ground and in trees [Hopson, 2001].

One study suggested that the earliest birds and their immediate ancestors did not climb trees. This study determined that the amount of toe claw curvature of early birds was more like that seen in modern ground-foraging birds than in perching birds [Glen and Bennett, 2007].

The Cursorial ("from the ground up") Theory

The cursorial theory is based on the evidence that *Archaeopteryx* was a strong, agile, biped, and from evidence that birds evolved from small, active, running predators. It has been suggested that feathers were used as nets to capture insects, but it has been shown that this action would cause the proto-bird to lose its balance by throwing off its angular momentum. As an alternative, it was suggested that insects were caught using the teeth with the arms held out laterally. An increased airfoil surface would increase lift and stability. This suggests that flight could have begun by running, leaping and sustaining short, leg-powered glides after prey or away from predators [Padian et al., 1998].

Criticisms of this include the problems of drag and needing to work against gravity. It is biomechanically easier to evolve flight from gliding than from the ground, although this does not indicate that it evolved this way. The other problem is the ground speed that would be required to reach a typical flight speed of 6-7 m/s. Modern lizards have been reported to reach theses speeds, but it is not known whether proto-birds could reach such speeds. There has been some doubt as to whether selection could improve both the forelimb and hindlimb at the same. However, decoupling of the fore- and hindlimbs was achieved in the coelurosaurs and the tail and hindlimb were progressively decoupled in the earliest birds [Padian et al., 1998].

Any model also needs to account for the evolution of the flight stroke. It has been shown that the immediate sister groups of birds, *Deinonychus* and the other dromaeosaurs, already had the sideways-flexing wrist joint that in birds is essential for the production of thrust. In dinosaurs, this feature was used as a prey-seizing stroke [Padian et al., 1998].

It would have only taken a slight adjustment of the angle of attack of this predatory stroke to create a suitable vortex wake. By running, leaping and a few such strokes, extension of the time in the air, and eventually flight from the ground up could have evolved. This idea requires no features not already known from fossils. It has been suggested that advantage may have been taken of any ridge or incline and so the model does have an element of the arboreal theory [Padian et al., 1998].

Overall, the evidence seems to point to a modified version of the cursorial theory of bird. Running leaps were aided by wings outstretched for balance; the wings were expanded at the distal ends for increased stability. The leaps were gradually extended by short flapping motions that elaborated the down and forward motion already present in the sister groups of the first birds. Running and leaping may have been enhanced by "ridge-gliding" or jumping from small heights [Padian et al., 1998].

The cursorial theory of the origin of flight was first proposed by Samuel Wendell Williston, and elaborated upon by Baron Nopcsa. This hypothesis proposes that some fast-running animals with long tails used their arms to keep their balance while running. Modern versions of this theory differ in many details from the Williston-Nopcsa version, mainly as a result of discoveries since Nopcsa's time.

Nopcsa theorized that increasing the surface area of the outstretched arms could have helped small cursorial predators keep their balance, and that the scales of the forearms elongated, evolving into feathers. The feathers could also have been used to trap insects or other prey. Progressively, the animals leapt for longer distances, helped by their evolving wings. Nopcsa also proposed three stages in the evolution of flight. First, animals developed passive flight, in which developing wing structures served as a sort of parachute. Second, they achieved active flight by flapping the wings. He used *Archaeopteryx* as an example of this second stage. Finally, birds gained the ability to soar [Terres, 1980].

Current thought is that feathers did not evolve from scales, as feathers are made of different proteins. More seriously, Nopcsa's theory assumes that feathers evolved as part of the evolution of flight, and recent discoveries prove that assumption is false

(Source: Wikipedia).

Figure 13. Reconstruction of *Rahonavis*, a ground-dwelling feathered dinosaur that some researchers think was well-equipped for flight.

Feathers are very common in coelurosaurian dinosaurs (including the early tyrannosauroid *Dilong*) [Prum and Brush, 2002]. Modern birds are classified as coelurosaurs by nearly all palaeontologists [Mayr et al., 2005] though not by a few ornithologists [Feduccia, 1999]. The modern version of the "from the ground up" hypothesis argues that birds' ancestors were small, *feathered*, ground-running predatory dinosaurs (rather like roadrunners in their hunting style) that used their forelimbs for balance while pursuing prey, and that the forelimbs and feathers later evolved in ways that provided gliding and then powered flight. The most widely suggested original functions of feathers include thermal insulation and competitive displays, as in modern birds [Burgers and Chiappe, 1999].

All of the *Archaeopteryx* fossils come from marine sediments, and it has been suggested that wings may have helped the birds run over water in the manner of the *Jesus Christ Lizard* (common basilisk) [Videler, 2005].

Most recent refutations of the "from the ground up" hypothesis attempt to refute the modern version's assumption that birds are modified coelurosaurian dinosaurs. The strongest attacks are based on embryological analyses that conclude that birds' wings are formed from digits 2, 3, and 4, corresponding to the index, middle, and ring fingers in humans. The first of a bird's three digits forms the alula, which they use to avoid stalling in low-speed flight—for example, when landing. The hands of coelurosaurs, however, are formed by digits 1, 2, and 3 (thumb and first two fingers in humans). However, these embryological analyses were immediately challenged on the embryological grounds that the "hand" often develops differently in clades that have lost some digits in the course of their evolution, and that birds' "hands" do develop from digits 1, 2, and 3 [Chatterjee, 1998; Vargas, 2004].

Wing-assisted Incline Running

The wing-assisted incline running (WAIR) hypothesis was prompted by observation of young chukar chicks, and proposes that wings developed their aerodynamic functions as a result of the need to run quickly up very steep slopes such as tree trunks, for example to escape from predators. This makes it a specialized type of cursorial ("from the ground up") theory. Note that in this scenario birds need *downforce* to give their feet increased grip [Bundle and Dial, 2003; Dial, 2003]. But early birds, including *Archaeopteryx*, lacked the shoulder mechanism by which modern birds' wings produce swift, powerful upstrokes. Since the downforce WAIR depends on upstrokes, it seems that early birds were incapable of WAIR [Senter, 2006]. Because WAIR is a behavioural trait without osteological specializations. The phylogenetic placement of the flight stroke before the divergence of Neornithes makes it impossible to determine if WAIR is ancestral to the avian flight stroke or derived from it [Dececchi, 2011].

Flight Capabilities of Archaeopteryx

The general consensus is that Archaeopteryx was a weak flier. This is supported by two arguments. The first is that *Archaeopteryx* lacks evidence of a supracoracoideus system, which in birds is the tendon that powers the upstroke. Experiments have shown that pigeons with a severed supracoracoideus tendon cannot take-off from ground level, cannot maintain level flight and cannot land safely. However, it is questionable whether Archaeopteryx can be compared with modern birds in this way [Padian et al., 1998].

The second argument is that the feathers of *Archaeopteryx* are asymmetrical, which suggests that it was capable of some flight. However, data shows that the asymmetry is less than that of modern fliers and gliders and this argument is only valid if it is assumed that fossil animals must look exactly like modern ones when performing the same function [Padian et al., 1998].

Generally it is thought that *Archaeopteryx* could glide as well as most modern birds and fly by flapping to some extent. The evidence for this is; 1) its wing planform and size are like those of some modern weakly flying birds, 2) the flight feathers are well-developed, 3) the sternum was a strong site of origin for flight muscles, and 4) its aerodynamic planform is unlike that of birds that only glide [Padian et al., 1998].

Flight Improvements after Archaeopteryx

Phylogenetic analysis has shown that many of the characteristics associated with the origin of flight were already present in non-avian theropod dinosaurs before birds evolved. Feathers evolved in non-avian coelurosaurs whose forelimbs were too short to bear functional wings. *Archaeopteryx* had a forelimb longer than the hindlimb, and flight feathers on the wings and long tail feathers. It is thought to have had a fully evolved flight stroke capable of generating thrust as well as lift. The presence of an alula in the early Cretaceous Eoalulavis shows that the wing mechanism that allowed flying at lower speeds and to manoeuvre like living birds evolved early in bird history. *Iberomesornis*, of the early Cretaceous, has features diagnostic of a perching ability [Padian et al., 1998].

5.17. DIMINISHED SIGNIFICANCE OF *ARCHAEOPTERYX*

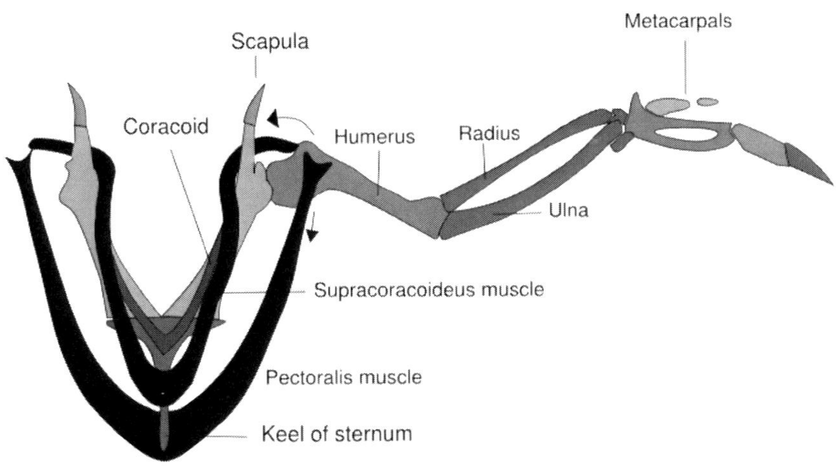

Figure 14. The supracoracoideus works using a pulley-like system to lift the wing while the pectorals provide the powerful downstroke.

Archaeopteryx was the first and for a long time the only known feathered Mesozoic animal. As a result, discussion of the evolution of birds and of bird flight centered on *Archaeopteryx* at least until the mid-1990s.

There has been debate about whether *Archaeopteryx* could really fly. It appears that *Archaeopteryx* had the brain structures and inner-ear balance sensors that birds use to control their flight [Alonso et al., 2004]. *Archaeopteryx* also had a wing feather arrangement like that of modern birds and similarly asymmetrical flight feathers on its wings and tail. But *Archaeopteryx* lacked the shoulder mechanism by which modern birds' wings produce swift, powerful upstrokes (see figure 14, above of supracoracoideus pulley); this may mean that it and other early birds were incapable of flapping flight and could only glide.

But the discovery since the early 1990s of many feathered dinosaurs means that *Archaeopteryx* is no longer the key figure in the evolution of bird flight. Other small, feathered coelurosaurs from the Cretaceous and Late Jurassic show possible precursors of avian flight. These include *Rahonavis* (Figure 13), a ground-runner with a *Velociraptor*-like raised sickle claw on the second toe, that some paleontologists assume to have been better adapted for flight than *Archaeopteryx*, *Scansoriopteryx*, an arboreal dinosaur that may support the "from the trees down" theory [Zhang et al., 2007], and *Microraptor*, an arboreal dinosaur possibly capable of powered flight but, if so, more like a biplane, as it had well-developed feathers on its legs [Chatterjee and Templin, 2007]. As early as 1915, some scientists argued that the evolution of bird flight may have gone through a four-winged (or*tetrapteryx*) stage [Beebe, 1915].

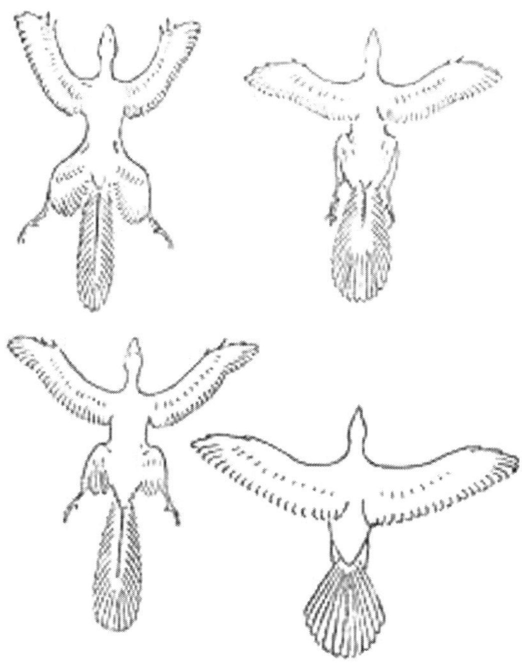

Figure 15. Proposed development of flight in a book from Lucas, Frederic, A. 1922: Tetrapteryx, *Archaeopteryx*, Hypothetical Stage, Modern Bird.

5.18. SECONDARY FLIGHTLESSNESS IN DINOSAURS

A hypothesis, credited to Gregory Paul and propounded in his books *Predatory Dinosaurs of the World* (1988) and *Dinosaurs of the Air* (2002), suggests that some groups of non-flying carnivorous dinosaurs—especially deinonychosaurs, and others such as oviraptorosaurs, therizinosaurs,alvarezsaurids and ornithomimosaurs—actually descend from birds. Paul also proposed that the bird ancestor of these groups were more advanced in its flight adaptations than *Archaeopteryx*. This would mean that *Archaeopteryx* is thus less closely related to extant birds than these dinosaurs are [Paul, 2002].

Paul's hypothesis received additional support when Mayr et al. (2005) analyzed a new, tenth specimen of *Archaeopteryx*, and concluded that *Archaeopteryx* was the sister clade to the Deinonychosauria, but that the more advanced bird *Confuciusornis* was within the Dromaeosauridae. This result supports Paul's hypothesis, suggesting that the Deinonychosauria and the Troodontidae are part of Aves, the bird lineage proper, and secondarily flightless [Mayr, 2006]. This paper, however, excluded all other birds and thus did not sample their character distributions. The paper was criticized by Corfe and Butler (2006) who found the authors could not support their conclusions statistically. Mayr et al. agreed that the statistical support was weak, but added that it is also weak for the alternative scenarios [Corfe and Butler, 2006].

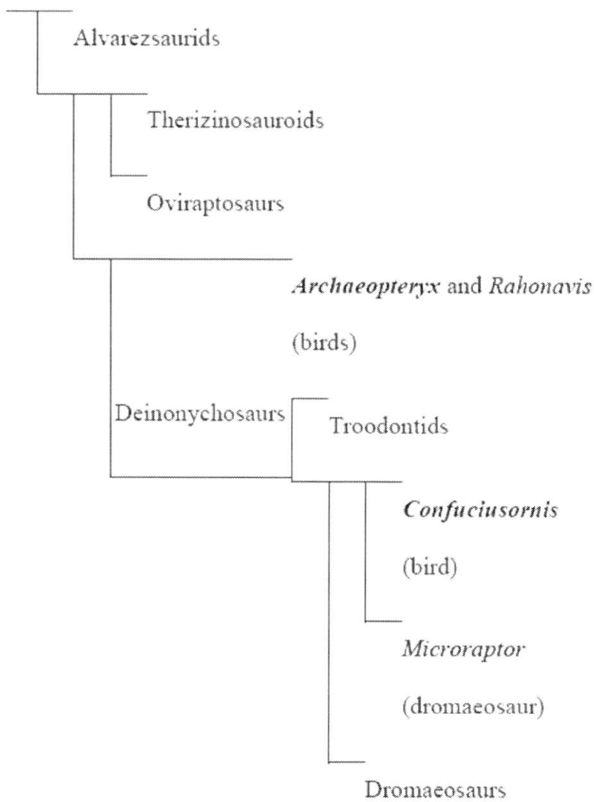

Figure16. Simplified cladogram from Mayr et al. (2005) Groups usually regarded as birds are in bold type.

Current cladistic analyses do not support Paul's hypothesis about the position of *Archaeopteryx*. Instead, they indicate that *Archaeopteryx* is closer to birds, within the clade *Avialae*, than it is to deinonychosaurs or oviraptorosaurs. However, some fossils support the version of this theory that holds that some non-flying carnivorous dinosaurs may have had flying ancestors. In particular—*Microraptor*, *Pedopenna*, and *Anchiornis* all have winged feet, share many features, and lie close to the base of the clade Paraves. This suggests that the ancestral paravian was a four-winged glider, and that larger Deinonychosaurs secondarily lost the ability to glide, while the bird lineage increased in aerodynamic ability as it progressed.

5.19. Digit Homology

There is a debate between embryologists and paleontologists whether the hands of theropod dinosaurs and birds are essentially different, based on phalangeal counts, a count of the number of phalanges (fingers) in the hand. This is an important and fiercely debated area of research because its results may challenge the consensus that birds are descendants of dinosaurs.

Embryologists and some paleontologists who oppose the bird-dinosaur link, have long numbered the digits of birds II-III-IV on the basis of multiple studies of the development in the egg [Chatterjee, 1998]. This is based on the fact that in most amniotes, the first digit to form in a 5-fingered hand is digit IV, which develops a primary axis. Therefore, embryologists have identified the primary axis in birds as digit IV, and the surviving digits as II-III-IV. The fossils of advanced theropod (Tetanurae) hands appear to have the digits I-II-III (some genera within Avetheropoda also have a reduced digit IV). If this is true, then the II-III-IV development of digits in birds is an indication against theropod (dinosaur) ancestry. However, with no ontogenical (developmental) basis to definitively state which digits are which on a theropod hand (because no non-avian theropods can be observed growing and developing today), the labelling of the theropod hand is not absolutely conclusive.

Paleontologists have traditionally identified avian digits as I-II-III. They argue that the digits of birds number I-II-III, just as those of theropod dinosaurs do, by the conserved phalangeal formula. The phalangeal count for archosaurs is 2-3-4-5-3; many archosaur lineages have a reduced number of digits, but have the same phalangeal formula in the digits that remain. In other words, paleontologists assert that archosaurs of different lineages tend to lose the same digits when digit loss occurs, from the outside to the inside. The three digits of dromaeosaurs, and *Archaeopteryx* have the same phalangeal formula of I-II-III as digits I-II-III of basal archosaurs. Therefore, the lost digits would be V and IV. If this is true, then modern birds would also possess digits I-II-III [Chatterjee, 1998]. Also, one research team has proposed a frame-shift in the digits of the theropod line leading to birds (thus making digit I into digit II, II to III, and so forth) [Wagner and Gauthier, 1999]. However, such frame shifts are rare in amniotes and—to be consistent with the theropod origin of birds—would have had to occur solely in the bird-theropod lineage forelimbs and not the hindlimbs (a condition unknown in any animal). This is called *Lateral Digit Reduction* (LDR) versus *Bilateral Digit Reduction* (BDR) (see also *Limusaurus*) [Vargas et al., 2009].

5.20. EVOLUTION OF BIRDS

The **evolution of birds** is thought to have begun in the Jurassic Period, with the earliest birds derived from a clade of theropoda dinosaurs named Paraves. Birds are categorized as a biological class, **Aves**. The earliest known species of class Aves is *Archaeopteryx lithographica*, from the Late Jurassic period, though *Archaeopteryx* is not commonly considered to have been a true bird. Modern phylogenies place birds in the dinosaur clade Theropoda. According to the current consensus, Aves and a sister group, the order Crocodilia, together are the sole living members of an unranked "reptile" clade, the Archosauria.

Phylogenetically, Aves is usually defined as all descendants of the most recent common ancestor of a specific modern bird species (such as the House Sparrow, *Passer domesticus*), and either *Archaeopteryx* [Padian and Chiappe, 1997], or some prehistoric species closer to Neornithes (to avoid the problems caused by the unclear relationships of *Archaeopteryx* to other theropods) [Gauthier, 1986]. If the latter classification is used then the larger group is termed Avialae. Currently, the relationship between dinosaurs, *Archaeopteryx*, and modern birds is still under debate.

There is significant evidence that birds emerged within theropod dinosaurs, specifically, that birds are members of Maniraptora, a group of theropods which includes dromaeosaurs and oviraptorids, among others [Hou et al., 1996]. As more non-avian theropods that are closely related to birds are discovered, the formerly clear distinction between non-birds and birds becomes less so. This was noted already in the 19th century, with Thomas Huxley writing:

> We have had to stretch the definition of the class of birds so as to include birds with teeth and birds with paw-like fore limbs and long tails. There is no evidence that Compsognathus possessed feathers; but, if it did, it would be hard indeed to say whether it should be called a reptilian bird or an avian reptile [Huxley, 1876].

Discoveries in northeast China (Liaoning Province) demonstrate that many small theropod dinosaurs did indeed have feathers, among them the compsognathid *Sinosauropteryx* and the microraptorian dromaeosaurid *Sinornithosaurus*. This has contributed to this ambiguity of where to draw the line between birds and reptiles [Norell and Ellison, 2005]. *Cryptovolans*, a dromaeosaurid found in 2002 (which may be a junior synonym of *Microraptor*) was capable of powered flight, possessed a sternal keel and had ribs with uncinate processes. *Cryptovolans* seems to make a better "bird" than Archaeopteryx which lacks some of these modern bird features. Because of this, some paleontologists have suggested that dromaeosaurs are actually basal birds whose larger members are secondarily flightless, i.e., that dromaeosaurs evolved from birds and not the other way around. Evidence for this theory is currently inconclusive, but digs continue to unearth fossils (especially in China) of feathered dromaeosaurs. At any rate, it is fairly certain that flight utilizing feathered wings existed in the mid-Jurassic theropods. The Cretaceous unenlagiine *Rahonavis* also possesses features suggesting it was at least partially capable of powered flight.

Although ornithischian (bird-hipped) dinosaurs share the same hip structure as birds, birds actually originated from the saurischian (lizard-hipped) dinosaurs if the dinosaurian origin theory is correct. They thus arrived at their hip structure condition independently. In fact, a bird-like hip structure also developed a third time among a peculiar group of theropods, the Therizinosauridae.

An alternate theory to the dinosaurian origin of birds, espoused by a few scientists, notably Larry Martin and Alan Feduccia, states that birds (including maniraptoran "dinosaurs") evolved from early archosaurs like Longisquama [Feduccia et al., 2005]. This theory is contested by most other paleontologists and experts in feather development and evolution [Prum, 2003].

5.21. Mesozoic Birds

The basal bird *Archaeopteryx*, from the Jurassic, is well known as one of the first "missing links" to be found in support of evolution in the late 19th century. Though it is not considered a direct ancestor of modern birds, it gives a fair representation of how flight evolved and how the very first bird might have looked. It may be predated by *Protoavis texensis*, though the fragmentary nature of this fossil leaves it open to considerable doubt whether this was a bird ancestor. The skeleton of all early bird candidates is basically that of a small theropod dinosaur with long, clawed hands, though the exquisite preservation of the Solnhofen Plattenkalk shows *Archaeopteryx* was covered in feathers and had wings [Huxley, 1876]. While *Archaeopteryx* and its relatives may not have been very good fliers, they would at least be competent gliders, setting the stage for the evolution of life on the wing.

The evolutionary trend among birds has been the reduction of anatomical elements to save weight. The first element to disappear was the bony tail, being reduced to a pygostyle and the tail function taken over by feathers. *Confuciusornis* is an example of their trend. While keeping the clawed fingers, perhaps for climbing, it had a pygostyle tail, though longer than in modern birds. A large group of birds, the Enantiornithes, evolved into ecological niches similar to those of modern birds and flourished throughout the Mesozoic. Though their wings resembled those of many modern bird groups, they retained the clawed wings and a snout with teeth rather than a beak in most forms.

The Cretaceous saw the rise of more modern birds with a more rigid ribcage with a carina and shoulders able to allow for a powerful upstroke, essential to sustained powered flight. They also had a more derived pygostyle, with a ploughshare-shaped end. An early example is *Yanornis*. Many were coastal birds, strikingly resembling modern shorebirds, like *Ichthyornis*, or ducks, like *Gansus*. Some evolved as swimming hunters, like the Hesperornithiformes – a group of flightless divers resembling grebes and loons. While modern in most respects, most of these birds retained typical reptilian-like teeth and sharp claws on the manus.

The modern toothless birds evolved from the toothed forefathers in the Cretaceous [Hope, 2002]. While the earlier primitive birds, particularly the Enantiornithes, continued to thrive and diversify alongside the pterosaurs, all but a few groups of the toothless Neornithes were cut short at the Chicxulub impact. The surviving lineages of birds were the comparatively primitive Paleognathae (ostrich and its allies), the aquatic duck lineage, the terrestrial fowl, and the highly volant Neoaves.

5.22. Adaptive Radiation of Modern Birds

Modern birds are classified in Neornithes, which are now known to have evolved into some basic lineages by the end of the Cretaceous (see *Vegavis*). The Neornithes are split into the paleognaths and neognaths.

The paleognaths include the tinamous (found only in Central and South America) and the ratites, which nowadays are found almost exclusively on the Southern Hemisphere. The ratites are large flightless birds, and include ostriches, rheas, cassowaries, kiwis and emus. A few scientists propose that the ratites represent an artificial grouping of birds which have independently lost the ability to fly in a number of unrelated lineages [Phillips, 2010]. In any case, the available data regarding their evolution is still very confusing, partly because there are no uncontroversial fossils from the Mesozoic.

The basal divergence from the remaining Neognathes was that of the Galloanserae, the superorder containing the Anseriformes (ducks,geese and swans), and the Galliformes (chickens, turkeys, pheasants, and their allies). The presence of basal anseriform fossils in the Mesozoic and likely some galliform fossils implies the presence of paleognaths at the same time, in spite of the absence of fossil evidence.

The dates for the splits are a matter of considerable debate amongst scientists. It is agreed that the Neornithes evolved in the Cretaceous and that the split between the Galloanserae and the other neognaths - the Neoaves - occurred before the Cretaceous–Paleogene extinction event, but there are different opinions about whether the radiation of the remaining neognaths occurred before or after the extinction of the other dinosaurs [Ericson et al., 2006]. This disagreement is in part caused by a divergence in the evidence, with molecular dating suggesting a Cretaceous radiation, a small and equivocal neoavian fossil record from Cretaceous, and most living families turning up during the Paleogene. Attempts made to reconcile the molecular and fossil evidence have proved controversial [Brown et al., 2007].

On the other hand, two factors must be considered: First, molecular clocks cannot be considered reliable in the absence of robust fossil calibration, whereas the fossil record is naturally incomplete. Second, in reconstructed phylogenetic trees, the time and pattern of lineage separation corresponds to the evolution of the *characters* (such as DNA sequences, morphological traits etc.) studied, *not* to the actual evolutionary pattern of the lineages; these ideally should not differ by much, but may well do so in practice.

Considering this, it is easy to see that fossil data, compared to molecular data, tends to be more accurate in general, but also to underestimate divergence times: morphological traits, being the product of entire developmental genetics networks, usually only start to diverge some time *after* a lineage split would become apparent in DNA sequence comparison - especially if the sequences used contain many silent mutations.

The Cretaceous Radiation

Avians radiated extensively in the Cretaceous, and a cladogram of the basal avian taxa with stratigraphic ranges is shown in figure 17.

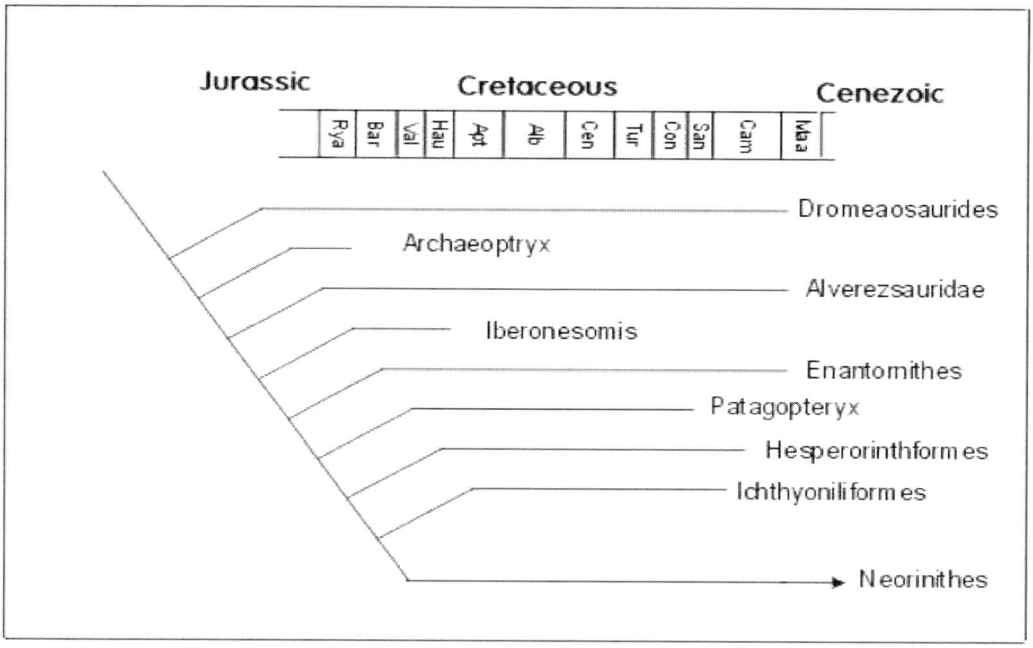

Figure 17. Cladogram of birds.

Mononykus

Mononykus, of the Alvarezsauridae, is the most recent dramatic find and occupies the most basal position on the avian family tree. It is the size of a turkey and has a skull with tiny teeth, an elongated neck, and very short and unusual forelimbs [Chiappe, 1995]. The phylogenetic position of *Mononykus* is supported by the presence of several characters diagnostic of Aves. These characters are concentrated in the braincase, the teeth and the tail (Figure 18). Cladistic analysis of dromaeosaurid theropods and birds places *Mononykus* as the sister taxon of all birds other than *Archaeopteryx* [Padian et al., 1998].

Figure 18. Skeleton of Mononykus, redrawn from Paddin (1998).

Mononykus is very unusual, especially its forelimb which is nothing like a flying wing. It is not understood why, so early in avian evolution, flying was abandoned. Also, the function of this strange forelimb is not understood. The rest of the skeleton is not dissimilar to the homologous elements of *Archaeopteryx* and other early birds. *Mononykus* is late Cretaceous in age and the split of its lineage from that of other birds must have taken place by the early Cretaceous, and since then autapomorphies may have evolved (such as the forelimb) that should not be used in phylogenetic analysis. It is unknown if *Mononykus* lost flight secondarily, as less derived members of the clade are not known [Padian et al., 1998].

Other early birds are providing a more complete picture of the early evolution of birds. The curved foot claws of the sparrow-sized *Iberomesornis*, of the early Cretaceous, provide evidence for an arboreal habitat, although the tail, pelvis and hindlimb are not much more advanced than in *Archaeopteryx* [Padian et al., 1998].

Enantiornithes

These are the most diverse and widely distributed group of early birds. They include about 15 taxa from the early Cretaceous, known from all continents except Antarctica. The anatomy of this group is highly variable. The best known skull is from *Cathayornis*, in which the teeth are unserrated and have constricted bases. This is the same as in all the toothed birds. The furcula is U-shaped and many of these birds have a grasping, perching foot suggestive of an arboreal lifestyle. Other enantiornithines have been interpreted as having wading and aquatic lifestyles. Most were small but some had wingspans of up to 1m across [Padian et al., 1998].

Most of the known Mesozoic avian diversity can be grouped within the Enantiornithes. This clade documents a major radiation of basal birds during the 80 million years of the Cretaceous [Padian et al., 1998].

Patagopteryx

The late Cretaceous *Patagopteryx* is another basal avian. It was chicken-sized, cursorial and flightless, with reduced wings. It has been interpreted to acquire flightlessness independently from the ratites and other modern flightless birds [Padian et al., 1998].

Ornithurae

The Ornithurae includes Neornithes and its sister groups Hesperornithiformes and Ichthyorniformes. *Hesperornis* is part of a radiation of secondarily flightless divers from the late Cretaceous. Ichthyornithiformes are known from marine late Cretaceous rocks, they were tern-like, but otherwise little is known about their anatomy [Padian et al., 1998].

Cretaceous Extinctions

None of the groups discussed so far are known to survive into the Tertiary. This has been used as evidence that birds underwent a dramatic Cretaceous-Tertiary bottleneck and then radiated in the early Tertiary. However, the fossil record of birds is so poor compared with other taxa that it is not really possible to distinguish whether extinction was massive or gradual [Padian et al., 1998].

5.23. Classification of Modern Species

The phylogenetic classification of birds is a contentious issue. Sibley & Ahlquist's *Phylogeny and Classification of Birds* (1990) is a landmark work on the classification of birds (although frequently debated and constantly revised). A preponderance of evidence suggests that most modern bird orders constitute good clades. However, scientists are not in agreement as to the precise relationships between the main clades. Evidence from modern bird anatomy, fossils and DNA has all been brought to bear on the problem but no strong consensus has emerged. As of the mid-2000s, new fossil and molecular data provide an increasingly clear picture of the evolution of modern bird orders, and their relationships. For example, the Charadriiformes seem to consititute an ancient and distinct lineage, while the Mirandornithes and Cypselomorphae are supported by a wealth of anatomical and molecular evidence. The understanding of the interrelationships of lower level taxa also continues to increase, particularly in the massively diverse perching bird group Passeriformes.

Bird classification and phylogenetic analysis is still under debate and requires more research. A recent 2008 study published in *Science* examined DNA sequences from 169 species of birds that represented all of the major extant groups. The findings may necessitate a wholesale restructuring of the avian phylogenetic tree. The findings also supported unestablished relationships between orders and confirmed disputes over particular groupings [Hackett, 2008].

5.24. Current Evolutionary Trends in Birds

Evolution generally occurs at a scale far too slow to be witnessed by humans. However, bird species are currently going extinct at a far greater rate than any possible speciation or other generation of new species. The disappearance of a population, subspecies, or species represents the permanent loss of a range of genes.

Another concern with evolutionary implications is a suspected increase in hybridization. This may arise from human alteration of habitats enabling related allopatric species to overlap. Forest fragmentation can create extensive open areas, connecting previously isolated patches of open habitat. Populations that were isolated for sufficient time to diverge significantly, but not sufficient to be incapable of producing fertile offspring may now be interbreeding so broadly that the integrity of the original species may be compromised. For example, the many hybrid hummingbirds found in northwest South America may represent a threat to the conservation of the distinct species involved [Fjeldså and Krabbe, 1990].

Several species of birds have been bred in captivity to create variations on wild species. In some birds this is limited to color variations, while others are bred for larger egg or meat production, for flightlessness or other characteristics.

CONCLUSION

There are several important points that have been raised which are as follows:

1. Birds evolved from the theropod dinosaurs. *Archaeopteryx* is the first known bird but was probably not the direct ancestor of modern birds.
2. Birds probably evolved flight from the ground up, possibly assisted by "ridge-gliding" or by jumping from small heights.
3. Feathers did not evolve for the purpose of flight. They were already present in coelurosaurian theropods not closely related to those that birds evolved from.
4. After *Archaeopteryx*, bird evolution focussed on the improvement of the flight apparatus. Other changes occurred later.
5. Secondary flightlessness evolved several times and arboreal lifestyles may have been adopted early in the evolution of birds.
6. The Tertiary bird record is very different to the Mesozoic one, suggesting a possible massive extinction.

To conclude, there is a lot that is still not known about the origin of birds. New fossils need to be found in order to fill the gaps in our knowledge. There are many puzzles, such as the function of *Mononykus'* unusual forelimb and much more work needs to be done before these puzzles can be solved.

The phylogeny of the early birds is constantly being reassessed and refined with every new discovery. The key to discovering the facts of avian origins and early evolution lies in this phylogenetic analysis.

REFERENCES

Alonso, P. D., Milner, A. C., Ketcham, R. A., Cokson, M. J. & Rowe, T. B. (2004). The avian nature of the brain and inner ear of Archaeopteryx. *Nature*, *430*, 666–9.

Asara, J. M., Schweitzer, M. H., Freimark, L. M., Phillips, M. Cantley, L. C. (2007). Protein Sequences from Mastodon and Tyrannosaurus Rex Revealed by Mass Spectrometry. *Science*, *316*, 280–5.

Bakker, R. T. (1972). Anatomical and ecological evidence of endothermy in dinosaurs. *Nature*, *238*, 81–85.

Barsbold, R. (1983). Carnivorous dinosaurs from the Cretaceous of Mongolia. *Trudy Soumestnaya Sovetsko-Mongol'skaya Paleontogicheskaya Ekspeditsiya* (in Russian), *19*, 1–117.

Beebe, C. W. A. (1915). Tetrapteryx stage in the ancestry of birds. *Zoologica*, *2*, 38–52.

Bhullar, B. A. S., Marugán-Lobón, J. S., Racimo, F., Bever, G. S., Rowe, T. B., Norell, M. A. & Abzhanov, A. (2012). Birds have paedomorphic dinosaur skulls. *Nature*, *487*, 223–226.

Brown, J. W., Payne, R. B. & Mindell, D. P. et al. (2007). Nuclear DNA does not reconcile 'rocks' and 'clocks' in Neoaves: a comment on Ericson et al. *Biol Lett*, *3*, 257–259.

Bundle, M. W. & Dial, K. P. (2003). Mechanics of wing-assisted incline running (WAIR). *J Exp Biol*, *206*, 4553–64.

Burgers, P. & Chiappe, L. M. (1999). The wing of Archaeopteryx as a primary thrust generator. *Nature*, *399*, 60–62.

Burke, A. C. & Feduccia, A. (1997). Developmental patterns and the identification of homologies in the avian hand. *Science*, *278*, 666–668.

Camp, C. L. (1936). A new type of small theropod dinosaur from the Navajo Sandstone of Arizona. *Bull Univ California Dept Geol Sci*, *24*, 39–65.

Carrano, M. R., Hutchinson, J. R. & Sampson, S. D. (2005). New information on *Segisaurus halli*, a small theropod dinosaur from the Early Jurassic of Arizona. *J Vert Paleontol*, *25*, 835–849.

Chambers, P. (2002). *Bones of Contention: The Archaeopteryx Scandals*. London: John Murray Ltd. 183–184.

Chang, B. S., Jönsson, K., Kazmi, M. A., Donoghue, M. J. & Sakmar, T. P. (2002). Recreating a Functional Ancestral Archosaur Visual Pigment. *Mol Biol Evol*, *19*, 1483–9.

Chatterjee, S. (1998). Counting the Fingers of Birds and Dinosaurs. *Science*, *280*, 355a–355.

Chatterjee, S. & Templin, R. J. (2004). Feathered coelurosaurs from China: new light on the arboreal origin of avian flight. 251-281. In Feathered Dragons: Studies on the Transition from Dinosaurs to Birds (Currie, P. J., Koppelhus, E. B., Shugar, M. A. & Wright, J.L. (eds.). Indiana University Press, Bloomington.

Chatterjee, S. & Templin, R. J. (2007). Biplane wing planform and flight performance of the feathered dinosaur Microraptor gui. *Proc Natl Acad Sci USA*, *104*, 1576–80.

Chatterjee, S. (1998). Counting the Fingers of Birds and Dinosaurs. *Science*, *280*, 355a.

Chen, P., Dong, Z. & Zhen, S. (1998). An exceptionally preserved theropod dinosaur from the Yixian Formation of China. *Nature*, *391*, 147–152.

Chiappe, L. M. (1997). "Aves". In Currie, P.J. & Padian, K. (eds.). *Encyclopedia of Dinosaurs*. San Diego: Academic Press. 45–50.

Chiappe, L. M. (2009). Downsized Dinosaurs: The Evolutionary Transition to Modern Birds. *Evol: Edu Outreach*, *2*, 248–256.

Chinsamy, A. & Hillenius, W. J. (2004). Physiology of nonavian dinosaurs. *The Dinosauria*, 2nd. 643–659.

Chipe, L., Norell, M. A. & Clark, J. M. (2002). The Cretaceous, short-armed Alvarezsauridae: *Mononykus* and its kin. In Chiappe, L.M. & Witmer, L.M. (eds.). *Mesozoic Birds: Above the Heads of Dinosaurs*. Berkeley: University of California Press. 87–119.

Cleland, T. P., Stoskopf, M. K. & Schweitzer, M. H. (2011). Histological, chemical, and morphological reexamination of the "heart" of a small Late *Cretaceous Thescelosaurus*. *Naturwissenschaften*, *98*, 203–211.

Corfe, I. J. & Butler, R. J. (2006). Comment on *A Well- Preserved Archaeopteryx Specimen With Theropod Features*. *Science*, *313*, 1238b.

Currie, P. J. & Chen, P. J. (2001). Anatomy of *Sinosauropteryx prima* from Liaoning, northeastern China. *Can J Earth Sci*, *38*, 705–727.

Dal Sasso, C. & Signore, M. (1998). Exceptional soft-tissue preservation in a theropod dinosaur from Italy. *Nature*, *392*, 383–387.

Darwin, C. R. (1859). *On the Origin of Species by Means of Natural Selection, or the Preservation of Favoured Races in the Struggle for Life*. London: John Murray. 502pp.

Dial, K. P. (2003). Wing-ssisted Incline Running and the Evolution of Flight. *Science*, *299*, 402–4.

Embery, G., Milner, A. C., Waddington, R. J., Hall, R. C., Langley, M. S. & Milan, A. M. (2003). Identification of proteinaceous material in the bone of the dinosaur Iguanodon. *Connect tiss res*, *44*, 41–6.

Ericson, P. G. P., Anderson, C. L., Britton, T., Elzanowski, A., Johansson, U. S., Kallersjo, M., Ohlson, J. I., Parsons, T. J. & Zuccon, D. et al. (2006). Diversification of Neoaves: integration of molecular sequence data and fossils. *Biol Lett*, *2*, 543–547.

Feduccia, A., Lingham-Soliar, T. & Hinchliffe, J. R. (2005). Do feathered dinosaurs exist? Testing the hypothesis on neontological and paleontological evidence. *J Morphol*, *266*, 125–166.

Feduccia, A. (1995). Explosive Evolution in Tertiary Birds and Mammals. *Science*, *267*, 637–8.

Feduccia, A. (1999). *The Origin and Evolution of Birds*. Yale University Press.

Feduccia, A. (2012). *Riddle of the Feathered Dragons: Hidden Birds of China*. Yale University Press.

Feduccia, A., Lingham-Soliar, T. & Hincliffe, J. R. (2005). Do feathered dinosaurs exist? Testing the hypothesis on neontological and paleontological evidence. *J Morpol*, *266*, 125–66.

Fisher, P. E., Russell, D. A., Stoskopf, M. K., Barrick, R. E., Hammer, M. & Kuzmitz, A. A. (2000). Cardiovascular evidence for an intermediate or higher metabolic rate in an ornithischian dinosaur. *Science*, *288*, 503–5.

Fjeldså, J. & Niels, K. (1990). *Birds of the High Andes: A Manual to the Birds of the Temperate Zone of the Andes and Patagonia, South America*. Apollo Books.

Forster, C. A., Sampson, S. D., Chiappe, L. M. & Krause, D. W. (1998). The theropod ancestry of birds: new evidence from the Late Cretaceous of Madagascar. *Science*, *279*, 1915–9.

Foth, C. (2012). On the identification of feather structures in stem-line representatives of birds: evidence from fossils and actuopalaeontology. *Paläontologische Zeitschrift*.

Gauthier, J. (1986). Saurischian Monophyly and the origin of birds. In Padian, K. *The Origin of Birds and the Evolution of Flight. Mem California Acad Sci*, *8*, 1–55.

Glen, C. L. & Bennett, M. B. (2007). Foraging modes of Mesozoic birds and non-avian theropods. *Curr Biol*, *17*, 911–2.

Hackett, S. J., Kimball, R. T., Reddy, S., Bowie, R. C. K., Braun, E. L., Braun, M. J., Chojnowski, J. L., Cox, W A. et al. (2008). A Phylogenomic Study of Birds Reveals Their Evolutionary History. *Science*, *320*, 1763–1768.

Heilmann, G. (1926). *The Origin of Birds*. London: Witherby. 208pp.

Hillenius, W. J. & Ruben, J. A. (2004). The evolution of endothermy in terrestrial vertebrates: Who? when? why? *Physiol Biochem Zool*, *77*, 1019–42.

Holtz, T. R. & Osmólska, H. (2004). "Saurischia". In Weishampel, D.B., Dodson, P. & Osmólska, H. (eds.). *The Dinosauria* (Second ed.). Berkeley: University of California Press. 21–24.

Hope, S. (2002). The Mesozoic Radiation of Neornithes. In Chiappe, L.M. & Witmer, L.M. *Mesozoic Birds: Above the Heads of Dinosaurs*. 339–388.

Hopson, J. A. (2001). Ecomorphology of avian and nonavian theropod phalangeal proportions: Implications for the arboreal versus terrestrial origin of bird flight. From New Perspectives on the Origin and Early Evolution of Birds: Proceedings of the International Symposium in Honor of Ostrom, J.H., Gauthier, J. & Gall, L.F. eds. New Haven: *Peabody Mus. Nat. Hist*., Yale Univ. Peabody Museum of Natural History, Yale University.

Horner, J. R. & Makela, R. (1979). Nest of juveniles provides evidence of family structure among dinosaurs. *Nature*, *282*, 296–298.

Hou, L. H., Zhou, Z., Martin, L. D. & Feduccia, A. (1995). A beaked bird from the Jurassic of China. *Nature*, *377*, 616–618.

Hou, L., Martin, M., Zhou, Z. & Feduccia, A. (1996). Early Adaptive Radiation of Birds: Evidence from Fossils from Northeastern China. *Science*, *274*, 1164–1167.

Hu, D., Hou, L. , Zhang, L. & Xu, X. (2009). A pre-*Archaeopteryx* troodontid theropod from China with long feathers on the metatarsus. *Nature*, *461*, 640–3.

Huxley, T. H. (1876). Lectures on Evolution. *New York Tribune*. Extra. no 36. In Collected Essays IV: 46-138

Huxley, T. H. (1868). On the animals which are most nearly intermediate between birds and reptiles. *Ann Mag Nat Hist*, *2*, 66–75.

Huxley, T. H. (1870). Further evidence of the affinity between the dinosaurian reptiles and birds. *Quar J Geol Soc London*, *26*, 12–31.

Ji, Q. & Ji, S. (1996). On the discovery of the earliest bird fossil in China and the origin of birds. *Chinese Geol*, *233*, 30–33.

Ji, Q., Philip, J. C., Philip, J., Norell, M. A. & Ji, S. (1998). Two feathered dinosaurs from northeastern China. *Nature*, *393*, 753–761.

Lee, A. H. & Werning, S. (2008). Sexual maturity in growing dinosaurs does not fit reptilian growth mdels. *Proc Nat Acad Sci*, *105*, 582–7.

Lingham-Soliar, T., Feduccia, A. & Wang, X. (2007). A new Chinese specimen indicates that 'protofeathers' in the Early Cretaceous theropod dinosaur Sinosauropteryx are degraded collagen fibres. *Proc R Soc B*, *274*, 1823–9.

Lipkin, C., Sereno, P. C. & Horner, J. R. (2007). The Furcula In Suchomimus Tenerensis And Tyrannosaurus Rex (Dinosauria: Theropoda: Tetanurae). *J Paleontol*, *81*, 1523–1527.

Makovicky, P. J., Pesteguía, S. & Agnolín, F. L. (2005). The earliest dromaeosaurid theropod from South America. *Nature*, *437*, 1007–11.

Martin, L. D. (2004). A basal archosaurian origin for birds. *Acta Zool Sinica*, *50*, 978–990.

Martinelli, Agustín G. & Vera, E. I. (2007). *Achillesaurus manazzonei*, a new alvarezsaurid theropod (Dinosauria) from the Late Cretaceous Bajo de la Carpa Formation, Río Negro Province, Argentina. *Zootaxa*, *1582*, 1–17.

Mayr, G. (2006). Response to Comment on *A Well-Preserved Archaeopteryx Specimen with Theropod Features*. *Science*, *313*, 1238c.

Mayr, G., Pohl, B. & Peters, D. S. (2005). A well-preserved*Archaeopteryx* specimen with theropod features. *Science*, *310*, 1483–6.

Mayr, G., Pohl, B. & Peters, D. S. (2005). A well-preserved *Archaeopteryx* specimen with theropod features. *Science*, *310*, 1483–6.

Mezentseva, N. V., Kumaratilake, J. S. & Newman, S. A. (2008). The brown adipocyte differentiation pathway in birds: An evolutionary road not taken. *BMC Biol*, *6*, 17.

Monastersky, R. (2000). All mixed up over birds and dinosaurs". *Sci News*, *157*, 38.

Newman, S. A. (2011). Thermogenesis, muscle hyperplasia, and the origin of birds. *BioEssays*, *33*, 653–656.

Newman, S. A., Mezentseva, N. V. & Badyaev, A. V. (2013). Gene loss, thermogenesis, and the origin of birds. *Ann Nw York Acad Sci*, *1289*, 36–47.

Nieuwland, I. J. J. (2004). Gerhard Heilmann and the artist's eye in science, 1912-1927. *J Vert Palaeontol*, *3*.

Nopcsa, F. (1907). Ideas on the origin of flight". *Proc Zool Soc London*, 223–238.

Norell, M. & Ellison, M. (2005). *Unearthing the Dragon, The Great Feathered Dinosaur Discovery* Pi Press, New York.

Norell, M. A., Clark, J. M., Dashzeveg, D., Barsbold, T., Chiappe, L. M., Davidson, A. R., McKenna, M. C. & Novacek, M. J. (1994). A theropod dinosaur embryo and the affinities of the Flaming Cliffs Dinosaur eggs. *Science*, *266*, 779–82.

Novas, F. E. & Pol, D. (2002). Alvarezsaurid relationships reconsidered". In Chiappe, L.M. & Witmer, L.M. (eds.). *Mesozoic Birds: Above the Heads of Dinosaurs*. Berkeley: University of California Press. 121–125.

O' Connor, P. M. & Claessens, L. P. A. M. (2005). Basic avian pulmonary design and flow-through ventilation in non-avian theropod dinosaurs. *Nature*, *436*, 253–6.

Osmólska, H., Maryańska, T. & Wolsan, M. (2002). Avialan status for Oviraptorosauria". *Acta Palaeontol Pol*, *47*, 97–116.

Ostrom, J. H. (1969). Osteology of *Deinonychus antirrhopus*, an unusual theropod from the Lower Cretaceous of Montana. *Bull Peabod Mus Nat Hist*, *30*, 1–165.

Ostrom, J. H. (1970). *Archaeopteryx*: Notice of a "new" specimen. *Science*, *170*, 537–8.

Ostrom, J. H. (1973). The ancestry of birds. *Nature*, *242*, 136–136.

Ostrom, J. H. (1975). The origin of birds. *Ann Rev Ear Planet Sci*, *3*, 55–77.

Ostrom, J. H. (1976). *Archaeopteryx* and the origin of birds. *Biol J Linn Soc*, *8*, 91–182.

Owen, R. (1863). On the *Archeopteryx* [sp] of von Meyer, with a description of the fossil remains of a long-tailed species, from the lithographic stone of Solenhofen. *Philos Trans R Soc London*, *153*, 33–47.

Padian, K. & Chiappe, L. M. (1997). Bird Origins. In Currie, P.J. & Padian, K. *Encyclopedia of Dinosaurs*. San Diego: Academic Press. 41–96.

Padian, K. (2004). Basal Avialae". In Weishampel, D.B., Dodson, P. & Osmólska, H. (eds.). *The Dinosauria* (Second ed.). Berkeley: University of California Press. pp. 210–231.

Paul, C. S., Ricardo, N. M., Jeffrey, A. W., David, J. V., Oscar, A. A. & Hans, C. E. L. (2008). Evidence for Avian Intrathoracic Air Sacs in a New Predatory Dinosaur from Argentina". In Kemp, Tom. *PLoS ONE*, *3*, e3303.

Paul, G. S. (2002). *Dinosaurs of the Air: The Evolution and Loss of Flight in Dinosaurs and Birds*. JHU Press.

Paul, G. S. (2002). Dinosaurs of the Air: The Evolution and Loss of Flight in Dinosaurs and Birds. Baltimore: Johns Hopkins University Press. pp 257

Paul, G. S. (2002). *Dinosaurs of the Air: The Evolution and Loss of Flight in Dinosaurs and Birds*. Baltimore: Johns Hopkins University Press. 472pp.

Perle, A., Norell, M. A., Chiappe, L. M. & Clark, J. M. (1993). Flightless bird from the Cretaceous of Mongolia. *Nature*, *362*, 623–626.

Phillips, M. J. et al. (2010). Tinamous and Moa Flock Together: Mitochondrial Genome Sequence Analysis Reveals Independent Losses of Flight among Ratites. *Syst Biol*, *59*, 90.

Prum, R. (2003). Are Current Critiques Of The Theropod Origin Of Birds Science? Rebuttal To Feduccia 2002. *Auk*, *120*, 550–561.

Prum, R. & Brush, A. H. (2002). The evolutionary origin and diversification of feathers. *Quart Rev Biol*, *77*, 261–95.

Rowe, T., McBride, E. F., Sereno, P. C., Russell, D. A., Fisher, P. E., Barrick, R. E. & Stoskopf, M. K. (2001). Dinosaur with a Heart of Stone. *Science*, *291*, 783.

Schweitzer, M. H., Marshall, M., Carron, K., Bohle, D. S., Busse, S. C., Arnold, E. V., Barnard, D., Horner, J. R. & Starkey, J. R. (1997). Heme compounds in dinosaur trabecular bone. *Proc Nat Acad Sci USA*, *94*, 6291–6.

Schweitzer, M. H., Wittmeyer, J. L. & Horner, J. R. (2005). Gender-specific reproductive tissue in ratites and *Tyrannosaurus rex*. *Science*, *308*, 1456–60.

Schweitzer, M. H., Wittmeyer, J. L., Horner, J. R. & Toporski, J. K. (2005). Soft-Tissue Vessels and Cellular Preservation in *Tyrannosaurus rex*. *Science*, *307*, 1952–5.

Schweitzer, M. H., Zheng, W., Organ C. L., Avci, R., Suo, Z., Freimark, L. M., Lebleu, V. S., Duncan, M. B., Vander, H. M. G., Neveu, J. M., Lane W. S., Cottrell, J. S., Horner, J. R., Cantley L. C., Kalluri, R. & Asara J. M. (2009). Biomolecular Characterization and Protein Sequences of the Campanian Hadrosaur B. canadensis. *Science*, *324*, 626–31.

Seeley, H. G. (1901). *Dragons of the Air: An Account of Extinct Flying Reptiles*. London: Methuen & Co., 239pp.

Senter, P. (2006). Scapular orientation in theropods and basal birds, and the origin of flapping flight. *Acta Palaeontol Pol*, *51*, 305–313.

Sereno, P. C. (1997). The origin and evolution of dinosaurs. *Ann Rev Earth Planet Sci*, *25*, 435–489.

Sereno, P. C. (1999). The evolution of dinosaurs. *Science*, *284*, 2137–47.

Sereno, P.C. & Rao, C. (1992). Early evolution of avian flight and perching: new evidence from the Lower Cretaceous of China. *Science*, *255*, 845–8.

Sloan, C. P. (1999). Feathers for *T. rex*? *Nat Geo*, *196*, 98–107.

Tarsitano, S. F., Russell, A. P., Hornel, F., Plummer, C. & Millerchip, K. (2000). On the Evolution of Feathers from an Aerodynamic and Constructional View Point. *Am Zool*, *40*, 676-686.

Terres, J. K. (1980). *The Audubon Society Encyclopedia of North American Birds*. New York, NY: Knopf. 398–401.

Turner, A. H., Pol, D., Clarke, J. A., Erickson, G. M. & Norell, M. A. (2007). A basal dromaeosaurid and size evolution preceding avian flight. *Science*, *317*, 1378–81.

Vargas, A. O. & Fallon, J. F. (2004). Birds have dinosaur wings: The molecular evidence. *J Exp Zool Part B: Mol Dev Evol*, *304B*, 86–90.

Videler, J. J. (2005). Avian Flight. Oxford University. Press, Oxford.

von Meyer, C. E. H. (1861). *Archaeopteryx lithographica* (Vogel-Feder) und *Pterodactylus* von Solnhofen. *Neues Jahrbuch für Mineralogie, Geologie und Paläontologie* (in German) 1861, 678–679.

Wagner, G. P. & Gautthier, J. A. (1999). 1,2,3 = 2,3,4: A solution to the problem of the homology of the digits in the avian hand. *Proc Natl Acad Sci USA*, *96*, 5111–6.

Walker, A. D. (1972). New light on the origin of birds and crocodiles. *Nature*, *237*, 257–263.

Wang, H. L., Yan, Z. Y. & Jin, D. Y. (1997). Reanalysis of published DNA sequence amplified from Cretaceous dinosaur egg fossil. *Mol Biol Evol*, *14*, 589–91.

Wellnhofer, P. (1988). Ein neuer Exemplar von *Archaeopteryx*. *Archaeopteryx*, *6*, 1–30.

Wings, O. (2007). A review of gastrolith function with implications for fossil vertebrates and a revised classification. *Palaeontol Pol*, *52*, 1–16.

Witmer, L. M. (2009). Feathered dinosaurs in a tangle. *Nature*, *461*, 601-602.

Xu, X. X., Norell, M. A., Kuang, X., Wang, X. Zhao, Q. & Jia, C. (2004). Basal tyrannosauroids from China and evidence for protofeathers in tyrannosauroids. *Nature*, *431*, 680–4.

Xu, X. X., Zhou, Z., Wang, X., Kuang, X., Zhang, F. & Du, X. (2003). Four-winged dinosaurs from China. *Nature*, *421*, 335–40.

Xu, X., Norell, M. A., Kuang, X., Wang, X., Zhao, Q. & Jia, C. (2004). Basal tyrannosauroids from China and evidence for protofeathers in tyrannosauroids. *Nature*, *431*, 680–4.

Xu, X. & Norell, M. A. (2004). A new troodontid dinosaur from China with avian-like sleeping posture. *Nature*, *431*, 838–41.

Xu, X., Tang, Z. & Wang, X. (1999). A therizinosaurid dinosaur with integumentary structures from China. *Nature*, *399*, 350–354.

Yates, A. M. & Vasconcelos, C. C. (2005). Furcula-like clavicles in the prosauropod dinosaur *Massospondylus*. *J Vert Paleontol*, *25*, 466–468.

Zhang, F., Kearns, S. L., Orr, P. J., Benton, M. J., Zhou, Z., Johnson, D., Xu, X. & Wang, X. (2010). Fossilized melanosomes and the colour of Cretaceous dinosaurs and birds. *Nature*, *463*, 1075–1078.

Zhang, F., Zhou, Z., Xu, X. & Wang, X. (2002). A juvenile coelurosaurian theropod from China indicates arboreal habits. *Naturwissenschaften*, *89*, 394–8.

Zhou, Z. & Zhang, F. (2002). A long-tailed, seed-eating bird from the Early Cretaceous of China. *Nature*, *418*, 405–9.

Chapter 6

BIRDS MIGRATION

6.1. BIRD MIGRATION: A BIRD EYE VIEW

Bird migration is the regular seasonal movement, often north and south along a flyway between breeding and wintering grounds, undertaken by many species of birds. Migration, which carries high costs in predation and mortality, including from hunting by humans, is driven primarily by availability of food. Migration occurs mainly in the Northern Hemisphere where birds are funnelled on to specific routes by natural barriers such as the Mediterranean Sea.

Historically, migration has been recorded as much as 3,000 years ago by Ancient Greek authors including Homer and Aristotle, and in the Book of Job, for species such as storks, Turtle Doves, and swallows. More recently, Johannes Leche began recording dates of arrivals of spring migrants in Finland in 1749, and scientific studies have used techniques including bird ringing and satellite tracking. Threats to migratory birds have grown with habitat destruction especially of stopover and wintering sites, as well as structures such as power lines and wind farms.

The Arctic Tern holds the long-distance migration record for birds, travelling between Arctic breeding grounds and the Antarctic each year. Some species of tubenoses (Procellariiformes) such as albatrosses circle the earth, flying over the southern oceans, while others such as Manx Shearwaters migrate 14,000 km (8,700 mi) between their northern breeding grounds and the southern ocean. Shorter migrations are common, including altitudinal migrations on mountains such as the Andes and Himalayas.

The timing of migration is controlled primarily by changes in day length. Migrating birds navigate using celestial cues from the sun and stars, the earth's magnetic field, and probably also mental maps. Migration has developed independently in different groups of birds and does not appear to require genetic change; some birds have acquired migratory behaviour since the last ice age.

Figure 1. Flocks of Common Starling Assembling before migration south wards.

Figure 2. Migrating waders in Roebuck Bay, Western Australia.

Migration is the regular seasonal movement, often north and south, undertaken by many species of birds. Bird movements include those made in response to changes in food availability, habitat, or weather. Sometimes, journeys are not termed "true migration" because they are irregular (nomadism, invasions, irruptions) or in only one direction (dispersal, movement of young away from natal area). Migration is marked by its annual seasonality [Berthold et al., 2001]. Non-migratory birds are said to be resident or sedentary. Approximately 1800 of the world's 10,000 bird species are long-distance migrants [Sekercioglu, 2007].

Many bird populations migrate long distances along a flyway. The most common pattern involves flying north in the spring to breed in the temperate or Arctic summer and returning in the autumn to wintering grounds in warmer regions to the south. Of course, in the Southern Hemisphere the directions are reversed, but there is less land area in the far South to support long-distance migration.

The primary motivation for migration appears to be food; for example, some hummingbirds choose not to migrate if fed through the winter. Also, the longer days of the northern summer provide extended time for breeding birds to feed their young. This helps diurnal birds to produce larger clutches than related non-migratory species that remain in the tropics. As the days shorten in autumn, the birds return to warmer regions where the available food supply varies little with the season.

These advantages offset the high stress, physical exertion costs, and other risks of the migration such as predation. Predation can be heightened during migration: the Eleonora's Falcon, which breeds on Mediterranean islands, has a very late breeding season, coordinated with the autumn passage of southbound passerine migrants, which it feeds to its young. A similar strategy is adopted by the Greater Noctule bat, which preys on nocturnal passerine migrants [Dondini and Vergari, 2000; Ibáñez et al., 2001; Popa-Lisseanu et al., 2007]. The higher concentrations of migrating birds at stopover sites make them prone to parasites and pathogens, which require a heightened immune response [Newton, 2008].

Within a species not all populations may be migratory; this is known as "partial migration". Partial migration is very common in the southern continents; in Australia, 44% of non-passerine birds and 32% of passerine species are partially migratory [Chan, 2001]. In some species, the population at higher latitudes tends to be migratory and will often winter at lower latitude. The migrating birds bypass the latitudes where other populations may be sedentary, where suitable wintering habitats may already be occupied. This is an example of *leap-frog migration* [Boland, 1990]. Many fully migratory species show leap-frog migration (birds that nest at higher latitudes spend the winter at lower latitudes), and many show the alternative, "chain migration" where populations 'slide' more evenly North and South without reversing order.

Within a population, it is common for different ages and/or sexes to have different patterns of timing and distance. Female Chaffinches in Eastern Fennoscandia migrate earlier in the autumn than males do [Panov and Ilya, 2011].

Most migrations begin with the birds starting off in a broad front. Often, this front narrows into one or more preferred routes termed flyways. These routes typically follow mountain ranges or coastlines, sometimes rivers, and may take advantage of updrafts and other wind patterns or avoid geographical barriers such as large stretches of open water. The specific routes may be genetically programmed or learned to varying degrees. The routes taken on forward and return migration are often different [Newton, 2008]. A common pattern in North America is clockwise migration, where birds flying North tend to be further West, and flying South tend to shift Eastwards.

Many, if not most, birds migrate in flocks. For larger birds, flying in flocks reduces the energy cost. Geese in a V-formation may conserve 12–20% of the energy they would need to fly alone [Hummel and Beukenberg, 1989; Cuts and Speakman, 1994]. Red Knots *Calidris canutus* and Dunlins *Calidris alpina* were found in radar studies to fly 5 km per hour faster in flocks than when they were flying alone [Newton, 2008].

Birds fly at varying altitudes during migration. An expedition to Mt. Everest found skeletons of Pintail and Black-tailed Godwit at 5000 m (16,400 ft) on the Khumbu Glacier [Geroudet, 1954]. Bar-headed Geese have been recorded by GPS flying at up to 6,540 metres while crossing the Himalayas, at the same time engaging in the highest rates of climb to altitude for any bird. Anecdotal reports of them flying much higher have yet to be corroborated with any direct evidence [Swan, 1970]. Seabirds fly low over water but gain altitude when crossing land, and the reverse pattern is seen in landbirds [Eastwood and Rider, 1965; Williams, 1950]. However, most bird migration is in the range of 150 m (500 ft) to 600 m (2000 ft). Bird-hit aviation records from the United States show most collisions occur below 600 m (2000 ft) and almost none above 1800 m (6000 ft) [Williams, 1950].

Bird migration is not limited to birds that can fly. Most species of penguin migrate by swimming. These routes can cover over 1000 km. Blue Grouse *Dendragapus obscures* perform altitudinal migration mostly by walking. Emus in Australia have been observed to undertake long-distance movements on foot during droughts [Newton, 2008].

6.2. HISTORICAL VIEWS

Records of bird migration were made 3,000 years ago by the Ancient Greek writers Hesiod, Homer, Herodotus and Aristotle. The Bible also notes migrations, as in the Book of Job (39:26), where the inquiry is made: *"Doth the hawk fly by Thy wisdom and stretch her wings toward the south?"* The author of Jeremiah (8:7) wrote: *"The stork in the heavens knoweth her appointed time; and the turtledove, and the crane, and the swallow, observe the time of their coming."*

Aristotle noted that cranes traveled from the steppes of Scythia to marshes at the headwaters of the Nile. Pliny the Elder, in his *Historia Naturalis*, repeats Aristotle's observations [Lincoln, 1979].

Swallow Migration versus Hibernation

Aristotle however suggested that swallows and other birds hibernated. This belief persisted as late as 1878, when Elliott Coues listed the titles of no less than 182 papers dealing with the hibernation of swallows [Cocker et al., 2005]. It was not until the end of the eighteenth century that migration as an explanation for the winter disappearance of birds from northern climes was accepted [Lincoln, 1979]. Thomas Bewick's *A History of British Birds* (Volume 1, 1797) mentions a report from "a very intelligent master of a vessel" who, "between the islands of Minorca and Majorca, saw great numbers of Swallows flying northward", and states the situation in Britain as follows:

Figure 3. Minoan fresco of swallows in springtime at Akrotiri, c. 1500 BC.

> Swallows frequently roost at night, after they begin to congregate, by the sides of rivers and pools, from which circumstance it has been erroneously supposed that they retire into the water.

—Bewick , Bewick (1797), then describes an experiment which succeeded in keeping swallows alive in Britain for several years, where they remained warm and dry through the winters. He concludes:

> These experiments have since been amply confirmed by ... M. Natterer, of Vienna ... and the result clearly proves, what is in fact now admitted on all hands, that Swallows do not in any material instance differ from other birds in their nature and propensities [for life in the air]; but that they leave us when this country can no longer furnish them with a supply of their proper and natural food...
>
> —Bewick (1797).

6.3. How Birds Migrate

A streamlined body shape and a lightweight skeleton composed of hollow bones minimize air resistance and reduce the amount of energy necessary to become and remain airborne. Well-developed pectoral muscles, which are attached to a uniquely avian structure called the furculum, power the flapping motion of the wings. The long feathers of the wings act as airfoils which help generate the lift necessary for flight [Berthold, 2001].

Birds have a large, four-chambered heart which proportionately weighs 6 times more than a human heart. This, combined with a rapid heartbeat (the resting heart rate of a small songbird is about 500 beats per minute; that of a hummingbird is about 1,000 beats per minute) satisfies the rigorous metabolic demands of flight. Unlike mammalian or reptilian lungs, the lungs of birds remain inflated at all times, with the air sacs acting as bellows to provide the lungs with a constant supply of fresh air [Berthold, 2001].

6.4. How High Do Birds Fly?

Most songbirds migrate at 500 to 2,000 meters, but some fly as high as 6,800 meters; swans have been recorded at 8,000 meters and Bar-headed Geese at 9,000 meters.

Flight = Mobility

Flight affords mobility and has made possible the evolution of avian migration as a means of access to distant food resources. It also means avoiding the physiological stress associated with cold weather. Variations in the patterns of migration are numerous. Some species move only a few kilometers up and down mountain slopes. Others will travel hundreds or even thousands of kilometers, some over vast bodies of water or tracts of inhospitable terrain [Berthold, 2001; Newton, 2008].

The Arctic Tern makes an annual round-trip of about 30,000 kilometers from the Arctic breeding grounds to Antarctic seas.

According to the Smithsonian Migratory Bird Center, a Blackpoll Warbler could boast of getting 720,000 miles to the gallon if it were burning gasoline instead of reserves of body fat.

The tiny Ruby-throated Hummingbird, weighing only about as much as a penny, makes the 1,000 km, 24-hour spring flight across the Gulf of Mexico from the Yucatán Peninsula to the southern coast of the United States [Newton, 2008].

6.5. Hazards of Migration

Death during migration takes a heavy toll. It is estimated that half of all migrants heading south for the winter will not return to breed in the spring.

Predation and bad weather are two natural causes of mortality during migration.

Collisions with tall buildings, windows, and other structures, being shot or trapped by hunters, and getting struck by automobiles are a few of the numerous human-made dangers. Unfortunately, many birds cannot distinguish the difference between real sky and a reflection of the sky in a window [Newton, 2008].

Dr. Dan Klem of Muhlenberg College estimates that each year, in the United States alone, during migration 98 to 976 million birds fly full tilt into windows and are fatally injured. Dr. Klem says we can minimize these collisions by breaking up the reflection on the outside of the window with a non-reflective window coating, window screens, flash tape and bird netting.

Planting trees and installing window awnings to block the sun from hitting the window may eliminate some reflection. Putting a bird feeder on or within a few feet of a window helps to slow birds down and lessen the effect of impact.

The continued loss and degradation of stopover habitat, however, is potentially the greatest threat of all.

As birds fly south, they may find fewer places to rest. Forests are being cleared and fragmented. Neo-tropical migrants have a hard time when they migrate and find fragments of forest or no forest at all. Clear cutting in forests has had a serious negative effect on neo-tropical birds. Agricultural practices can also have negative effects on neo-tropical migrants.

6.6. Importance of Stopover Sites

A migrating bird faced with the dilemma of a stopover site having disappeared may not have any viable options. Without places along the way that provide an adequate food supply for the quick replenishment of fat reserves, shelter from predators, and water, these birds are probably not going to make it [Boland, 1990; Chan, 2001].

Declines in the numbers of many Neotropical migratory bird species have been detected over the past several decades. There are two main causes: fragmentation of breeding habitat and destruction of tropical forests on the wintering grounds.

As the birds move north in waves, they fan out across the eastern U.S., feeding on the all-you-can-eat buffet of insects that hatch out in the spring. This broad-front movement pattern means that songbird stopover sites are widely dispersed across the wooded areas of the eastern U.S. in the spring [Boland, 1990; Chan, 2001].

For those birds heading south in the fall across the Gulf of Mexico, the Gulf coast is an important area for stopovers.

In the fall, higher elevation sites--especially mountain meadows--become important because of the abundant populations of insects that peak late in the season. Also in the fall, at lower elevations, foothill riparian areas (rivers, streams and creeks) provide important fruit bearing plants for birds such as tanagers and grosbeaks.

Loss and degradation of stopover habitat not only can result in more birds dying while on migration, but it can also have serious repercussions in terms of nesting success. For example, birds heading north are already constrained by the relatively short amount of time available to get to the breeding grounds, establish a territory, pair with a mate, and get on with the further demands of raising young. Late arrival, or arrival in poor condition, on the breeding grounds because of inadequate food and rest en route, is likely to jeopardize a bird's ability to reproduce [Boland, 1990; Chan, 2001].

6.7. Conservation = Economic Sense

The conservation of migratory birds depends on conservation of habitats, but parks and reserves alone will not provide adequate space for protection. The fate of migratory birds and other wide-ranging species depends upon the quality of human-managed habitats.

The health of temperate and tropical ecosystems is bound together by the migration of billions of birds each year--and shaded coffee plantations play a key role. This form of land use may itself be on the way to becoming an endangered species [Berthold, 2001; Boland, 1990; Chan, 2001; Newton, 2008].

6.8. Protect and Create Habitat

You can create and restore habitat at home, at your workplace and in your local park or wildlife refuge. You can make any property attractive to birds by offering water, shelter, food and nesting habitat [Newton, 2008].

6.9. Why Do Birds Migrate?

Birds migrate to move from areas of low or decreasing resources to areas of high or increasing resources. The two primary resources being sought are food and nesting locations. Birds that nest in the northern hemisphere tend to migrate northward in the spring to take advantage of burgeoning insect populations, budding plants and an abundance of nesting locations. As winter approaches, and the availability of insects and other food resources drops, the birds move south again.

Escaping the cold is a motivating factor but many species, including hummingbirds, can withstand freezing temperatures as long as an adequate supply of food is available.

Why do some birds fly thousands of miles back and forth between breeding and non-breeding areas every year whereas others never travel at all? The pressure to migrate comes from seasonal food scarcity [Dorst, 1963].

To figure out the underlying pressures that drive some birds to leave home for the season, the team wanted to examine a related set of species and compare their size, food type, habitat, migratory behavior and whether they fed in flocks.

Boyle and Conway focused on 379 species of New World flycatchers from the suborder Tyranni. One of the largest groups of New World birds, the Tyranni includes kingbirds, flycatchers, phoebes and such southern Arizona birdwatchers' delights as vermillion flycatchers and rose-throated becards. Tropical members include manakins and cotingas [Dorst, 1963].

Food scarcity was the number one issue that predicted a species' migratory behavior, "Food availability is the underlying process, not diet and habitat."

One strategy for dealing seasonal changes in food availability is migration. Species that forage in flocks are less likely to migrate.

> "If you are faced with food scarcity, you have two options," Boyle said. "You can either forage with other birds or you can migrate."

When birds band together to search for food, the group is more likely to find a new patch of food than is one lone individual, she said. "Flocking can be an alternative way to deal with food shortages."

A universal assumption about bird migration has been that short-distance migration is an evolutionary stepping stone to long-distance migration. The team's work contradicts that idea by showing that short-distance migrants are inherently different from their globe-trotting cousins.

Older Cranes Help Keep Younger Birds on the Right Track

A study of whooping cranes has shed new light on how much of a bird's migration route is learnt, and how much is innate in its genetics.

Scientists found that learning from older, more experienced birds was a crucial factor in the migratory habits of this long-lived, social species. The US-German team of researchers also found that the cranes' migratory performance improved over time [Lincoln, 1979].

For a bird to journey hundreds or thousands of miles between its breeding and non-breeding ranges is a difficult, perilous journey, one that not all birds survive. So why do birds migrate? What reasons send millions of birds into the skies every spring and fall? There is more than one single reason for birds to migrate, but it all comes down to survival [Lincoln, 1979].

If No Birds Migrated

Without a reason to migrate, birds would have even more challenging lives than making these excruciating journeys. If no birds migrated, food supplies in breeding areas would be rapidly depleted, and many chicks would starve. Competition for nesting sites would be fierce, and predators would be attracted to the high concentrations of breeding birds and easy meals of nestlings. It is for those two reasons – food and breeding – that birds migrate, but those reasons are far more complicated than they seem [Lincoln, 1979; Sprague et al., 2008].

Migrating for a Meal

For all birds, one of the principle driving forces behind migration is food scarcity. If all birds were to stay in the same tropical regions year-round, food would become scarce and breeding would be less successful. But as food sources are regenerating in the north each spring, millions of birds migrate to those areas to take advantage of the abundance. As the food supplies then dwindle in the fall, they return to replenished tropical regions.

This pattern of migrating for a meal is true not only for neotropical migrants, but also short-range migratory birds that may move only short distances to pursue a food source. Bird irruptions are also the result of changes in the food supply, with greater irruptions occurring in years when food supplies are low for northern birds, forcing them to seek adequate food further south, well outside their typical range [Sprague et al., 2008].

Migrating for Family

Over millennia, birds have evolved different migration patterns, timing and destinations to disperse around the world to breed, taking advantage of a wide variety of suitable conditions to raising their young. Those conditions can vary for every bird species, and may involve specific food sources, habitats that provide adequate shelter and breeding colonies that offer greater protection than a single pair of bird parents [Walraff, 2005].

It may seem contradictory to argue that birds migrate to help their offspring survive when many of those same bird parents quickly abandon their young as they mature, leaving the young birds to make their dangerous first migration without adult guidance. It is exactly because the birds have raised their chicks in a relatively rich, safe environment, however, that gives them the advantage of being prepared to make that journey.

More Reasons Birds Migrate

Food may be the key to a regular migration, but birds migrate for other reasons related to helping their offspring survive, including...

- **Climate**: Birds have evolved different types of plumage to survive different climates, and changes in those climates can affect migration. Many birds leave the Arctic breeding grounds, for example, when temperatures begin to dip and they need more temperate habitat. Similarly, the hottest tropical regions can be a harsh environment for raising chicks, and it is advantageous to lay eggs further north.

- **Predators**: Habitats that have abundant food sources year-round also attract a greater number of predators that can threaten nests. Birds that migrate to different habitats can avoid that onslaught of predators, giving their young a better chance of reaching maturity.
- **Disease**: Any large group of birds crammed in one type of habitat is susceptible to parasites and diseases that can decimate thousands of birds in a short period of time. While diseases can and do occasionally devastate breeding colonies, birds that disperse to different locations have less chance of spreading a disease to their entire population, including their new offspring.

In the end, the reasons why birds migrate all come down to survival – not of the migrating birds themselves, but of the chicks they will raise. Finding richer food sources, seeking safer habitats and avoiding predators are all migration behaviors designed to ensure breeding success, allowing the birds to survive for another generation and allowing birders the pleasure of witnessing another year's migration.

It is very possible this applies to other bird species that are long-lived and are social.
In other bird species, genetics has been shown to play a significant role in the timing and direction of migration.

Back from the Edge

After being pushed to the brink of extinction by unregulated hunting and loss of habitat, whooping cranes are starting to spread their wings across the skies of North America.

Nine years ago, the Necedah National Wildlife Refuge in the US state of Wisconsin took a group of captive-bred whooping cranes and taught them their ancestors' migration route.

The birds followed the dulcet sounds of crane calls, which were emitted from a piloted ultra-light aircraft across nearly 1,500 miles of sky from Wisconsin to Florida [Helm and Gwinner, 2006].

After the initial training, the whooping cranes began to migrate to and from their breeding grounds freely in groups.

Researchers examined the data from these birds' migration paths and found that the presence of older, more experienced birds helped the migratory group stay on a straight-line course to their breeding grounds.

Moreover, the more frequently a crane flew the route, the straighter the line the crane took in following years [Helm and Gwinner, 2006].

The research shows that learning of migration happens over many years and that social learning plays an important role as younger birds are able to learn from older birds [Helm and Gwinner, 2006].

The researchers were not able to detect a genetic component to the cranes' migratory habits in this study. But they say that previous evidence suggests that innate behaviour must play a role in some aspects of crane migration [Gill et al., 2005].

The team says their results have important implications for conservation and recovery efforts focused on whooping cranes, and potentially other long-lived, social birds such as storks and geese as well.

6.10. Types of Migration

The term *migration* is used to describe movements of populations of birds (or other animals). One way to look at migration is to consider the distances traveled.

- **Short-distance migrants:** May move only a short distance, as from higher to lower elevations on a mountainside.
- **Medium-distance migrants:** Some species may cover distances that span from one to several states.
- **Long-distance migrants:** Birds that typically have ranges that extend from the United States and Canada in the summer to Mexico and further south in the winter.

The pattern of migration can vary within each category, but is most variable in short and medium distance migrants [Berthold, 2001].

6.11. Origins of Migration

The origin of migration is related to the distance traveled. For short-distance migrants it is as simple as a search for food. The origins of long-distant migration patterns are more complex and include the development of the genetic make-up of the bird [Berthold, 2001].

6.12. Migration Triggers

The mechanisms initiating migratory behavior vary and are not always completely understood. Migration can be triggered by a combination of changes in day length, lower temperatures, changes in food supplies, and genetic predisposition. Different species of birds and even segments of the population within the same species may follow different migratory patterns Newton, 2008].

6.13. Navigation

Migrating birds can cover thousands of miles in their annual travels, often traveling the same course year after year with little deviation in the path followed. First year birds may migrate unescorted to a winter home they have never before seen and return the following spring to the area in which they were born [Wiltschko, 2006].

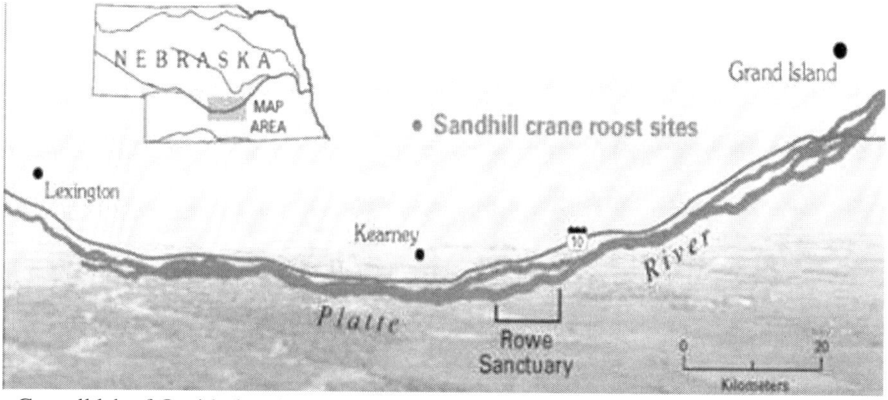

(Source: Cornell lab of Ornithology).

Figure 4. Migration pathway of Sandhill Crane.

The secrets of their amazing navigational skills remain largely hidden. Birds appear to navigate using a variety of techniques, including navigation by the stars, sensing changes in the earth's magnetic field, and even smell [Wiltschko, 2006].

Some species follow preferred pathways on their annual migrations. These pathways are often related to important stopover locations that provide food supplies critical to the birds' survival [Wiltschko, 2006].

Each spring approximately 500,000 Sandhill Cranes and some endangered Whooping Cranes use the Central Platte River Valley in Nebraska as a staging habitat during their migration north to breeding and nesting grounds in Canada, Alaska, and the Siberian Arctic (Figure 4).

6.14. Migration Hazards

Taking a journey that can stretch to a round-trip distance of several thousand miles is a dangerous and arduous undertaking. It is an effort that tests both the birds' physical and mental capabilities. The physical stress of the trip, lack of adequate food supplies along the way, bad weather, and increased exposure to predators all add to the hazards of the journey.

In recent years long-distant migrants have been facing a growing threat from communication towers and tall buildings. Many species are attracted to the lights of tall buildings and millions are killed each year in collisions with the structures [Hedenström, 2008].

6.15. Migrant Traps

Migrating birds can sometimes be found in certain areas in larger than normal numbers. This may be the result of local weather conditions, an abundance of food or the local topography. For example, small songbirds may migrate north in the spring, flying directly over the Gulf of Mexico to the coastlines of Texas and other Gulf Coast states. Under favorable weather conditions they may continue inland for many miles before stopping to rest and feed. However, storms and headwinds can often leave the birds near exhaustion when they reach land. In such cases they will head for the nearest location offering food and cover. Many of the motts along the gulf coast can provide these needs and become a temporary home to large numbers of birds in a very short time. These migration traps have become very popular with birders, even earning international reputations [Hedenström, 2008].

Spring migration is an especially good time for those that feed birds in their backyard to attract species they normally do not see. Offering a variety of food sources, water, and adding natural food sources to the landscape can make a backyard attractive to migrating songbirds.

6.16. Range Maps

Many birders use the range maps in their field guides to help determine if and when a particular species might be in their area. Range maps are especially useful when working with non-resident species. Range maps can be confusing and have limitations. Ranges of birds can vary year-to-year, as with several of the irruptive species such as redpolls. The range of some species can expand or contract fairly rapidly, with changes occurring in time periods shorter than the republication time of a field guide.

6.17. Migration and the Spread of Disease

Wild birds have historically played a very limited role in the spread of disease to humans. West Nile virus has had an impact on bird populations, with Corvid species such as crows and jays being the most susceptible. The mosquito is responsible for the transfer of West Nile virus to humans. Bird or avian flu can also be carried by migrating birds and is often fatal to the bird. To date there are no confirmed cases of the spread of avian flu from wild birds directly to humans [Rappole et al., 2000, 2006].

Figure 5. Introduced into the Bahamas in the mid-1970s, the Eurasian Collared-Dove is now established throughout the southeastern United States. Its range has rapidly spread north and west, all the way to the west coast of the United States. It prefers an urban or suburban habitat [Shimazaki, 2004].

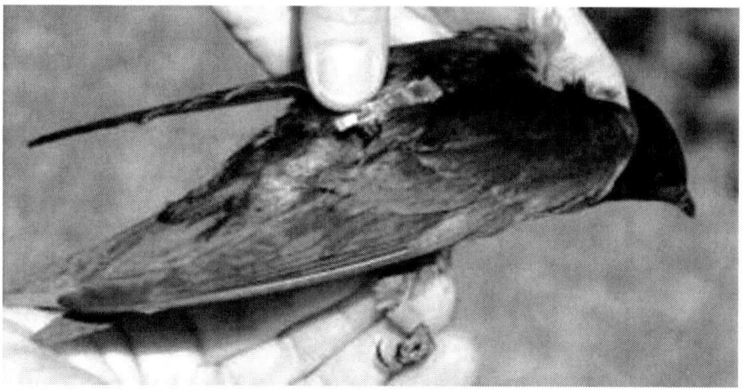

Figure 6. Studying migration patterns in Purple martins.

6.18. Additional Resources

Migration is a fascinating study and there is much yet to learn. The Lab of Ornithology's Miyoko Chu has published a book titled **Songbird Journeys** that explorers many aspects of migration in an interesting and easy-to-read style. The Cornell Lab of Ornithology's Handbook of Bird Biology provides even more information on the amazing phenomenon of bird migration.

Some songbirds can't alter their migration schedule to take into account earlier, warmer springs as a result of climate change, which could be deadly.

The researchers, led by ecologist Kevin Fraser at York University in Toronto, reached their conclusion after tracking purple martins, members of the swallow family that range as far north as southern Canada, all along their spring migration route of more than 7,000 kilometres [Schmaljohann, 2007].The results of the study were published in the journal PLoS ONE.

Fraser and his colleagues found that two populations of birds that their research team was tracking took off from their wintering grounds in Brazil's Amazon Basin for their breeding sites in the northeastern U.S. at the same time every year for five years in a row, no matter what the temperature was like at both their summer and winter homes [Schmaljohann, 2007].

Record high spring temperatures at their breeding grounds in 2012 didn't break the pattern.

Sometimes Martins may arrive at their breeding grounds "too late," after the spring population of the insects they eat has peaked and started to decline, and the reduced food supply could hamper their ability to breed successfully.

Songbird Populations Falling

Populations of many songbirds have fallen drastically in North America since the 1960s — more than 75 per cent in the case of some species, such as the rusty blackbird, the olive-sided flycatcher, the Canada warbler, the American black duck and the evening grosbeak. A variety of factors may play a role, including habitat destruction and climate change [Schmaljohann, 2007].

Currently, purple martin populations are declining by three to four per cent per year in the northern part of their range, around the Great Lakes. Some scientists have suggested that an increasing mismatch between the birds' migration times and the start of spring may be to blame. European studies show insect populations have been peaking earlier and earlier each year, and that places where the peak happens earliest are where populations of migratory insect-eating birds called pied flycatchers were declining the most. The flycatchers did not appear to migrate earlier to compensate [Schmaljohann, 2007]. There is some evidence that certain weather patterns in the Amazon and the northeastern U.S. are correlated. Individual birds could respond to these cues, or perhaps change their migration speed in response to conditions along their route [Schmaljohann, 2007].

Tracking Backpack

Migration pattern of purple martins at their breeding sites was studied by attaching tiny tracking devices to their backs using a little "backpack," starting in 2008. The devices log light levels, providing data about latitude from the timing of sunrise and sunset, and data about longitude from the timing of the solar noon, which varies with longitude. Studies show that the backpacks do not appear to affect the birds' ability to fly, breed or survive.

In 2012, the researchers managed to recapture 52 purple martins that had been tagged in Erie, Penn., and Woodbridge, Va., and to retrieve the tracking devices, with their data. They then matched the tracking data with data from weather stations along the route to find out what conditions were like as the birds flew by, and compared it to data from previous years.

(Source: Shyamal, L., The Migration Ecology of Birds, Academic Press).

Figure 7. A flock of Barnacle Geese during autumn migration.

Researchers were surprised to find that the birds did not seem to respond to temperatures or rainfall. The findings suggest that individual purple martins can't compensate for short-term changes in the timing of spring, unlike species with shorter migrations, such as tree swallows.

Purple martins and the dozen or so other songbirds in North America that migrate to very distant wintering sites might be particularly at risk with changes in climate because they're further away and less likely to get the signals they might need of changes up north. Scientists expect that natural selection will eventually cause the birds to shift their migration times, but that's happening too slowly [Schmaljohann, 2007].

6.19. Long Distance Migration

The typical image of migration is of northern landbirds, such as swallows and birds of prey, making long flights to the tropics. However, many Holarctic wildfowl and finch species winter in the North Temperate Zone, but in regions with milder winters than their summer breeding grounds. For example, the pink-footed goose migrates from Iceland to Britain and neighbouring countries, whilst the Dark-Eyed Junco migrates from subarctic and arctic climates to the contiguous United States and the American Goldfinch from taiga to wintering grounds extending from the American South northwestward to Western Oregon. Migratory routes and wintering grounds are traditional and learned by young during their first migration with their parents. Some ducks, such as the Garganey, move completely or partially into the tropics. The European pied flycatcher also follows this migratory trend, breeding in Asia and Europe and wintering in Africa [Weber, 2009].

Figure 8. Swainson's Thrush.

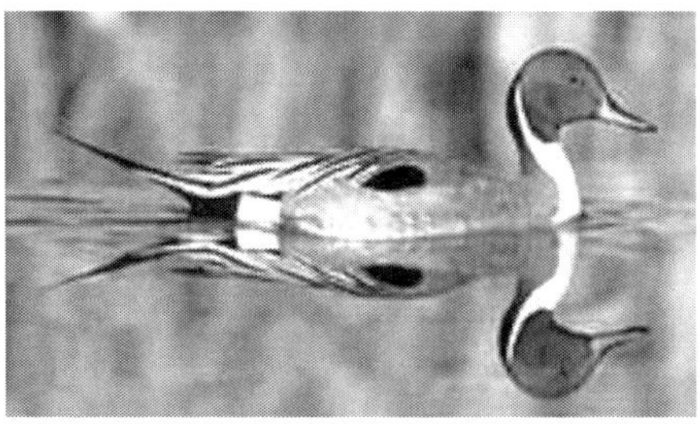

Figure 9. Northern Pintail.

The same considerations about barriers and detours that apply to long-distance land-bird migration apply to water birds, but in reverse: a large area of land without bodies of water that offer feeding sites is a barrier to many may also be a barrier to a bird that feeds in coastal waters. Detours avoiding such barriers are observed: for example, Brent Geese migrating from the Taymyr Peninsula to the Wadden Sea travel via the White Sea coast and the Baltic Sea rather than directly across the Arctic Ocean and northern Scandinavia [Weber, 2009].

A similar situation occurs with waders (called *shorebirds* in North America). Many species, such as Dunlin and Western Sandpiper, undertake long movements from their Arctic breeding grounds to warmer locations in the same hemisphere, but others such as Semipalmated Sandpiper travel longer distances to the tropics in the Southern Hemisphere. Like the large and powerful wildfowl, the waders are strong fliers. This means that birds wintering in temperate regions have the capacity to make further shorter movements in the event of particularly inclement weather [Sprague et al., 2008].

For some species of waders, migration success depends on the availability of certain key food resources at stopover points along the migration route. This gives the migrants an opportunity to refuel for the next leg of the voyage. Some examples of important stopover locations are the Bay of Fundy and Delaware Bay [Kathleen et al., 2013].

Some Bar-tailed Godwits have the longest known non-stop flight of any migrant, flying 11,000 km from Alaska to their New Zealand non-breeding areas [Gill et al., 2005]. Prior to migration, 55 percent of their bodyweight is stored fat to fuel this uninterrupted journey.

Figure 10. Bar-tailed Godwit.

Figure 11. Arctic Terns.

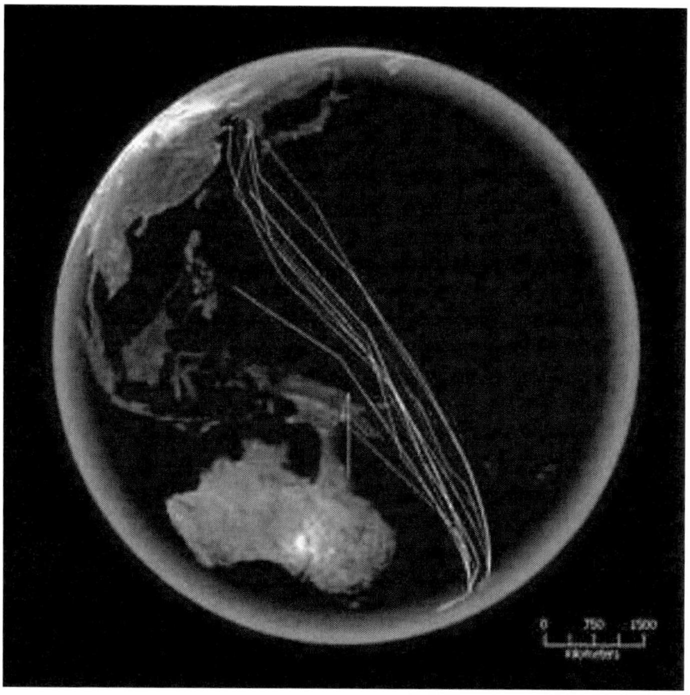

Figure 12. The routes of satellite tagged Bar-tailed Godwits migrating north from New Zealand. This species has the longest known non-stop migration of any species, up to 10,200 km (6,300 miles).

Seabird migration is similar in pattern to those of the waders and waterfowl. Some, such as the Black Guillemot and some gulls, are quite sedentary; others, such as most terns and auks breeding in the temperate northern hemisphere, move varying distances south in the northern winter. The Arctic Tern has the longest-distance migration of any bird, and sees more daylight than any other, moving from its Arctic breeding grounds to the Antarctic non-breeding areas (Figure 12). One Arctic Tern, ringed (banded) as a chick on the Farne Islands off the British east coast, reached Melbourne, Australia in just three months from fledging, a sea journey of over 22,000 km (14,000 miles). A few seabirds, such as Wilson's Petrel and Great Shearwater, breed in the southern hemisphere and migrate north in the southern winter. Seabirds have the additional advantage of being able to feed during migration over open waters.

The most pelagic species, mainly in the 'tubenose' order Procellariiformes, are great wanderers, and the albatrosses of the southern oceans may circle the globe as they ride the "roaring forties" outside the breeding season. The tubenoses spread widely over large areas of open ocean, but congregate when food becomes available. Many are also among the longest-distance migrants; Sooty Shearwaters nesting on the Falkland Islands migrate 14,000 km (8,700 mi) between the breeding colony and the North Atlantic Ocean off Norway. Some Manx Shearwaters do this same journey in reverse. As they are long-lived birds, they may cover enormous distances during their lives; one record-breaking Manx Shearwater is calculated to have flown 8 million km (5 million miles) during its over-50 year lifespan [Anon, 2007].

Figure 13. Griffon Vulture soaring.

Some large broad-winged birds rely on thermal columns of rising hot air to enable them to soar. These include many birds of prey such as vultures, eagles, and buzzards, but also storks. These birds migrate in the daytime. Migratory species in these groups have great difficulty crossing large bodies of water, since thermals only form over land, and these birds cannot maintain active flight for long distances. Mediterranean and other seas present a major obstacle to soaring birds, which must cross at the narrowest points. Massive numbers of large raptors and storks pass through areas such as Gibraltar, Falsterbo, and the Bosphorus at migration times. More common species, such as the European Honey Buzzard, can be counted in hundreds of thousands in autumn. Other barriers, such as mountain ranges, can also cause funnelling, particularly of large diurnal migrants. This is a notable factor in the Central American migratory bottleneck. Batumi bottleneck in the Caucasus is one of the heaviest migratory funnels on earth. Avoiding flying over the Black Sea surface and across high mountains, hundreds of thousands of soaring birds funnel through an area around the city of Batumi, Georgia [Maanen et al., 2001]. It has been suggested [Gensbol, 1984]. Birds of prey (such as Honey Buzzards) which migrate using thermals may benefit from losing 10 to 20% of their weight and this may explain why they forage less on migration than do smaller birds of prey with more active flight such as Falcons, Hawks and Harriers (Figure 13).

Figure 14. Ruby-throated Hummingbird.

Many of the smaller insectivorous birds including the warblers, hummingbirds and flycatchers migrate large distances, usually at night (Figure 14). They land in the morning and may feed for a few days before resuming their migration. The birds are referred to as *passage migrants* in the regions where they occur for short durations between the origin and destination [Schmaljohann, 2007].

Nocturnal migrants minimize predation, avoid overheating, and feed during the day. One cost of nocturnal migration is the loss of sleep. Migrants may be able to alter their quality of sleep to compensate for the loss [Rattenborg, 2004].

6.20 Short Distance and Altitudinal Migration

Many long-distance migrants appear to be genetically programmed to respond to changing day length. Species that move short distances, however, may not need such a timing mechanism, and may move in response to local weather conditions. Thus mountain and moorland breeders, such as Wall creeper and White-throated Dipper, may move only altitudinally to escape the cold higher ground. Other species such as Merlin and Skylark will move further to the coast or to a more southerly region. Species like the Common Chaffinch are not migratory in Britain, but move south or to Ireland in very cold weather [Kreft, 2004].

Short-distance passerine migrants have two evolutionary origins. Those that have long-distance migrants in the same family, such as the Common Chiffchaff, are species of southern hemisphere origins that have progressively shortened their return migration to stay in the northern hemisphere.

Species that have no long-distance migratory relatives, such as the waxwings, are effectively moving in response to winter weather and the loss of their usual winter food, rather than enhanced breeding opportunities.

Figure 15. Cedar Waxwing.

Figure 16. Woodland Kingfisher.

In the tropics there is little variation in the length of day throughout the year, and it is always warm enough for a food supply (although because of competition, there may not be enough food for every bird). Migration within the tropics has been far less studied than in the temperate zones. It was once assumed that tropical birds were mostly sedentary; however, altitudinal migration and other within-tropics movements appear to be surprisingly common [Helm and Gwinner, 2006]. Many tropical regions have wet and dry seasons, inducing some birds to migrate or wander widely to find food. Indeed, the monsoons of India are preceded by the arrival of the Jacobin Cuckoo, the "harbinger of the monsoon". Other examples include the Woodland Kingfisher of west Africa and many Australian birds (Figure 16) [Helm and Gwinner, 2006].

There are a few species, notably cuckoos, which are genuine long-distance migrants within the tropics. An example is the Lesser Cuckoo, which breeds in India and spends the non-breeding season in Africa. Such examples help make the case that food supplies, not weather per se, drive migration patterns.

Altitudinal migration is common on mountains worldwide, such as in the Himalayas and the Andes [Kreft, 2004]. Quite often, altitudinal migration is combined with distanct migration; for example, the Himalayan Kashmir Flycatcher and Pied Thrush both move far south as the highlands of Sri Lanka. Altitudinal migration may even be important to birds living on relatively small islands, such as the Hawaiian Islands, which have high mountains [Kreft, 2004].

6.21. Irruptions and Dispersal

Sometimes circumstances such as a good breeding season followed by a food source failure the following year lead to irruptions in which large numbers of a species move far beyond the normal range. Bohemian Waxwings will show this unpredictable variation in annual numbers, with five major arrivals in Britain during the nineteenth century, but eighteen between the years 1937 and 2000. Red Crossbills too are irruptive, with invasions noted in 1251, 1593, 1757, and 1791.

Figure 17. Rainbow Bee-eater.

The temperate zones of the southern continents have extensive arid areas, particularly in Australia and western southern Africa, and weather-driven movements are common but not always predictable. A couple of weeks of heavy rain in one part or another of the usually dry centre of Australia, for example, causes dramatic plant and invertebrate growth, attracting birds from all directions. This can happen at any time of year, and, in any given area; may not happen again for a decade or more, depending on the frequency of El Niño and La Niña periods.

Bird migration is primarily, but not entirely, a Northern Hemisphere phenomenon. In the Southern Hemisphere, seasonal migration tends to be much less obvious. There are several reasons for this.

First, the largely uninterrupted expanses of land mass or ocean tend not to funnel migrations into narrow and obvious pathways, making them less obvious to the human observer. Second, at least for terrestrial birds, climatic regions tend to fade into one another over a long distance rather than be entirely separate: this means that rather than make long trips over unsuitable habitat to reach particular destinations, migrant species can usually travel at a relaxed pace, feeding as they go. Without banding studies it is often not obvious that the birds seen in any particular locality as the seasons change are in fact different members of the same species passing through, gradually working their way north or south [Rubolini et al., 2004].

Many species do in fact breed in the temperate southern hemisphere regions and winter further north in the tropics. The southern African Greater Striped Swallow, and the Australian Satin Flycatcher, Dollarbird, and Rainbow Bee-eater for example, winters well north of their breeding range (Figure 16) [Edwards et al., 2004].

6.22. Physiology and Control

The control of migration, its timing and response are genetically controlled and appear to be a primitive trait that is present even in non-migratory species of birds. The ability to navigate and orient themselves during migration is a much more complex phenomenon that may include both endogenous programs as well as learning [Helm and Gwinner, 2006].

6.23. TIMING

The primary physiological cue for migration is the changes in the day length. These changes are also related to hormonal changes in the birds. In the period before migration, many birds display higher activity or Zugunruhe (German: *migratory restlessness*) as well as physiological changes such as increased fat deposition. The occurrence of Zugunruhe even in cage-raised birds with no environmental cues (e.g., shortening of day and falling temperature) has pointed to the role of circannual endogenous programs in controlling bird migrations. Caged birds display a preferential flight direction that corresponds with the migratory direction they would take in nature, even changing their preferential direction at roughly the same time their wild conspecifics change course.

In species where there is polygyny and with considerable sexual dimorphism, there is a tendency for males to return earlier to the breeding sites than their females. This is termed as protandry [Rubolini et al., 2004; Edwards et al., 2007].

6.24. ORIENTATION AND NAVIGATION

Navigation is based on a variety of senses. Many birds have been shown to use a sun compass. Using the sun for direction involves the need for making compensation based on the time. Navigation has also been shown to be based on a combination of other abilities including the ability to detect magnetic fields (magnetoception), use visual landmarks as well as olfactory cues [Walraff, 2005].

Long distance migrants are believed to disperse as young birds and form attachments to potential breeding sites and to favourite wintering sites. Once the site attachment is made they show high site-fidelity, visiting the same wintering sites year after year [Ketterson and Nolan, 1990].

The ability of birds to navigate during migrations cannot be fully explained by endogenous programming, even with the help of responses to environmental cues. The ability to successfully perform long-distance migrations can probably only be fully explained with an accounting for the cognitive ability of the birds to recognize habitats and form mental maps. Satellite tracking of day migrating raptors such as Ospreys and Honey Buzzards has shown that older individuals are better at making corrections for wind drift [Thorup et al., 2003].

As the circannual patterns indicate, there is a strong genetic component to migration in terms of timing and route, but this may be modified by environmental influences. An interesting example where a change of migration route has occurred because of such a geographical barrier is the trend for some Blackcaps in central Europe to migrate west and winter in Britain rather than cross the Alps [Thorup et al., 2003].

Migratory birds may use two electromagnetic tools to find their destinations: one that is entirely innate and another that relies on experience. A young bird on its first migration flies in the correct direction according to the Earth's magnetic field, but does not know how far the journey will be. It does this through a radical pair mechanism whereby chemical reactions in special photo pigments sensitive to long wavelengths are affected by the field. Although this only works during daylight hours, it does not use the position of the sun in any way. At this stage the bird is similar to a boy scout with a compass but no map, until it grows accustomed to the journey and can put its other facilities to use. With experience they learn various landmarks and this "mapping" is done by magnetites in the trigeminal system, which tell the bird how strong the field is. Because birds migrate between northern and southern regions, the magnetic field strengths at different latitudes let it interpret the radical pair mechanism more accurately and let it know when it has reached its destination [Wiltschko, 2006]. More recent research has found a neural connection between the eye and "Cluster N", the part of the forebrain that is active during migrational orientation, suggesting that birds may actually be able to *see* the magnetic field of the earth [Deutschlander et al., 1999; Heyers et al., 2006]

6.25. Vagrancy

Migrating birds can lose their way and appear outside their normal ranges. This can be due to flying past their destinations as in the "spring overshoot" in which birds returning to their breeding areas overshoot and end up further north than intended. Reverse migration, where the genetic programming of young birds fails to work properly, can lead to great rarities turning up as vagrants thousands of kilometres out of range. Certain areas, because of their location, have become famous as watchpoints for migrating birds. Examples are the Point Pelee National Park in Canada, and Spurn in England. Drift migration of birds blown off course by the wind can result in "falls" of large numbers of migrants at coastal sites [Guillemain, 2005].

A related phenomenon called "abmigration" involves birds from one region joining similar birds from a different breeding region in the common winter grounds and then migrating back along with the new population. This is especially common in some waterfowl, which shift from one flyway to another [Guillemain, 2005].

6.26. Migration Conditioning

It has been possible to teach a migration route to a flock of birds, for example in re-introduction schemes. After a trial with Canada Geese, microlight aircraft were used in the US to teach safe migration routes to reintroduced Whooping Cranes [Rohwer et al., 2005].

Adaptation

Birds need to alter their metabolism in order to meet the demands of migration. The storage of energy through the accumulation of fat and the control of sleep in nocturnal migrants require special physiological adaptations. In addition, the feathers of a bird suffer from wear-and-tear and require to be molted. The timing of this molt - usually once a year but sometimes two - varies with some species molting prior to moving to their winter grounds and others molting prior to returning to their breeding grounds [Hedenstrom, 2008]. Apart from physiological adaptations, migration sometimes requires behavioural changes such as flying in flocks to reduce the energy used in migration or the risk of predation [Weber, 2009].

6.27. Evolutionary and Ecological Factors

Migration in birds is highly labile and is believed to have developed independently in many avian lineages [Pulido, 2007]. While it is agreed that the behavioral and physiological adaptations necessary for migration are under genetic control, some authors have argued that no genetic change is necessary for migratory behavior to develop in a sedentary species because the genetic framework for migratory behavior exists in nearly all avian lineages [Rappole et al., 2003]. This explains the rapid appearance of migratory behavior after the most recent glacial maximum [Mila et al., 2006].

Whether a particular species migrates depends on a number of factors. The climate of the breeding area is important, and few species can cope with the harsh winters of inland Canada or northern Eurasia. Thus, the partially migratory Blackbird *Turdus merula* is migratory in Scandinavia, but not in the milder climate of southern Europe. The nature of the staple food is also significant. Most specialist insect eaters outside the tropics are long-distance migrants, and have little choice but to head south in winter [Jenny and Kery, 2003].

Sometimes the factors are finely balanced. The Whinchat *Saxicola rubetra* of Europe and the Siberian Stonechat *Saxicola maura* of Asia are long-distance migrants wintering in the tropics, whereas their close relative, the European Stonechat *Saxicola rubicola* is a resident bird in most of its range, and moves only short distances from the colder north and east. A possible factor here is that the resident species can often raise an extra brood [Both et al., 2006].

Recent research suggests that long-distance passerine migrants are of South American and African, rather than northern hemisphere, evolutionary origins. They are effectively southern species coming north to breed rather than northern species going south to winter [Both et al., 2006].

Theoretical analyses, summarized by Alerstam (2001), show that detours that increase flight distance by up to 20% will often be adaptive on aerodynamic grounds - a bird that loads itself with food to cross a long barrier flies less efficiently. However, some species show circuitous migratory routes that reflect historical range expansions and are far from optimal in ecological terms. An example is the migration of continental populations of Swainson's Thrush, which fly far east across North America before turning south via Florida to reach northern South America; this route is believed to be the consequence of a range expansion that occurred about 10,000 years ago. Detours may also be caused by differential wind conditions, predation risk, or other factors [Wormworth and Mallon, 2006].

Climate Change

Large scale climatic changes, as have been experienced in the past, are expected to have an effect on the timing of migration. Studies have shown a variety of effects including timing changes in migration, breeding [Jenny and Kery, 2003] as well as population variations [Both et al., 2006; Wormworth and Mallon, 2006].

Ecological Effects

The migration of birds also aids the movement of other species, including those of ectoparasites such as ticks and lice [Smith et al., 1996], which in turn may carry micro-organisms including those of concern to human health. Considerable interest has been taken due to the global spread of avian influenza, however migrant birds have not been found to be a special risk, with import of pet and domestic birds being a greater threat [Rappole et al., 2006]. Some viruses that are maintained in birds without lethal effects, such as the West Nile Virus may however be spread by migrating birds [Rappole et al., 2000]. Birds may also have a role in the dispersal of propagules of plants and plankton [Cruden, 1966; Figuerola and Green, 2002].

Some predators take advantage of the concentration of birds during migration. Greater Noctule bats feed on nocturnal migrating passerines. Some birds of prey specialize on migrating waders [Ronald et al., 2004].

6.28. Studying Migration

Scientists use several techniques in studying migration, from banding to satellite tracking. One of the goals is to locate important stopover and wintering locations. Once identified, steps can be taken to protect and save these key locations.

Study Techniques

Early studies on the timing of migration began in 1749 in Finland, with Johannes Leche of Turku collecting the dates of arrivals of spring migrants.

Bird migration routes have been studied by a variety of techniques including the oldest, marking. Swans have been marked with a nick on the beak since about 1560 in England. Scientific ringing was pioneered by Mortensen in 1899 [Greenwood, 2007]. Other techniques include radar and satellite tracking.

Stable isotopes of hydrogen, oxygen, carbon, nitrogen, and sulphur can establish avian migratory connectivity between wintering sites and breeding grounds. Stable isotopic methods to establish migratory linkage rely on spatial isotopic differences in bird diet that are incorporated into inert tissues like feathers, or into growing tissues such as claws and muscle or blood [Hobson and Wassenaar, 1997; Bowen et al., 2005].

Figure 18. Emlen funnel.

An approach to identify migration intensity makes use of upward pointing microphones to record the nocturnal contact calls of flocks flying overhead. These are then analyzed in a laboratory to measure time, frequency and species [Farnsworth et al., 2004].

An older technique to quantify migration involves observing the face of the moon towards full moon and counting the silhouettes of flocks of birds as they fly at night [Lowery, 1951; Liechti, 1996].

Orientation behaviour studies have been traditionally carried out using variants of a setup known as the **Emlen funnel**, which consists of a circular cage with the top covered by glass or wire-screen so that either the sky is visible or the setup is placed in a planetarium or with other controls on environmental cues (Figure 18). The orientation behaviour of the bird inside the cage is studied quantitatively using the distribution of marks that the bird leaves on the walls of the cage [Emlen and Emlen 1966]. Other approaches used in pigeon homing studies make use of the direction in which the bird vanishes on the horizon.

6.29. Threats and Conservation

Human activities have threatened many migratory bird species. The distances involved in bird migration mean that they often cross political boundaries of countries and conservation measures require international cooperation. Several international treaties have been signed to protect migratory species including the Migratory Bird Treaty Act of 1918 of the US and the African-Eurasian Migratory Waterbird Agreement.

The concentration of birds during migration can put species at risk. Some spectacular migrants have already gone extinct, the most notable being the Passenger Pigeon (*Ectopistes migratorius*). During migration the enormous flocks were a mile (1.6 km) wide, darkening the sky and 300 miles (480 km) long, taking several days to pass.

Other significant areas include stop-over sites between the wintering and breeding territories [Shimazaki et al., 2004]. A capture-recapture study of passerine migrants with high fidelity for breeding and wintering sites did not show similar strict association with stop-over sites [Catry et al., 2004].

Hunting along the migratory route can also take a heavy toll. The populations of Siberian Cranes that wintered in India declined due to hunting along the route, particularly in Afghanistan and Central Asia. Birds were last seen in their favourite wintering grounds in Keoladeo National Park in 2002.

Structures such as power lines, wind farms and offshore oil-rigs have also been known to affect migratory birds. Habitat destruction by land use changes is the biggest threat, and shallow wetlands that are stopover and wintering sites for migratory birds are particularly threatened by draining and reclamation for human use.

6.30. Some Points to Help Birds

- You can start by evaluating your property. It is important to remember that a diversity of habitat encourages a larger variety of birds.
- Dead trees and brush piles provide shelter, nest sites and food (insects) for migrating birds.
- Providing water can be as simple as putting out a bird bath or as complicated as installing a pond with a creek and waterfall.
- Herbicides, fungicides and pesticides - can be lethal to birds.
- Put out bird feeders, seed, fruit and nectar feeders, and fruit.
- Landscape the yard with native evergreen and fruit bearing trees, shrubs, grasses and vines. Design the garden so that plants flower and fruit throughout the spring, summer and fall.
- Having a source of water (especially a moving source) can help attract more migratory birds.
- Getting involved in migratory bird conservation here in the United States or in Latin America can be as simple as writing a check, donating equipment or picking up a shovel.

References

Boland, J. M. (1990). Leapfrog migration in North American shorebirds: intra- and interspecific examples. *The Condor*, *92*, 284–290.

Both, C., Sandra, B., Lessells, C. M. & Marcel, E. V. (2006). Climate change and population declines in a long-distance migratory bird. *Nature*, *441*, 81–83.

Catry, P., Encarnacao, V., Araujo, A., Fearon, P., Fearon, A., Armelin, M. & Delaloye, P. (2004). Are long-distance migrant passerines faithful to their stopover sites? *J Av Biol*, *35*, 170–181.

Chan, K. (2001). Partial migration in Australian landbirds: a review. *Emu*, *101*, 281–292.

Cocker, M. & Mabey, R. (2005). *Birds Britannica*. Chatto & Windus. p. 315.

Cruden, R. W. (1966). Birds as Agents of Long-Distance Dispersal for Disjunct Plant Groups of the Temperate Western Hemisphere. *Evolution*, *20*, 517–532.

Cutts, C. J. & Speakman, J. R. (1994). Energy savings in formation flight of Pink-footed Geese. *J Exp Biol*, *189*, 251–261.

Deutschlander, M. E., Phillips, J. B. & Borland, S. C. (1999). The case for light-dependent magnetic orientation in animals. *J Exp Biol*, *202*, 891–908.

Diego, R., Fernando, S. & Nicola, S. (2004). Protandry and sexual dimorphism in trans-Saharan migratory birds. *Behav Ecol*, *15*, 592–601.

Dondini, G. & Vergari, S. (2000). Carnivory in the greater noctule bat (*Nyctalus lasiopterus*) in Italy . *J Zool*, *251*, 233–236.

Dorst, J. (1963). *The migration of birds*. Houghton Mifflin Co., Boston. p 476.

Eastwood, E. & Rider, G. C. (1965). Some radar measurements of the altitude of bird flight. *British Birds*, *58*, 393–426.

Edwards, D. B. & Forbes, M. R. (2007). Absence of protandry in the spring migration of a population of Song Sparrows Melospiza melodia. *Ibis*, *149*, 715–720.

Emlen, S. T. & Emlen, J. T. (1966). A technique for recording migratory orientation of captive birds. *Auk*, *83*, 361–367.

Farnsworth, A., Gauthreaux, S. A. & van Blaricom, D. G. (2004). A comparison of nocturnal call counts of migrating birds and reflectivity measurements on Doppler radar. *J Av Biol*, *35*, 365–369.

Figuerola, O. & Green, O. J. (2002). Dispersal of aquatic organisms by waterbirds: a review of past research and priorities for future studies. *Fres Biol*, *47*, 483–494.

Gabriel, B., Leonard, W. & Keith, H. (2005). Global application of stable hydrogen and oxygen isotopes to wildlife forensics. *Oecologia*, *143*, 337–348.

Gensbol, B. (1984). Collins Guide to the Birds of Prey of Britain and Europe, p 28

Geroudet, P. (1954). Des oiseaux migrateurs trouvés sur la glacier de Khumbu dans l'Himalaya. *Nos Oiseaux*, *22*, 254.

Gill, R. E. Jr, Theunis, P., Gary, H., Rene, S. & Adrian, R. (2005). Crossing the ultimate ecological barrier: evidence for an 11,000 km-long nonstop flight from Alaska to New Zealand and Eastern Australia by Bar-tailed Godwits. *The Condor*, *107*, 1–20.

Greenwood, J. J. D. (2007). Citizens, science and bird conservation. *J Ornithol*, *148*, 77–124.

Guillemain, M., Sadoul, N. & Simon, G. (2005). European flyway permeability and abmigration in Teal Anas crecca, an analysis based on ringing recoveries. *Ibis*, *147*, 688–696.

Hedenström, A. (2008). Adaptations to migration in birds: behavioural strategies, morphology and scaling effects. *Phil Trans R Soc B*, *363*, 287–299.

Helm, B. & Gwinner, E. (2006). Migratory Restlessness in an Equatorial Nonmigratory Bird. *PLoS Biol*, *4*, e110.

Heyers, D., Manns, M., Luksch, H., Güntürkün, O. & Mouritsen, H. (2007). A Visual Pathway Links Brain Structures Active during Magnetic Compass Orientation in Migratory Birds. In Iwaniuk, Andrew. *PLoS ONE*, *2*, e937.

Hummel, D. & Beukenberg, M. (1989). Aerodynamische Interferenzeffekte beim Formationsfl ug von Vogeln. *J Ornithol*, *130*, 15–24.

Ibáñez, C., Juste, J., García-Mudarra, J. L. & Agirre-Mendi, P. T. (2001). Bat predation on nocturnally migrating birds. *Proc Natl Acad Sci USA*, *98*, 9700–9702.

Jenni, L. & Kery, M. (2003). Timing of autumn bird migration under climate change: advances in long-distance migrants, delays in short-distance migrants. *Proc R Soc Lon B*, *270*, 1467.

Kathleen , E. C., Lawrence, J. N. & Joanna, B. (2013). Abundance and Distribution of Migrant Shorebirds in Delaware Bay. *The Condor*, *95*, 694–705.

Keith, H. & Leonard, W. (1997). Linking breeding and wintering grounds of neotropical migrant songbirds using stable hydrogen isotopic analysis of feathers. *Oecologia*, *109*, 142–148.

Ketterson, E. D. & Nolan, V. Jr (1990). Site attachment and site fidelity in migratory birds: experimental evidence from the field and analogies from neurobiology. In E. Gwinner. *Bird Migration*. Springer Verlag. 117–129.

Kreft, S. (2004). The Fourth Dimension: An Overview of Altitudinal Migration. 25th Annual Bonn Convention, Berlin.

Liechti, F. (1996). *Instructions to count nocturnal bird migration by watching the full moon*. Schweizerische Vogelwarte, CH-6204 Sempach, Switzerland.

Lincoln, F. C. (1979). *Migration of Birds.*. Fish and Wildlife Service. *Circular*, *16*.

Lowery, G. H. (1951). A quantitative study of the nocturnal migration of birds. *University Kan Pub Mus Nat Hist*, *3*, 361–472.

Maanen, E. V., Goradze, I., Gavashelishvili, A. & Goradze, R. (2001). Opinion: Trapping and hunting of migratory raptors in western Georgia. *Bird Conser Int*, *11*, 77–92.

Mila, B., Smith, T. & Wayne, R. (2006). Postglacial population expansion drives the evolution of long-distance avian migration in a songbird. *Evolution*, *60*, 2403–2409.

Newton, I. (2008). *The Migration Ecology of Birds*. Elselvier.

Panov, I. N. (2011). Overlap between moult and autumn migration in passerines in northern taiga zone of Eastern Fennoscandia. *Av Ecol Behav*, *19*, 33–64.

Peter, B., Hans-Günther, B. & Valerie, W. (2001). *Bird Migration: A General Survey*. Oxford: Oxford University Press.

Popa-Lisseanu, A. G., Delgado-Huertas, A., Forero, M. G., Rodriguez, A., Arlettaz, R. & Ibanez, C. (2007). Bats' Conquest of a Formidable Foraging Niche: The Myriads of Nocturnally Migrating Songbirds. In Rands, S. *PLoS ONE*, *2*, e205.

Pulido, F. (2007). The genetics and evolution of avian migration. *BioSci*, *57*, 165–174.

Rappole, J. H. & Hubálek, Z. (2006). Birds and Influenza H5N1 Virus Movement to and within North America. *Emerg Infect Dis*, *12*, 10.

Rappole, J. H., Derrickson, S. R. & Hubalek, Z. (2000). Migratory birds and spread of West Nile virus in the Western Hemisphere. *Emerg Infect Dis*, *6*, 319–328.

Rappole, J., Helm, B. & Ramos, M. (2003). An integrative framework for understanding the origin and evolution of avian migration. *J Av Biol*, *34*, 125.

Rattenborg, N. C., Mandt, B. H., Obermeyer, W. H., Winsauer, P. J. & Huber, R. (2004). Migratory Sleeplessness in the White-Crowned Sparrow (Zonotrichia leucophrys gambelii) . *PLoS Biol.*, *2*, e212.

Rohwer, S., Butler, L. K. & Froehlich, D. R. (2005). Ecology and Demography of East-West Differences in Molt Scheduling of Neotropical Migrant Passerines. In Greenberg, R. & Marra, P.P. *Birds of two worlds: the ecology and evolution of migration*. Johns Hopkins University Press. p. 94.

Ronald, C. Y., Robert, W. B., David, B. L., Barry, D. S. & John, I. (2004). Western sandpipers have altered migration tactics as peregrine falcon populations have recovered. *Proc R Soc Lond B*, *271*, 1263–1269 1263.

Schmaljohann, H., Felix, L. & Bruno, B. (2007). Songbird migration across the Sahara: the non-stop hypothesis rejected. *Proc Biol Sci*, *274*, 735–739.

Sekercioglu, C.H. (2007). Conservation ecology: area trumps mobility in fragment bird extinctions. *Curr Biol*, *17*, 283–286.

Shimazaki, H., Masayuki, T. & Hiroyoshi, H. (2004). Migration routes and important stopover sites of endangered oriental white storks (*Ciconia boyciana*) as revealed by satellite tracking. *Mem Natl Inst Polar Res Spec Issue*, *58*, 162–178.

Smith, R. P. Jr, Rand, P. W., Lacombe, E. H., Morris, S. R., Holmes, D. W. & Caporale, D. A. (1996). Role of bird migration in the long-distance dispersal of *Ixodes dammini*, the vector of Lyme disease. *J Infect Dis*, *174*, 221–4.

Spencer, R. (1985). *Marking*. In: Campbell, B. & Lack, E. (1985). *A dictionary of birds*. British Ornithologists' Union. London, 338–341.

Sprague, A. J., Hamilton, D. J. & Diamond, A. W. (2008). Site Safety and Food Affect Movements of Semipalmated Sandpipers (Calidris pusilla) Migrating Through the Upper Bay of Fundy . *Av Conser Ecol* 3, 4. Online: http://www.ace-eco.org/vol3/iss2/art4/

Swan, L. W. (1970). Goose of the Himalayas. *Nat Hist*, *79*, 68–75.

Thorup, K., Thomas, A., Mikael, H. & Nils, K. (2003). Bird orientation: compensation for wind drift in migrating raptors is age dependent. *Proc R Soc Lond B (Suppl)*, *270*, S8–S11.

Walraff, H. G. (2005). *Avian Navigation: Pigeon Homing as a Paradigm*. Springer.

Weber, J. M. (2009). The physiology of long-distance migration: extending the limits of endurance metabolism. *J Exp Biol*, *212*, 593–597.

Williams, G. G. (1950). Weather and spring migration. *Auk*, *67*, 52–65.

Wiltschko, W., Munro, U., Ford, H. & Wiltschko, R. (2006). Bird navigation: what type of information does the magnetite-based receptor provide? *Proc R Soc B*, *273*, 2815–20.

Wormworth, J. & Mallon, K. (2006). *Bird Species and Climate Change: The Global Status Report version 1.0.*. WWF.

Chapter 7

ORIGIN AND EVOLUTION OF MAMMALS

7.1. HISTORY

The evolution of mammals has passed through many stages since the first appearance of their synapsid ancestors in the late Carboniferous period. By the mid-Triassic, there were many synapsid species that looked like mammals. The lineage leading to today's mammals split up in the Jurassic. Synapsids from this period include *Dryolestes*, more closely related to extant placentals and marsupials than to monotremes, as well as *Ambondro*, more closely related to monotremes [Rougier et al., 2007]. Later on, the eutherian and metatherian lineages separated; the metatherians are the animals more closely related to the marsupials, while the eutherians are those more closely related to the placentals. Since *Juramaia*, the earliest known eutherian, lived 160 million years ago in the Jurassic, this divergence must have occurred in the same period.

After the Cretaceous–Paleogene extinction event wiped out the non-avian dinosaurs (birds are generally regarded as the surviving dinosaurs) and several other mammalian groups, placental and marsupial mammals diversified into many new forms and ecological niches throughout the Paleogene and Neogene, by the end of which all modern orders had appeared.

Mammals are the only living synapsids. The synapsid lineage became distinct from the sauropsid lineage in the late Carboniferous period, between 320 and 315 million years ago. The sauropsids are today's reptiles and birds along with all the extinct animals more closely related to them than to mammals [Waggoner, 1997]. This does not include the mammal-like reptiles, a group more closely related to the mammals.

Throughout the Permian period, the synapsids included the dominant carnivores and several important herbivores. In the subsequent Triassic period, however, a previously obscure group of sauropsids, the archosaurs, became the dominant vertebrates. The mammalian forms appeared during this period; their superior sense of smell, backed up by a large brain, facilitated entry into nocturnal niches with less exposure to archosaur predation. The nocturnal lifestyle may have contributed greatly to the development of mammalian traits such as endothermy and hair. Later in the Mesozoic, after theropod dinosaurs replaced rauisuchians as the dominant carnivores, mammals spread into other ecological niches. For example, some became aquatic, some were gliders, and some even fed on juvenile dinosaurs.

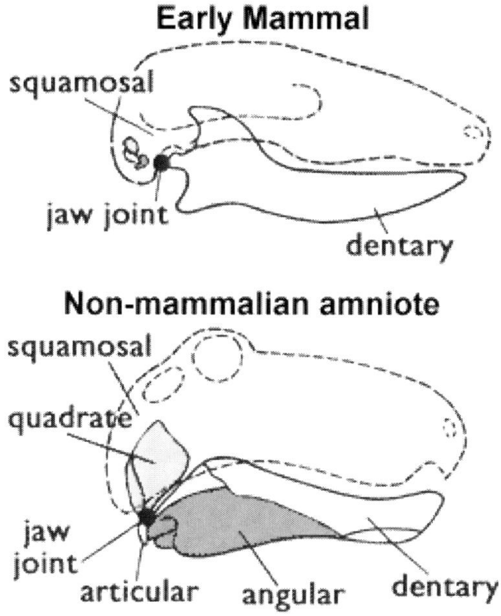

Figure 1. Mammalian and non-mammalian jaws.

Most of the evidence consists of fossils. For many years, fossils of Mesozoic mammals and their immediate ancestors were very rare and fragmentary; but, since the mid-1990s, there have been many important new finds, especially in China. The relatively new techniques of molecular phylogenetics have also shed light on some aspects of mammalian evolution by estimating the timing of important divergence points for modern species. When used carefully, these techniques often, but not always, agree with the fossil record.

Although mammary glands are a signature feature of modern mammals, little is known about the evolution of lactation. This is because these soft tissues are not often preserved in the fossil record. Most study of the evolution of mammals centers around the shapes of the teeth, the hardest parts of the tetrapod body. Other much-studied aspects include the evolution of the middle ear bones, erect limb posture, a bony secondary palate, fur and hair, and warm-bloodedness.

7.2. Mammals

In the mammalian configuration, the quadrate and articular bones are much smaller and form part of the middle ear. Note that in mammals, the lower jaw consists only of the dentary bone (Figure 1). Living mammal species can be identified by the presence of milk-producing mammary glands in females. Other features are required when classifying fossils, because mammary glands and other soft-tissue features are not visible in fossils.

One such feature available for paleontology, shared by all living mammals (including monotremes), but not present in any of the early Triassic therapsids, are jaw bones. The earliest amniotes had a jaw joint composed of the articular (a small bone at the back of the lower jaw) and the quadrate (a small bone at the back of the upper jaw). All non-mammalian tetrapods use this system including amphibians,turtles, lizards, snakes, crocodilians, dinosaurs (and their descendants the birds), and therapsids. But mammals have a different jaw joint, composed only of the dentary (the lower jaw bone, which carries the teeth) and the squamosal (another small skull bone). In the Jurassic, their quadrate and articular bones evolved into the incus and malleus bones in the middle ear [Cowen, 2000]. Mammals also have a double occipital condyle; they have two knobs at the base of the skull that fit into the topmost neck vertebra, while other tetrapods have a single occipital condyle.

In a 1981 article, Kenneth A. Kermack and his co-authors argued for drawing the line between mammals and earlier synapsids at the point where the mammalian pattern of molar occlusion was being acquired and the dentary-squamosal joint had appeared. The criterion chosen, they noted, is merely a matter of convenience; their choice was based on the fact that "the lower jaw is the most likely skeletal element of a Mesozoic mammal to be preserved." Today, most paleontologists consider the animals satisfying this criterion to be mammals [Kemp, 2005].

7.3. The Ancestary of Mammals

Amniotes

The first fully terrestrial vertebrates were amniotes — their eggs had internal membranes that allowed the developing embryo to breathe but kept water in. This allowed amniotes to lay eggs on dry land, while amphibians generally need to lay their eggs in water (a few amphibians, such as the Surinam toad, have evolved other ways of getting around this limitation). The first amniotes apparently arose in the middle Carboniferous from the ancestral reptiliomorphs [Carroll, 1991].

Within a few million years, two important amniote lineages became distinct: mammals' synapsid ancestors and the sauropsids, from which lizards, snakes, crocodilians, dinosaurs and birds are descended. The earliest known fossils of synapsids and sauropsids (such as *Archaeothyris* and *Hylonomus*, respectively) date from about 320 to 315 million years ago. Unfortunately, it is difficult to be sure about when each of them evolved, since vertebrate fossils from the late Carboniferous are very rare, and therefore the actual first occurrences of each of these types of animal might have been considerably earlier.

Synapsids

Synapsid skulls are identified by the distinctive pattern of the holes behind each eye, which served the following purposes:

- made the skull lighter without sacrificing strength.
- saved energy by using less bone.
- probably provided attachment points for jaw muscles. Having attachment points further away from the jaw made it possible for the muscles to be longer and therefore to exert a strong pull over a wide range of jaw movement without being stretched or contracted beyond their optimum range.

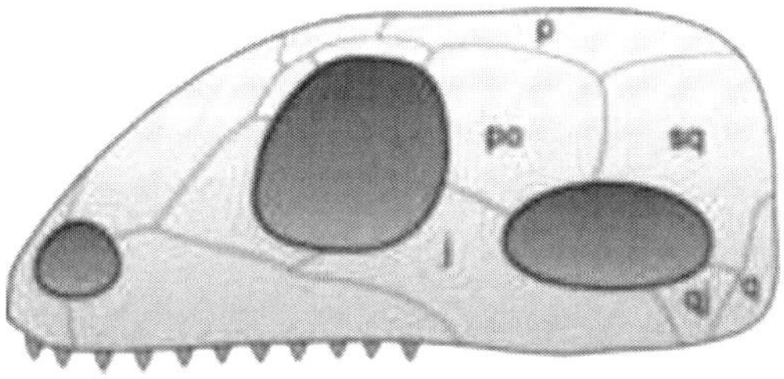

Figure 2. The original synapsid skull structure has one hole behind each eye, in a fairly low position on the skull (lower right in this image).

The synapsid pelycosaurs included the largest land vertebrates of the Early Permian such as the 6 m (20 ft) long *Cotylorhynchus hancocki*. Among the other large pelycosaurs were *Dimetrodon grandis* and *Edaphosaurus cruciger* (Figure 2).

Therapsids

Therapsids descended from pelycosaurs in the middle Permian and took over their position as the dominant land vertebrates. They differ from pelycosaurs in several features of the skull and jaws, including larger temporal fenestrae and incisors that are equal in size.

The therapsid lineage that led to mammals went through a series of stages, beginning with animals that were very like their pelycosaur ancestors and ending with some that could easily be mistaken for mammals [Kermack and Kermack, 1984].

- Gradual development of a bony secondary palate. Most books and articles interpret this as a prequisite for the evolution of mammals' high metabolic rate, because it enabled these animals to eat and breathe at the same time. But some scientists point out that some modern ectotherms use a fleshy secondary palate to separate the mouth from the airway, and that a *bony* palate provides a surface on which the tongue can manipulate food, facilitating chewing rather than breathing [Bennett and Ruben, 1986]. The interpretation of the bony secondary palate as an aid to chewing also suggests the development of a faster metabolism, since chewing makes it possible to digest food more quickly. In mammals, the palate is formed by two specific bones, but various Permian therapsids had other combinations of bones in the right places to function as a palate.
- the dentary gradually becomes the main bone of the lower jaw.
- Progress towards an erect limb posture, which would increase the animals' stamina by avoiding Carrier's constraint. But this process was erratic and very slow — for example: all herbivorous therapsids retained sprawling limbs (some late forms may have had semi-erect hind limbs); Permian carnivorous therapsids had sprawling forelimbs, and some late Permian ones also had semi-sprawling hindlimbs. In fact, modern monotremes still have semi-sprawling limbs.
- Progress towards the mammalian jaw and middle ear in the Triassic.
- Some scientists have argued that some Triassic therapsids show signs of lactation .

Only the dicynodonts, therocephalians and cynodonts survived into the Triassic.

Biarmosuchia

The Biarmosuchia were the most primitive and pelycosaur-like of the therapsids.

Dinocephalians

Dinocephalians ("terrible heads") included both carnivores and herbivores. They were large; *Anteosaurus* was up to 6 m (20ft) long. Some of the carnivores had semi-erect hindlimbs, but all dinocephalians had sprawling forelimbs. In many ways they were very primitive therapsids; for example, they had no secondary palate and their jaws were rather "reptilian".

Anomodonts

The anomodonts ("anomalous teeth") were among the most successful of the herbivorous therapsids — one sub-group, the dicynodonts (Figure 3), survived almost to the end of the Triassic. But anomodonts were very different from modern herbivorous mammals, as their only teeth were a pair of fangs in the upper jaw and it is generally agreed that they had beaks like those of birds or ceratopsians.

Theriodonts

The theriodonts ("beast teeth") and their descendants had jaw joints in which the lower jaw's articular bone tightly gripped the skull's very small quadrate bone. This allowed a much wider gape, and one group, the carnivorous gorgonopsians ("gorgon faces"), took advantage of this to develop "sabre teeth". But the theriodont's jaw hinge had long term significance — the much reduced size of the quadrate bone was an important step in the development of the mammalian jaw joint and middle ear.

Figure 3. *Lystrosaurus*, one of the few genera of dicynodonts that survived the Permian-Triassic extinction event.

The gorgonopsians still had some primitive features: no bony secondary palate (but other bones in the right places to perform the same functions); sprawling forelimbs; hindlimbs that could operate in both sprawling and erect postures. But the therocephalians ("beast heads"), which appear to have arisen at about the same time as the gorgonopsians, had additional mammal-like features, e.g., their finger and toe bones had the same number of phalanges (segments) as in early mammals (and the same number that primates have, including humans).

Cynodonts

The cynodonts, a theriodont group that also arose in the late Permian, include the ancestors of all mammals. Cynodonts' mammal-like features include further reduction in the number of bones in the lower jaw, a secondary bony palate, cheek teeth with a complex pattern in the crowns, and a brain which filled the endocranial cavity (Figure 4).

Multi-chambered burrows have been found, containing as many as 20 skeletons of the Early Triassic cynodont *Trirachodon*; the animals are thought to have been drowned by a flash flood. The extensive shared burrows indicate that these animals were capable of complex social behaviors [Groenewald et al., 2001].

7.4. Triassic Takeover

The catastrophic Permian-Triassic mass extinction slightly more than 250 million years ago killed off about 70 percent of terrestrial vertebrate species and the majority of land plants.

As a result, ecosystems and food chains collapsed, and the establishment of new stable ecosystems took about 30 million years. With the disappearance of the gorgonopsians, who were dominant predators in the late Permian [Benton, 2004], the cynodonts' principal competitors for dominance of the carnivorous niches were a previously obscure sauropsid group, the archosaurs, including the ancestors of crocodilians, dinosaurs, and birds.

The archosaurs quickly became the dominant carnivores [Benton, 2004] a development often called the "Triassic takeover." Their success may have been due to the fact that the early Triassic was predominantly arid and therefore archosaurs' superior water conservation gave them a decisive advantage. All known archosaurs have glandless skins and eliminate nitrogenous waste in a uric acid paste containing little water, while the cynodonts probably excreted most such waste in a solution of urea, as mammals do today; considerable water is required to keep urea dissolved [Campbell, 1979].

But the Triassic takeover may have been a vital factor in the evolution of the mammals. Two groups stemming from the early cynodonts were successful in niches with minimal competition from the archosaurs: the tritylodonts, who were herbivores, and the mammals, most of whom were small nocturnal insectivores although some, like *Sinoconodon*, were carnivores that fed on vertebrate prey, while still others were herbivores or omnivores. As a result:

Figure 4. Artist's conception of the cynodont *Trirachodon* within a burrow.

- The therapsids trend towards differentiated teeth with precise occlusion accelerated, because of the need to hold captured arthropods and crush their exoskeletons.
- As the body length of the mammals' ancestors fell below 50 mm (2 inches), advances in thermal insulation and temperature regulation may have become necessary for nocturnal life [Ruben and Jones, 2000].
- Acute senses of hearing and smell became vital.
- This accelerated the development of the mammalian middle ear.
- The increase in the size of the olfactory lobes of the brain increased brain weight as a percentage of total body weight [Rowe et al., 2011]. Brain tissue requires a disproportionate amount of energy [Raichle and Gusnard, 2002]. The need for more food to support the enlarged brains increased the pressures for improvements in insulation, temperature regulation and feeding.
- Probably as a side-effect of the nocturnal life, mammals lost two of the four cone opsins, photoreceptors in the retina present in the eyes of the earliest amniotes. Paradoxically, this may have improved their ability to discriminate colors in dim light [Vorobyev, 2006].

7.5. Fossil Record

Mesozoic synapsids that had evolved to the point of having a jaw joint composed of the dentary and squamosal bones are preserved in few good fossils, mainly because they were mostly smaller than rats:

- They were largely restricted to environments that are less likely to provide good fossils. Floodplains as the best terrestrial environments for fossilization provide few mammal fossils, because they are dominated by medium to large animals, and the mammals could not compete with archosaurs in the medium to large size range.
- Their delicate bones were vulnerable to being destroyed before they could be fossilized — by scavengers (including fungi and bacteria) and by being trodden upon.
- Small fossils are harder to spot and more vulnerable to being destroyed by weathering and other natural stresses before they are discovered.

In the past 40 years, however, the number of Mesozoic fossil mammals has increased decisively; only 116 genera were known in 1979, for example, but about 310 in 2007, with an increase in quality such that "at least 18 Mesozoic mammals are represented by nearly complete skeletons" [Luo, 2007].

7.6. Mammals or Mammalia Forms?

Some writers restrict the term "mammal" to the crown group mammals, the group consisting of the most recent common ancestor of monotremes, marsupials, and placentals together with all the descendants of that ancestor. To accommodate some related taxa falling outside the crown group, Mammaliaformes were described as comprising "the last common ancestor of Morganucodontidae and Mammalia and all its descendants." Besides Morganucodontidae, the newly defined taxon includes Docodonta and Kuehneotheriidae. Though haramiyids have been referred to the mammals since the 1860s, Rowe excluded them from the Mammaliaformes as falling outside his definition, putting them in a larger clade, the Mammaliamorpha [Rowe, 1988].

Some writers have adopted this terminology noting, to avoid misunderstanding, that they have done so. Most paleontologists, however, still think that animals with the dentary-squamosal jaw joint and the sort of molars characteristic of modern mammals should formally be members of Mammalia.

Where the ambiguity in the term "mammal" may be confusing, this article uses "mammaliaform" and "crown mammal".

7.7. Family Tree — Cynodonts to Crown Group Mammals

Morganucodontidae

The Morganucodontidae first appeared in the late Triassic, about 205M years ago. They are an excellent example of transitional fossils, since they have both the dentary-squamosal and articular-quadrate jaw joints. They were also one of the first discovered and most thoroughly studied of the mammaliaforms, since an unusually large number of morganucodont fossils have been found (Figure 5).

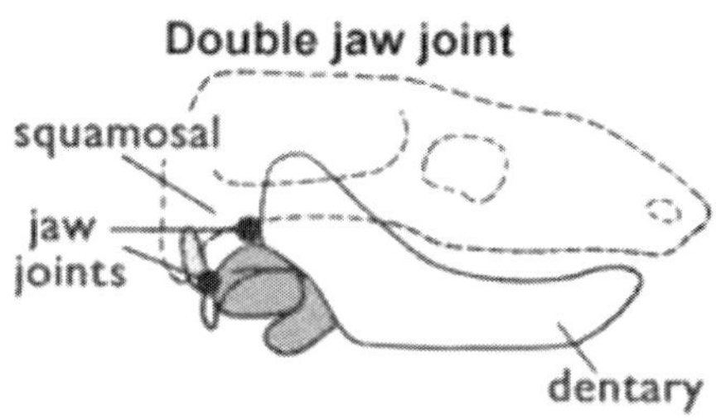

Figure 5. Morganucodontidae and other transitional forms had both types of jaw joint: dentary-squamosal (front) andarticular-quadrate (rear).

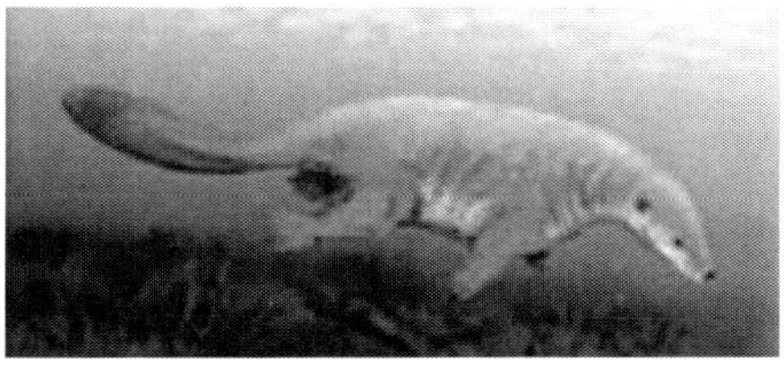

Figure 6. Reconstruction of *Castorocauda*. Note the fur and the adaptations for swimming (broad, flat tail; webbed feet) and for digging (robust limbs and claws).

Docodonts

Docodonts, among the most common Jurassic mammaliaforms, are noted for the sophistication of their molars. Though they were generally herbivorous and insectivorous, an exception is the fish-eating *Castorocauda* ("beaver tail"), which lived in the mid Jurassic about 164M years ago and was first discovered in 2004 and described in 2006. *Castorocauda* was not a crown group mammal, but it is extremely important in the study of the evolution of mammals because the first find was an almost complete skeleton (a real luxury in paleontology) and it breaks the "small nocturnal insectivore" stereotype (Figure 6) [Ji et al., 200] :

- It was noticeably larger than most Mesozoic mammaliaform fossils — about 17 in (43 cm) from its nose to the tip of its 5-inch (130 mm) tail, and may have weighed 500–800 g (18–28 oz).
- It provides the earliest absolutely certain evidence of hair and fur. Previously the earliest was *Eomaia*, a crown group mammal from about 125M years ago.
- It had aquatic adaptations including flattened tail bones and remnants of soft tissue between the toes of the back feet, suggesting that they were webbed. Previously the earliest known semi-aquatic mammaliaforms were from the Eocene, about 110M years later.
- *Castorocauda*'s powerful forelimbs look adapted for digging. This feature and the spurs on its ankles make it resemble the platypus, which also swims and digs.
- Its teeth look adapted for eating fish: the first two molars had cusps in a straight row, which made them more suitable for gripping and slicing than for grinding; and these molars are curved backwards, to help in grasping slippery prey.

Hadrocodium

The mammalian family tree show *Hadrocodium* as an "aunt" of crown mammals. This mammaliaform, dated about 195M years ago in the very early Jurassic, exhibits some important features [Luo et al., 2001]:

- The jaw joint consists only of the squamosal and dentary bones, and the jaw contains no smaller bones to the rear of the dentary, unlike the therapsid design.

- In therapsids and early mammaliaforms the eardrum may have stretched over a trough at the rear of the lower jaw. But *Hadrocodium* had no such trough, which suggests its ear was part of the cranium, as it is in crown-group mammals — and hence the former articular and quadrate had migrated to the middle ear and become the malleus and incus. On the other hand, the dentary has a "bay" at the rear that mammals lack. This suggests that *Hadrocodium's* dentary bone retained the same shape that it would have had if the articular and quadrate had remained part of the jaw joint, and therefore *Hadrocodium* or a very close ancestor may have been the first to have a full mammalian middle ear.
- Therapsids and earlier mammaliaforms had their jaw joints very far back in the skull, partly because the ear was at the rear end of the jaw but also had to be close to the brain. This arrangement limited the size of the braincase, because it forced the jaw muscles to run round and over it. *Hadrocodium's* braincase and jaws were no longer bound to each other by the need to support the ear, and its jaw joint was further forward. In its descendants or those of animals with a similar arrangement, the brain case was free to expand without being constrained by the jaw and the jaw was free to change without being constrained by the need to keep the ear near the brain — in other words it now became possible for mammaliaforms to develop large brains and to adapt their jaws and teeth in ways that were purely specialized for eating.

The Earliest Crown Mammals

The extant mammals and their relatives back to their last common ancestor constitute the earliest crown mammals. Since this group has living members, DNA analysis can be applied in an attempt to explain the evolution of features that do not appear in fossils. This endeavor often involves molecular phylogenetics, a technique that has become popular since the mid-1980s.

7.8. Color Vision

Early amniotes had four opsins in the cones of their retinas to use for distinguishing colors: one sensitive to red, one to green, and two corresponding to different shades of blue [Bailes et al., 2007; Jacobs, 2009]. The green opsin was not inherited by any crown mammals, but all normal individuals did inherit the red one. Early crown mammals thus had three cone opsins, the red one and both of the blues [Jacobs, 2009]. All their extant descendants have lost one of the blue-sensitive opsins but not always the same one: marsupials and placentals (except for cetaceans) retain one blue-sensitive opsin while monotremes retain the other [Collins, 2010]. Some placentals and marsupials, including humans, subsequently evolved green-sensitive opsins; like early crown mammals, therefore, their vision is trichromatic [Dulai, 1999; Arrese et al., 2007].

7.9. AUSTRALOSPHENIDA AND AUSKTRIBOSPHENIDAE

Ausktribosphenidae is a group name that has been given to some rather puzzling finds that:

- appear to have tribosphenic molars, a type of tooth that is otherwise known only in placentals and marsupials [Jacobs et al., 1989].
- come from mid Cretaceous deposits in Australia — but Australia was connected only to Antarctica, and placentals originated in the northern hemisphere and were confined to it until continental drift formed land connections from North America to South America, from Asia to Africa and from Asia to India.
- are represented only by teeth and jaw fragments, which is not very helpful.

Australosphenida is a group that has been defined in order to include the Ausktribosphenidae and monotremes. *Asfaltomylos* (mid- to late Jurassic, from Patagonia) has been interpreted as a basal australosphenid (animal that has features shared with both Ausktribosphenidae and monotremes; lacks features that are peculiar to Ausktribosphenidae or monotremes; also lacks features that are absent in Ausktribosphenidae and monotremes) and as showing that australosphenids were widespread throughout Gondwanaland (the old Southern hemisphere super-continent) [Rauhut et al., 2002].

Recent analysis of *Teinolophos*, which lived somewhere between 121 and 112.5 million years ago, suggests that it was a "crown group" (advanced and relatively specialised) monotreme. This was taken as evidence that the basal (most primitive) monotremes must have appeared considerably earlier, but this has been disputed. The study also indicated that some alleged Australosphenids were also "crown group" monotremes (e.g., *Steropodon*) and that other alleged Australosphenids (e.g.*Ausktribosphenos*, *Bishops*, *Ambondro*, *Asfaltomylos*) are more closely related to and possibly members of the Therian mammals (group that includes marsupials and placentals [Rowe et al., 2008].

7.10. MONOTREMES

Teinolophos, from Australia, is the earliest known monotreme. A 2007 study suggests that it was not a basal (primitive, ancestral) monotreme but a full-fledged platypus, and therefore that the platypus and echidna lineages diverged considerably earlier [Rowe et al., 2008]. A more recent study (2009), however, has suggested that while *Teinolophos* was a type of platypus, it also was a basal monotreme and predated the radiation of modern monotremes. The semi-aquatic lifestyle of platypuses prevented them from being outcompeted by the marsupials that migrated to Australia millions of years ago, since joeys need to keep attached to their mothers and would drown if the mothers ventured into water (though there are exceptions like the water opossum and the lutrine opossum). Genetic evidence has determined that echidnas diverged from the platypus lineage as recently as 19-48M when they made their transition from semi-aquatic to terrestrial lifestyle.

Monotremes have some features that may be inherited from the cynodont ancestors:

- like lizards and birds, they use the same orifice to urinate, defecate and reproduce ("monotreme" means "one hole").
- they lay eggs that are leathery and uncalcified, like those of lizards, turtles and crocodilians.

Unlike other mammals, female monotremes do not have nipples and feed their young by "sweating" milk from patches on their bellies.

Of course these features are not visible in fossils, and the main characteristics from paleontologists' point of view are:

- a slender dentary bone in which the coronoid process is small or non-existent.
- the external opening of the ear lies at the posterior base of the jaw.
- the jugal bone is small or non-existent.
- a primitive pectoral girdle with strong ventral elements: coracoids, clavicles and interclavicle.
  Note: therian mammals have no interclavicle.
- sprawling or semi-sprawling forelimbs.

7.11. Multituberculates

Multituberculates (named for the multiple tubercles on their "molars") are often called the "rodents of the Mesozoic" but this is an example of convergent evolution rather than meaning that they are closely related to the Rodentia (Figure 7). They existed for approximately one hundred twenty million years—the longest fossil history of any mammal lineage—but were eventually outcompeted by rodents, becoming extinct during the early Oligocene.

Some authors have challenged the phylogeny represented by the cladogram above. They exclude the multituberculates from the mammalian crown group, holding that multituberculates are more distantly related to extant mammals than even the Morganucodontidae [Butler, 2000; White, 2005]. Multituberculates are like undisputed crown mammals in that their jaw joints consist of only the dentary and squamosal bones and the quadrate and articular bones are part of the middle ear; their teeth are differentiated, occlude, and have mammal-like cusps; they have a zygomatic arch; and the structure of the pelvis suggests that they gave birth to tiny helpless young, like modern marsupials. On the other hand, they differ from modern mammals:

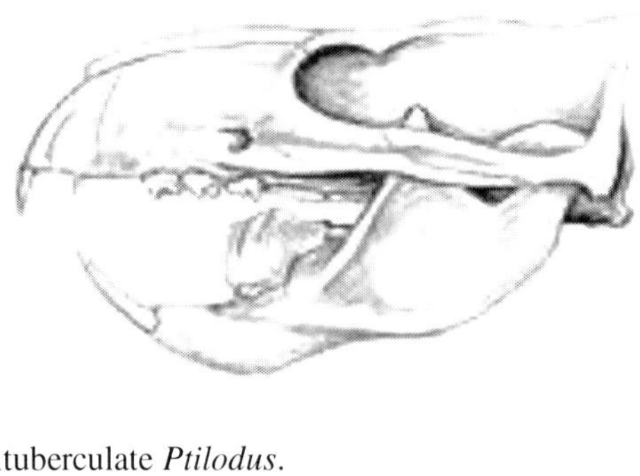

Figure 7. Skull of the multituberculate *Ptilodus*.

- Their "molars" have two parallel rows of tubercles, unlike the tribosphenic (three-peaked) molars of uncontested early crown mammals.
- The chewing action differs in that undisputed crown mammals chew with a side-to-side grinding action, which means that usually the molars occlude on only one side at a time, while multituberculates' jaws were incapable of side-to-side movement—they chewed, rather, by dragging the lower teeth backwards against the upper ones as the jaw closed.
- The anterior (forward) part of the zygomatic arch mostly consists of the maxilla (upper jawbone) rather than the jugal, a small bone in a little slot in the maxillary process (extension).
- The squamosal does not form part of the braincase.
- The rostrum (snout) is unlike that of undisputed crown mammals; in fact it looks more like that of a pelycosaur such as *Dimetrodon*. The multituberculate rostrum is box-like, with the large flat maxillae forming the sides, the nasal the top, and the tall premaxilla at the front.

7.12. METATHERIA

The living Metatheria are all marsupials ("animals with pouches"). A few fossil genera, such as the Mongolian late Cretaceous *Asiatherium*, may be marsupials or members of some other metatherian group(s) [Szalay and Trofimov, 1986].

The oldest known metatherian is *Sinodelphys*, found in 125M-year-old early Cretaceous shale in China's northeastern Liaoning Province. The fossil is nearly complete and includes tufts of fur and imprints of soft tissues.

Didelphimorphia (common opossums of the Western Hemisphere) first appeared in the late Cretaceous and still have living representatives, probably because they are mostly semi-arboreal unspecialized omnivores.

The best-known feature of marsupials is their method of reproduction:

- The mother develops a kind of yolk sack in her womb that delivers nutrients to the embryo. Embryos of bandicoots, koalas and wombats additionally form placenta-like organs that connect them to the uterine wall, although the placenta-like organs are smaller than in placental mammals and it is not certain that they transfer nutrients from the mother to the embryo.
- Pregnancy is very short, typically four to five weeks. The embryo is born at a very young age of development, and is usually less than 2 inches (5.1 cm) long at birth. It has been suggested that the short pregnancy is necessary to reduce the risk that the mother's immune system will attack the embryo.
- The newborn marsupial uses its forelimbs (with relatively strong hands) to climb to the nipple, which is usually in a pouch on the mother's belly. The mother feeds the baby by contracting muscles over her mammary glands, as the baby is too weak to suck. The newborn marsupial's need to use its forelimbs in climbing to the nipple has prevented the forelimbs from evolving into paddles or wings and has therefore

prevented the appearance of aquatic or truly flying marsupials (although there are several marsupial gliders).

Although some marsupials look like some placentals (the thylacine or "marsupial wolf" is a good example), marsupial skeletons have some features that distinguish them from placentals:

- Some, including the thylacine, have 4 molars. No placentals have more than 3.
- All have a pair of palatal fenestrae, window-like openings on the bottom of the skull (in addition to the smaller nostril openings).
- Marsupials also have a pair of marsupial bones (sometimes called "epipubic bones"), which support the pouch in females. But these are not unique to marsupials, since they have been found in fossils of multituberculates, monotremes, and even eutherians — so they are probably a common ancestral feature that disappeared at some point after the ancestry of living placental mammals diverged from that of marsupials ([Novacek et al., 1997]. Some researchers think the epipubic bones' original function was to assist locomotion by supporting some of the muscles that pull the thigh forwards [White, 1989].

7.14. Eutheria

The time of appearance of the earliest eutherians has been a matter of controversy. On one hand, recently discovered fossils of *Juramaia*, have been dated to 160 million years ago and classified as eutherian [Luo et al., 2011]. Fossils of *Eomaia* from 125 million years ago in the Early Cretaceous have also been classified as eutherian. A recent analysis of phenomic characters, however, classified *Eomaia* as pre-eutherian and reported that the earliest clearly eutherian specimens came from Maelestes, dated to 91 million years ago [O'Leary et al., 2013]. That study also reported that eutherians did not significantly diversify until after the catastrophic extinction at the Cretaceous–Paleogene boundary, about 66 million years ago.

Eomaia (Figure 8) was found to have some features that are more like those of marsupials and earlier metatherians:

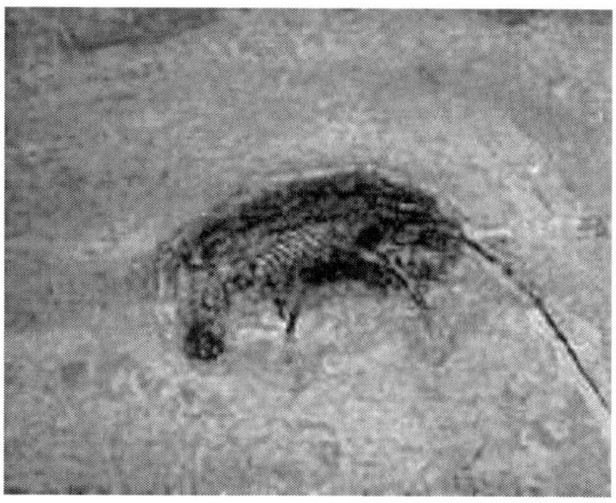

Figure 8. Fossil of *Eomaia* in the Hong Kong Science Museum.

- Epipubic bones extending forwards from the pelvis, which are not found in any modern placental, but are found in all other mammals — early mammaliaforms, non-placental eutherians, marsupials, and monotremes — as well as in the cynodont therapsids that are closest to mammals. Their function is to stiffen the body during locomotion [Reilly and White, 2003]. This stiffening would be harmful in pregnant placentals, whose abdomens need to expand [Novacek et al., 1997].
- A narrow pelvic outlet, which indicates that the young were very small at birth and therefore pregnancy was short, as in modern marsupials. This suggests that the placenta was a later development.
- 5 incisors in each side of the upper jaw. This number is typical of metatherians, and the maximum number in modern placentals is 3, except for homodonts such as the armadillo. But *Eomaia's* molar to premolar ratio (it has more pre-molars than molars) is typical of eutherians, including placentals, and not normal in marsupials.

Eomaia also has a Meckelian groove, a primitive feature of the lower jaw that is not found in modern placental mammals.

These intermediate features are consistent with molecular phylogenetics estimates that the placentals diversified about 110M years ago, 15M years after the date of the *Eomaia* fossil.

Eomaia also has many features that strongly suggest it was a climber, including several features of the feet and toes; well-developed attachment points for muscles that are used a lot in climbing; and a tail that is twice as long as the rest of the spine.

Placentals' best-known feature is their method of reproduction:

The embryo attaches itself to the uterus via a large placenta through which the mother supplies food and oxygen and removes waste products.

Pregnancy is relatively long and the young are fairly well-developed at birth. In some species (especially herbivores living on plains) the young can walk and even run within an hour of birth.

It has been suggested that the evolution of placental reproduction was made possible by retroviruses that: make the interface between the placenta and uterus into a syncytium, i.e., a thin layer of cells with a shared external membrane. This allows the passage of oxygen, nutrients and waste products but prevents the passage of blood and other cells, which would cause the mother's immune system to attack the fetus, which is good for the foetus but makes the mother more vulnerable to infections [Ji et al., 2002].

7.15. Theria

Theria ("beasts"), is the clade originating with the last common ancestor of the Eutheria (including placentals) and Metatheria (including marsupials). Common features include:

- no interclavicle.
- coracoid bones non-existent or fused with the shoulder blades to form coracoid processes.

- a type of crurotarsal ankle joint in which: the main joint is between the tibia and astragalus; the calcaneum has no contact with the tibia but forms a heel to which muscles can attach (Figure 9). (The other well-known type of crurotarsal ankle is seen in crocodilians and works differently — most of the bending at the ankle is between the calcaneum and astragalus).
- tribosphenic molars.

7.16. Expansion of Ecological Niches in Mesozoic

There is still some truth in the "small, nocturnal insectivores" stereotype, but recent finds, mainly in China, show that some mammalia forms and crown group mammals were larger and had a variety of lifestyles. For example:

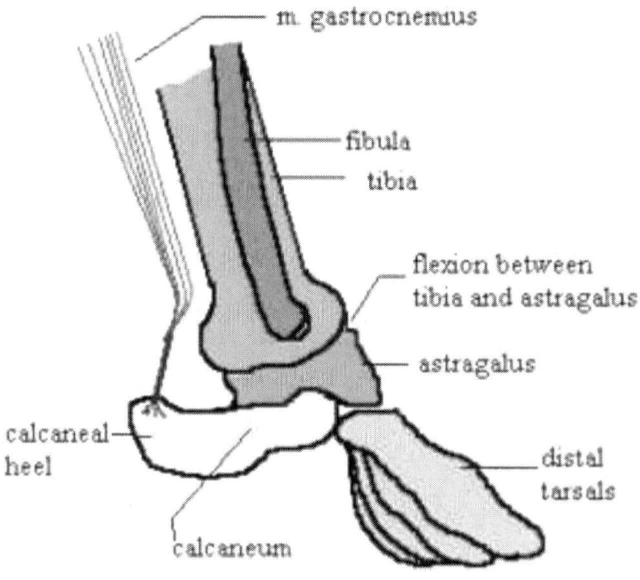

Figure 9. Therian form of crurotarsal ankle. Adapted with permission from Palaeos.

Figure 10. *Repenomamus* sometimes preyed on young dinosaurs.

- *Castorocauda*, a member of Docodonta which lived in the middle Jurassic about 164 million years, was about 42.5 cm (16.7 in) long, weighed 500–800 g (18–28 oz), had limbs that were adapted for swimming and digging and teeth adapted for eating fish.
- *Multituberculates* are allotherians that survived for over 125 million years (from mid Jurassic, about 160M years ago, to late Eocene, about 35M years ago) are often called the "rodents of the Mesozoic". As noted above, they may have given birth to tiny live neonates rather than laying eggs.
- *Fruitafossor*, from the late Jurassic period about 150 million years ago, was about the size of a chipmunk and its teeth, forelimbs and back suggest that it broke open the nest of social insects to prey on them (probably termites, as ants had not yet appeared) [Luo and Wible, 2005].
- *Volaticotherium*, from the boundary of the early Cretaceous about 125M years ago, is the earliest-known gliding mammal and had a gliding membrane that stretched out between its limbs, rather like that of a modern flying squirrel. This also suggests it was active mainly during the day [Meng et al., 2006].
- *Repenomamus*, a eutriconodont from the early Cretaceous 130 million years ago, was a stocky, badger-like predator that sometimes preyed on young dinosaurs (Figure 10). Two species have been recognized, one more than 1 m (39 inches) long and weighing about 12–14 kg (26–31 lb), the other less than 0.5 m (20 inches) long and weighing 4–6 kg (8.8–13 lb) [Li et al., 2000; Hu et al., 2005].

7.17. Evolution of Major Groups of Living Mammals

There are currently vigorous debates between traditional paleontologists and molecular phylogeneticists about how and when the modern groups of mammals diversified, especially the placentals. Generally the traditional paelontologists date the appearance of a particular group by the earliest known fossil whose features make it likely to be a member of that group, while the molecular phylogeneticists suggest that each lineage diverged earlier (usually in the Cretaceous) and that the earliest members of each group were anatomically very similar to early members of other groups and differed only in their genetics. These debates extend to the definition of and relationships between the major groups of placentals — the controversy about Afrotheria is a good example.

7.18. Fossil-Based Family Tree of Placental Mammals

Here is a very simplified version of a typical family tree based on fossils, based on Cladogram of Mammalia. It tries to show the nearest thing there is at present to a consensus view, but some paleontologists have very different views, for example [Wible et al., 2007]:

- The most common view is that placentals originated in the southern hemisphere, but some paleontologists argue that they first appeared in Laurasia (old supercontinent containing modern Asia, N. America and Europe).

- Paleontologists differ about when the first placentals appeared, with estimates ranging from 20M years before the end of the Cretaceous to just after the end of the Cretaceous. Molecular biologists argue for a much earlier origin, even suggesting appearance in the Middle Jurassic [Waddell et al., 1999].
- Most paleontologists suggest that placentals should be divided into Xenarthra and the rest, but a few think these animals diverged later.

For the sake of brevity and simplicity, the diagram omits some extinct groups in order to focus on the ancestry of well-known modern groups of placentals. This also conveys the following:

- the age of the oldest known fossils in many groups, since one of the major debates between traditional paleontologists and molecular phylogeneticists is about when various groups first became distinct.

This family tree contains some surprises and puzzles. For example:

The closest living relatives of cetaceans (whales, dolphins, porpoises) are artiodactyls, hoofed animals, which are almost all pure vegetarians.

Bats are fairly close relatives of primates.

The closest living relatives of elephants are the aquatic sirenians, while their next relatives are hyraxes, which look more like well-fed guinea pigs.

- There is little correspondence between the structure of the family (what was descended from what) and the dates of the earliest fossils of each group. For example the earliest fossils of perissodactyls (the living members of which are horses, rhinos and tapirs) date from the late Paleocene but the earliest fossils of their "sister group" the Tubulidentata date from the early Miocene, nearly 50M years later.

7.19. Family Tree of Placental Mammals According to Molecular Phylogenetics

Molecular phylogenetics uses features of organisms' genes to work out family trees in much the same way as paleontologists do with features of fossils — if two organisms' genes are more similar to each other than to those of a third organism, the two organisms are more closely related to each other than to the third.

Molecular phylogeneticists have proposed a family tree that is very different from the one with which paleontologists are familiar. Like paleontologists, molecular phylogeneticists have different ideas about various details [Murphy et al., 2001; Kriegs et al., 2006]:

Here are the most significant of the many differences between family tree and the one familiar to paleontologists:

- The top-level division is between Atlantogenata and Boreoeutheria, instead of between Xenarthra and the rest. However, analysis of transposable element insertions supports a 3-way top-level split between Xenarthra, Afrotheria and Boreoeutheria

[Churakov et al., 2009; Nishihara et al., 2009] and the Atlantogenata clade does not receive significant support in recent distance-based molecular phylogenetics [Meredith et al., 2011].

- Afrotheria contains several groups that are only distantly related according to the paleontologists' version: Afroinsectiphilia ("African insectivores"), Tubulidentata (aardvarks, which paleontologists regard as much closer to odd-toed ungulates than to other members of Afrotheria), Macroscelidea (elephant shrews, usually regarded as close to rabbits and rodents). The only members of Afrotheria that paleontologists would regard as closely related are Hyracoidea (hyraxes), Proboscidea (elephants) and Sirenia (manatees, dugongs).
- Insectivores are split into 3 groups: one is part of Afrotheria and the other two are distinct sub-groups within Boreoeutheria.
- Bats are closer to Carnivora and odd-toed ungulates than to primates and Dermoptera (colugos).
- Perissodactyla (odd-toed ungulates) are closer to Carnivora and bats than to Artiodactyla (even-toed ungulates).

The grouping together of the Afrotheria has some geological justification. All surviving members of the Afrotheria originate from South American or (mainly) African lineages — even the Indian elephant, which diverged from an African lineage about 7.6 million years ago. As Pangaea broke up, Africa and South America separated from the other continents less than 150M years ago, and from each other between 100M and 80M years ago. So it would not be surprising if the earliest eutherian immigrants into Africa and South America were isolated there and radiated into all the available ecological niches.

Nevertheless these proposals have been controversial. Paleontologists naturally insist that fossil evidence must take priority over deductions from samples of the DNA of modern animals. More surprisingly, these new family trees have been criticised by other molecular phylogeneticists, sometimes quite harshly:

- Mitochondrial DNA's mutation rate in mammals varies from region to region — some parts hardly ever change and some change extremely quickly and even show large variations between individuals within the same species [Springer et al., 1995, 1996].
- Mammalian mitochondrial DNA mutates so fast that it causes a problem called "saturation", where random noise drowns out any information that may be present. If a particular piece of mitochondrial DNA mutates randomly every few million years, it will have changed several times in the 60 to 75M years since the major groups of placental mammals diverged [Li, 1997].

7.20. Timing of Placental Evolution

Recent molecular phylogenetic studies suggest that most placental orders diverged late in the Cretaceous period, about 100 to 85 million years ago, but that modern families first appeared later, in the late Eocene and early Miocene epochs of the Cenozoic period [Bininda et al., 2007]. Fossil-based analyses, on the contrary, limit the placentals to the Cenozoic [Wible et al., 2007]. Many Cretaceous fossil sites contain well-preserved lizards, salamanders, birds, and mammals, but not the modern forms of mammals. It is likely that they simply did not exist, and that the molecular clock runs fast during major evolutionary radiations [Benton, 1999]. On the other hand there is fossil evidence from 85 million years ago of hoofed animals that may be ancestors of modern ungulates [Archibald, 1996].

Fossils of the earliest members of most modern groups date from the Paleocene, a few dates from later and very few from the Cretaceous, before the extinction of the dinosaurs. But some paleontologists, influenced by molecular phylogenetic studies, have used statistical methods to extrapolate *backwards* from fossils of members of modern groups and concluded that primates arose in the late Cretaceous [Martin et al., 2007]. However, statistical studies of the fossil record confirm that mammals were restricted in size and diversity right to the end of the Cretaceous, and rapidly grew in size and diversity during the Early Paleocene [Alroy, 1999; Archibald and Deutschman, 2001].

7.21. Evolution of Mammalian Features

Jaws and Middle Ears

Hadrocodium, whose fossils date from the early Jurassic, provides the first clear evidence of fully mammalian jaw joints and middle ears, in which the jaw joint is formed by the dentary and squamosal bones while the articular and quadrate move to the middle ear, where they are known as the incus and malleus.

One analysis of the monotreme *Teinolophos* suggested that this animal had a pre-mammalian jaw joint formed by the angular and quadrate bones and that the definitive mammalian middle ear evolved twice independently, in monotremes and in therian mammals, but this idea has been disputed [Rich et al., 2005].

Milk Production (Lactation)

It has been suggested that lactation's original function was to keep eggs moist. Much of the argument is based on monotremes (egg-laying mammals) [Oftedal, 2002]:

- While the amniote egg is usually described as able to evolve away from water, most reptile eggs actually need moisture if they are not to dry out.
- Monotremes do not have nipples but secrete milk from a hairy patch on their bellies.
- During incubation, monotreme eggs are covered in a sticky substance whose origin is not known. Before the eggs are laid, their shells have only three layers. Afterwards, a fourth layer appears with a composition different from that of the original three. The sticky substance and the fourth layer may be produced by the mammary glands.
- If so, that may explain why the patches from which monotremes secrete milk are hairy. It is easier to spread moisture and other substances over the egg from a broad, hairy area than from a small, bare nipple.

Later research demonstrated that caseins already appeared in the common mammalian ancestor approximately 200–310 million years ago [Border and Robertson, 1993]. The question of whether secretion of a substance to keep egg moist translated into actual lactation in therapsids is open. A small mammaliomorph called *Sinocodon*, generally assumed to be the sister group of all later mammals, had front teeth in even the smallest individuals. Combined with a poorly ossified jaw, they very probably did not suckle. Thus suckling may have evolved right at the pre-mammal/mammal transition.

Hair and Fur

The first clear evidence of hair or fur is in fossils of *Castorocauda*, from 164M years ago in the mid Jurassic. As both extant mammals and *Castorocauda* have a double coat of hair, with both guard hairs and an undercoat, it may be assumed that their last common ancestor did as well. This animal must have been Triassic as it was an ancestor of the Triassic *Tikitherium*.

In the mid-1950s, some scientists interpreted the foramina (passages) in the maxillae (upper jaws) and premaxillae (small bones in front of the maxillae) of cynodonts as channels that supplied blood vessels and nerves to vibrissae (whiskers) and suggested that this was evidence of hair or fur [Brink, 1955; Kemp, 1982]. It was soon pointed out, however, that foramina do not necessarily show that an animal had vibrissae; the modern lizard *Tupinambis* has foramina that are almost identical to those found in the non-mammalian cynodont *Thrinaxodon* [Estes, 1961]. Popular sources, nevertheless, continue to attribute whiskers to *Thrinaxodon*. A trace fossil from the Lower Triassic had been erroneously regarded as a cynodont footprint showing hair [Ellenberger, 1976], but this interpretation has been refuted [Olsen, 2012].

Ruben & Jones [2000] note that the Harderian glands that secrete lipids for coating the fur were present in the earliest mammals like *Morganucodon* but were absent in near-mammalian therapsids like *Thrinaxodon*.

Erect Limbs

The evolution of erect limbs in mammals is incomplete — living and fossil monotremes have sprawling limbs. Some scientists think that the parasagittal (non-sprawling) limb posture is limited to the Boreosphenida, a group that contains the therians but not, for example, the multituberculates. In particular, they attribute a parasagittal stance to the therians *Sinodelphys* and *Eomaia*, which means that the stance had arisen by 125 million years ago, in the Early Cretaceous [Kielan–Jaworowska and Hurum, 2006].

Warm-bloodedness

"Warm-bloodedness" is a complex and rather ambiguous term, because it includes some or all of the following:

Endothermy, the ability to generate heat internally rather than via behaviors such as basking or muscular activity.

Homeothermy, maintaining a fairly constant body temperature. Most enzymes have an optimum operating temperature; efficiency drops rapidly outside the preferred range. A homeothermic organism needs only to possess enzymes that function well in a small range of temperatures.

Tachymetabolism, maintaining a high metabolic rate, particularly when at rest. This requires a fairly high and stable body temperature because of the Q_{10} effect: biochemical processes run about half as fast if an animal's temperature drops by 10 °C.

Since scientists cannot know much about the internal mechanisms of extinct creatures, most discussion focuses on homeothermy and tachymetabolism.

Modern monotremes have a low body temperature compared to marsupials and placental mammals, around 32 °C (90 °F) [Paul, 1988]. Phylogenetic bracketing suggests that the body temperatures of early crown-group mammals were not less than that of extant monotremes. There is cytological evidence that the low metabolism of monotremes is a secondarily evolved trait [Watson and Graves, 1988].

Respiratory Turbinates

Modern mammals have respiratory turbinates, convoluted structures of thin bone in the nasal cavity. These are lined with mucous membranes that warm and moisten inhaled air and extract heat and moisture from exhaled air. An animal with respiratory turbinates can maintain a high rate of breathing without the danger of drying its lungs out, and therefore may have a fast metabolism. Unfortunately these bones are very delicate and therefore have not yet been found in fossils. But rudimentary ridges like those that support respiratory turbinates have been found in advanced Triassic cynodonts such as *Thrinaxodon* and *Diademodon*, which suggests that they may have had fairly high metabolic rates [Hillenius, 1992, Ruben, 1995].

Bony Secondary Palate

Mammals have a secondary bony palate, which separates the respiratory passage from the mouth, allowing them to eat and breathe at the same time. Secondary bony palates have been found in the more advanced cynodonts and have been used as evidence of high metabolic rates [McNab, 1978]. But some cold-blooded vertebrates have secondary bony palates (crocodilians and some lizards), while birds, which are warm-blooded, do not have them.

Diaphragm

A muscular diaphragm helps mammals to breathe, especially during strenuous activity. For a diaphragm to work, the ribs must not restrict the abdomen, so that expansion of the chest can be compensated for by reduction in the volume of the abdomen and *vice versa*. The advanced cynodonts have very mammal-like rib cages, with greatly reduced lumbar ribs. This suggests that these animals had diaphragms, were capable of strenuous activity for fairly long periods and therefore had high metabolic rates. On the other hand these mammal-like rib cages may have evolved to increase agility. The movement of even advanced therapsids was however "like a wheelbarrow", with the hindlimbs providing all the thrust while the forelimbs only steered the animal. In other words, advanced therapsids were not as agile as either modern mammals or the early dinosaurs [Ccowen, 2000]. So the idea that the main function of these mammal-like rib cages was to increase agility is doubtful.

Limb Posture

The therapsids had sprawling forelimbs and semi-erect hindlimbs [Jenkins, 1971]. This suggests that Carrier's constraint would have made it rather difficult for them to move and breathe at the same time, but not as difficult as it is for animals such as lizards, which have completely sprawling limbs [Pough et al., 1996]. Advanced therapsids may therefore have been significantly less active than modern mammals of similar size and so may have had slower metabolisms overall or else been bradymetabolic (lower metabolism when at rest).

Insulation (Hair and Fur)

Insulation is the "cheapest" way to maintain a fairly constant body temperature, without consuming energy to produce more body heat. Therefore, possession of hair or fur would be good evidence of homeothermy, but would not be such strong evidence of a high metabolic rate [Schmidt, 1975; Whithers, 1992].

The first clear evidence of hair or fur is in fossils of *Castorocauda*, from 164M years ago in the mid Jurassic.

Mammalian Brain

Mammals are noted for their large brain size relative to body size, compared to other animal groups. Recent findings suggest that the first brain area to expand was the brain area involved in smell. Scientists scanned the skulls of early mammal species dating back to 190-200 million years ago and compared the brain case shapes to earlier pre-mammal species and found that the brain area involved in the sense of smell was the first to enlarge. This change may have allowed these early mammals to hunt insects at night when dinosaurs were not active.

REFERENCES

Alroy J. (1999). The fossil record of North American mammals: evidence for a Paleocene evolutionary radiation. *Syst Biol*, *48*, 107–18.

Archibald, J. D. & Deutschman, D. H. (2001). Quantitativ Analysis of the Timing of the Origin and Diversification of Extant Placental Orders. *J Mamm Evol*, *8*, 107–124.

Archibald, J. D. (1996). Fossil Evidence for a Late Cretaceous Origin of "Hoofed" Mammals. *Science*, *272*, 1150–1153.

Arrese, C. A., Hart, N. S., Thomas, N., Beazley, L. D. & Shand, J. (2002). Trichromacy in Australian marsupials. *Curr Biol*, *12*, 657–660.

Bailes, H. J., Davies, W. L., Trezise, A. E. O. & Collin, S. P. (2007). Visual pigments in a living fossil, the Australian lungfish *Neoceratodus forsteri*. *BMC Evol Biol*, *7*, 200–207.

Ben, W. (1997). Introduction to the Synapsida". University of California Museum of Paleontology.

Bennett, A. F. & Ruben, J. A. (1986). The metabolic and thermoregulatory status of therapsids". In Hotton III, N., MacLean, P.D. & Roth, J.J. et al.. *The ecology and biology of mammal-like reptiles*. Washington: Smithsonian Institution Press, Washington. 207–218.

Benton, M. J. (1999). Early origins of modern birds and mammals: molecules vs. morphology. *BioEss*, *21*, 1043–1051.

Benton, M. J. (2004). *Vertebrate Palaeontology* (3rd ed.). Oxford: Blackwell Science.

Bininda-Emonds, O. R. P., Cardillo, M., Jones, K. E., Ross D. E., Robin M. D., Grenyer, R., Price, S. A. & Vos, R. A. (2007). The delayed rise of present-day mammals. *Nature*, *446*, 507–511.

Borders, R. & Robertson, W. (1993). Washington's pesticide panel: process and product. *Vet hum toxicol*, *35*, 258–259.

Brink, A. S. (1955). A study on the skeleton of *Diademodon*. *Palaeontol Afr*, *3*, 3–39.

Butler, P. M. (2000). Review of the early allotherian mammals. *Acta Palaeontol*, *45*, 317–342.

Campbell, J. W. (1979). Prosser, C.L. ed. *Comparative Animal Physiology* (3rd ed.). W. B. Sauders. 279–316.

Carroll, R. L. (1991). The origin of reptiles. In: Schultze, H.P. & Trueb, L. (ed) *Origins of the higher groups of tetrapods — controversy and consensus*. Ithaca: Cornell University Press, 331-353.

Ccowen, R. (2000). *History of Life*. Oxford: Blackwel Science. p. 432.

Churakov, G., Kriegs, J.O., Baertsch, R., Zemann, A., Brosius, J. & Schmitz, J. (2009). Mosaic retroposon insertion patterns in placental mammals. *Gen Res*, *19*, 868–75.

Collin, S. P. (2010). Evolution and Ecology of Retinal Photoreception in Early Vertebrates. *Br Behav Evol*, *75*, 174–185.

Cowen, R. (2000). *History of Life*. Oxford: Blackwell Science. p. 432.

Dulai, K. S., von Dornum, M., Mollon, J. D. & Hunt, D. M. (1999). The Evolution of Trichromatic Color Vision by Opsin Gene Duplication in New World and Old World Primates. *Gen Res*, *9*, 629–638.

Ellenberger, P. (1976). Une piste avec traces de soies épaisses dans le Trias inférieur a moyen de Lodève (Hérault, France): *Cynodontipus polythrix nov. gen., nov. sp.* les Cynodontes en France". *Geobios*, *9*, 769–787.

Estes, R. (1961). Cranial anatomy of the cynodont reptile *Thrinaxodon liorhinus*". *Bull Mus Comp Zool*, *125*, 165–180.

Groenewald, G. H., Welman, J. & MacEachern, J. A. (2001). Vertebrate Burrow Complexes from the Early Triassic Cynognathus Zone (Driekoppen Formation, Beaufort Group) of the Karoo Basin, South Africa. *PALAIOS*, *16*, 148–160.

Hillenius, W. H. (1992). The evolution of nasal turbinates and mammalian endothermy. *Paleobiol*, *18*, 17–29.

Hu, Y., Meng, J., Wang, Y. & Li, C. (2005). Large Mesozoimammals fed on young dinosaurs. *Nature*, *433*, 149–152.

Jacobs, G. H. (2009). Evolution of colour vision in mammals. *Philos Trans R Soc B*, *364*, 2957–67.

Jacobs, L. L., Winkler, D. A. & Murry P. A. 1989). Modern Mammal Origins: Evolutionary Grades in the Early Cretaceous of North America. *Proc Nat Acad Sci USA*, *86*, 4992–4995.

Jenkins, F. A., Jr (1971). The postcranial skeleton of African cynodonts. *Bull Pea Mus Nat Hist*, *36*, 1–216.

Ji, Q., Luo, Z. X., Yuan, C. X. & Tabrum, A. R. (2006). A Swimming Mammaliaform from the Middle Jurassic and Ecomorphological Diversification of Early Mammals. *Science*, *311*, 1123–7.

Ji, Q., Luo, Z. X., Yuan, C. X., Wible, J. R., Zhang, J. P. & Georgi, J. A. (2002). The earliest known eutherian mammal. *Nature*, *416*, 816–822.

Kemp, T. S. (1982). *Mammal-like reptiles and the origin of mammals*. London: Academic Press. p. 363.

Kemp, T. S. (2005). *The Origin and Evolution of Mammals*. Oxford University Press. p. 3.

Kermack, D. M. & Kermack, K. A. (1984). *The evolution of mammalian characters*. Croom Helm.

Kielan−Jaworowska, Z. & Hurum, J. H. (2006). Limb posture in early mammals: Sprawling or parasagittal. *Acta Palaeontol Pol*, *51*, 10237–10239.

Kriegs, J. O., Churakov, G. & Kiefmann, M. (2006). Retroposed Elements as Archives for the Evolutionary History of Placental Mammals. *PLoS Biol*, *4*, e91.

Li, J., Wang, Y., Wang, Y. & Li, C. (2000). A new family of primitive mammal from the Mesozoic of western Liaoning, China. *Chin Sci Bull*, *46*, 782–785.

Li, W. H. (1997). *Molecular Evolution*. Sinauer Associates.

Luo, Z. X. & Wible, J. R. (2005). A Late Jurassic Digging Mammal and Early Mammal Diversification. *Science*, *308*, 103–107.

Luo, Z. X. (2007). Transformation and diversification in early mammal evolution. *Nature*, *450*, 1011–1019.

Luo, Z. X., Crompton, A. W. & Sun, A. L. (2001). A New Mammalia form from the Early Jurassic and Evolution of Mammalian Characteristics. *Science*, *292*, 1535–1540.

Luo, Z. X., Yuan, C. X., Meng, Q. J. & Qiang, J. (2011). A Jurassic eutherian mammal and divergence of marsupials and placentals. *Nature*, *476*, 442–445.

Martin, R. D., Soligo, C. & Tavaré, S. (2007). Primate Origins: Implications of a Cretaceous Ancestry. *Folia Primatol*, *78*, 277–296.

McNab, B. K. (1978). The evolution of endothermy in the phylogeny of mammals. *.Am Nat*, *112*, 1–21.

Meng, J., Hu, Y., Wang, Y., Wang, X. & Li, C. (2006). A Mesozoic gliding mammal from northeastern China. *Nature*, *444*, 889–893.

Meredith, R. W., Janecka, J. E., Gatesy, J., Ryder, O. A., Fisher, C. A., Teeling, E. C., Goodbla, A. & Eizirik, E. (2011). Impacts of the Cretaceous Terrestrial Revolution and KPg Extinction on Mammal Diversification. *Science*, *334*, 521–4.

Murphy, W. J., Eizirik, E. & Springer, M. S. (2001). Resolution of the Early Placental Mammal Radiation Using Bayesian Phylogenetics. *Science*, *294*, 2348–2351.

Nishihara, H., Maruyama, S. & Okada, N. (2009). Retroposon analysis and recent geological data suggest near-simultaneous divergence of the three superorders of mammals. *Proc Nat Acad Sci USA*, *106*, 5235–40.

Novacek, M. J., Rougier, G. W., Wible, J. R., McKenna, M. C., Dashzeveg, D. & Horovitz, I. (1997). Epipubic bones in eutherian mammals from the late Cretaceous of Mongolia. *Nature*, *389*, 440–441.

O'Leary, M. A., Bloch, J. I., Flynn, J. J., Gaudin, T. J., Giallombardo, A., Giannini, N. P., Goldberg, S. L. & Kraatz, B. P. (2013). The placental mammal ancestor and the post-K-Pg radiation of placentals. *Science*, *339*, 662–7.

Olsen, P. E. (2012). *Cynodontipus*: A procolophonid burrow - not a hairy cynodont track (Middle-Late Triassic: Europe, Morocco, Eastern North America)). *Program for Northeastern Section - 47th Annual Meeting (18–20 March 2012) Hartford, Connecticut*, the Geological society of America

Paul, G. S. (1988). *Predatory Dinosaurs of the World*. NeYork: Simon and Schuster. p. 464.

Pough, F. H., Heiser, J. B. & McFarland, W. N. (1996). *Vertebrate Life*. New Jersey: Prentice-Hall. p. 798.

Raichle, M. E. & Gusnard, D. A. (2002). Appraising the brain's energy budget. *Proc Nat Acad Sci USA*, *99*, 10237–10239.

Rauhut, O. W. M., Martin, T., Ortiz-Jaureguizar, E. & Puerta, P. (2002). A Jurassic mammal from South America. *Nature*, *416*, 165–168.

Reilly, S. M. & White, T. D. (2003). Hypaxial Motor Patterns and the Function of Epipubic Bones in Primitive Mammals. *Science*, *299*, 400–402.

Rich, T. H., Hopson, J. A., Musser, A. M., Flannery., T. G. F. & Vickers-Rich, P. (2005). Independent Origins of Middle Ear Bones in Monotremes and Therians. *Science*, *307*, 910–914.

Rougier, G. W., Martinelli, A. G., Forasiepi, A. M. & Novacek, M. J. (2007). New Jurassic mammals from Patagonia, Argentina: A reappraisal of australosphenidan morphology and interrelationships. *Am Mus Nov*, *3566*, 1–54.

Rowe, T. (1988). Definition, diagnosis, and origin of Mammalia. *J Vert Paleontol*, *8*, 241–264.

Rowe, T. B., Macrini, T. E. & Luo, Z. X. (2011). Fossil evidence on origin of the mammalian brain. *Science*, *332*, 955–957.

Rowe, T., Rich, T. H., Vickers-Rich, P., Springer, M. & Woodburne, M. O. (2008). The oldest platypus and its bearing on divergence timing of the platypus and echidna clades. *Proc Nat Acad Sci USA*, *105*, 1238–1242.

Ruben, J. (1995). The evolution of endothermy in mammals and birds: from physiology to fossils. *Ann Rev Physiol*, *57*, 69–95.

Ruben, J. A. & Jones, T. D. (2000). Selective Factors Associated with the Origin of Fur and Feathers. *Am. Zool*, *40*, 585–596.

Ruben, J. A. & Jones, T. D. (2000). Selective Factors Associated with the Origin of Fur and Feathers. *Am Zool*, *40*, 585–596.

Schmidt-Nielsen, K. (1975). *Animal physiology: Adaptation and environment*. Cambridge: Cambridge University Press. p. 699.

Springer, M. S. & Douzery, E. (1996). Secondary Structure and patterns of evolution among mammalian mitochondrial 12S rRNA molecules. *J Mol Evol*, *43*, 357–373.

Springer, M. S., Hollar, L. J. & Burk, A. (1995). Compensatory substitutions and the evolution of the mitochondrial 12S rRNA gene in mammals. *Mol Biol Evol*, *12*, 1138–1150.

Szalay, F. S. & Trofimov, B. A. (1996). The Mongolian Late Cretaceous Asiatherium, and the early phylogeny and paleobiogeography of Metatheria. *J Vert Paleontol*, *16*, 474–509.

Vorobyev, M. (2006). Evolution of colour vision: The story of lost visual pigments. *Perception* (ECVP Abstract Supplement) 35.

Waddell, P. J., Cao, Y., Hasegawa, M. & Mindell, D. P. (1999). Assessing the Cretaceous superordinal divergence times within birds and placental mammals by using whole mitochondrial protein sequences and an extended statistical framework. *Syst Biol*, *48*, 119–137.

Watson, J. M. & Graves, J. A. M. (1988). Monotreme Cell-Cycles and the Evolution of Homeothermy. *Aust J Zool*, *36*, 573–584.

White, A. T. (2005). Mammaliaformes—Palaeos.

White, T. D. (1989). An analysis of epipubic bone function in mammals using scaling theory. *J Theor Biol*, *139*, 343–57.

Wible, J. R., Rougier, G. W., Novacek, M. J. & Asher, R. J. (2007). Cretaceous eutherians and Laurasian origin for placental mammals near the K/T boundary. *Nature*, *447*, 1003–1006.

Wible, J. R., Rougier, G. W., Novacek, M. J. & Asher, R. J. (2007). Cretaceous eutherians and Laurasian origin for placental mammals near the K/T boundary. *Nature*, *447*, 1003–1006.

Withers, P. C. (1992). *Comparative Animal Physiology*. Fort Worth: Saunders College. p. 949.

Chapter 8

DIVERSITY OF MAMMALS

8.1. MONOTREMES (EGG-LAYING MAMMALS)

Monotremes (from the Greek "single" + "hole", referring to the cloaca) are mammals that lay eggs (Prototheria) instead of giving birth to live young like marsupials (Metatheria) and placental mammals (Eutheria). The only surviving examples of monotremes are all indigenous to Australia and New Guinea, although there is evidence that they were once more widespread. Among living mammals they include the platypus and four species of echidnas (or spiny anteaters); there is debate regarding monotreme taxonomy [Mckenna et al.,1997].

There are only five living monotreme species: the duck-billed platypus and four species of echidna (also known as spiny anteaters) (Figure 1-2). All of them are found only in Australia and New Guinea. Monotremes are not a very diverse group today, and there has not been much fossil information known until rather recently.

In some ways, monotremes are very primitive for mammals because, like reptiles and birds, they lay eggs rather than having live birth. In a number of other respects, monotremes are rather derived, having highly modified snouts or beaks, and modern adult monotremes have no teeth. Like other mammals, however, monotremes have a single bone in their lower jaw, three middle ear bones, are warm-blooded with high metabolic rates, hair, and they produce milk to nourish the young [Groves, 2005].

8.2. GENERAL CHARACTERISTICS

In common with reptiles and marsupials, monotremes lack the connective structure (corpus callosum), which in placental mammals is the primary communication route between the right and left brain hemispheres. The anterior commissure provide an alternate communication route between the two hemispheres. In monotremes and marsupials, it carries all the commissural fibers arising from the neocortex, whereas in placental mammals, the anterior commissure carries only some of these fibers [Hugall et al. 2007].

The key anatomical difference between monotremes and other mammals is the one that gave them their name; *monotreme* means 'single opening' in Greek and comes from the fact that their urinary, defecatory, and reproductive systems all open into a single duct, the cloaca. This structure is very similar to the one found in reptiles. Monotremes and marsupials have a single cloaca (though marsupials also have a separate genital tract), while most placental mammal females have separate openings for reproduction, urination, and defecation: the vagina, the urethra, and the anus (Figure 1).

Monotremes lay eggs. However, the egg is retained for some time within the mother, which actively provides the egg with nutrients. Monotremes also lactate, but have no defined nipples, excreting the milk from their mammary glands via openings in their skin. All species are long-lived, with low rates of reproduction and relatively prolonged parental care of infants (Figure 2).

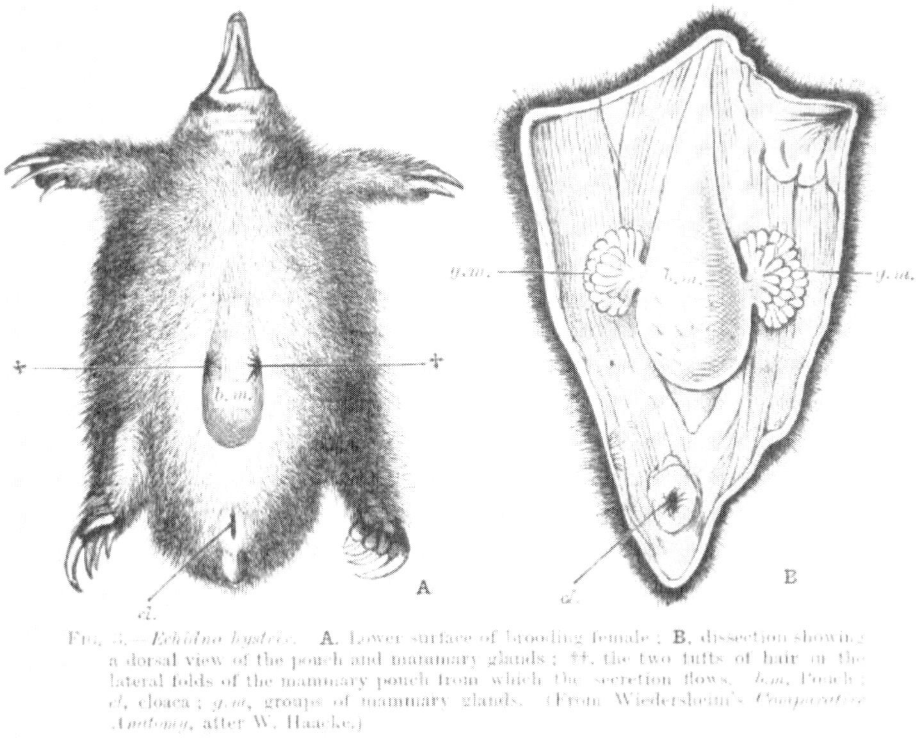

Figure 1. Monotreme female reproductive organs (Source: Beddard, Frank E. (Frank Evers), 1858-1925).

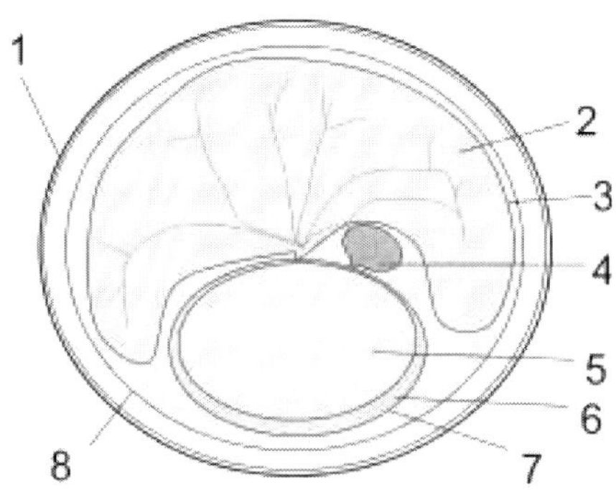

Figure 2. Diagram of a Monotreme Egg. 1) Shell; 2) Yolk; 3) Yolk Sac; 4) Allantois; 5) Embryo; 6) Amniotic Fluid; 7) Amniotic Membrane; and 8) Membrane.

Extant monotremes lack teeth as adults. Fossil forms and modern platypus youngs have "tribosphenic" form of molars (occlusal surface formed by three cusps arranged in a triangle), which is one of the hallmarks of extant mammals. Some recent work suggests monotremes acquired this form of molar independently of placental mammals and marsupials [Butler and Hodos, 2005], although this is not well established. The jaw of monotremes is constructed somewhat differently from those of other mammals, and the jaw opening muscle is different. As in all true mammals, the tiny bones that conduct sound to the inner ear are fully incorporated into the skull, rather than lying in the jaw as in cynodonts and other premammalian synapsids. This feature, too, is now claimed to have evolved independently in monotremes and therians, although, as with the analogous evolution of the tribosphenic molar, this is disputed. The external opening of the ear still lies at the base of the jaw. The sequencing of the platypus genome has also provided insight into the evolution of a number of monotreme traits, such as venom and electroreception, as well as showing some new unique features, such as the fact that monotremes possess 10 sex chromosomes and that their X chromosome resembles the sex chromosome of birds, suggesting that the two sex chromosomes of marsupial and placental mammals evolved more recently than the split from the monotreme lineage. This feature, along with some other genetic similarities with birds, such as shared genes related to egg-laying, is thought to provide some insight into the most recent common ancestor of the synapsid lineage leading to mammals and the sauropsid lineage leading to birds and modern reptiles, which are believed to have split about 315 million years ago [Rich et al. 2005].

The monotremes also have extra bones in the shoulder girdle, including an interclavicle and coracoid, which are not found in other mammals. Monotremes retain a reptile-like gait, with legs on their sides rather than underneath their bodies. The monotreme leg bears a spur in the ankle region; the spur is not functional in echidnas, but contains powerful venom in the male platypus.

Monotreme metabolic rate is remarkably low by mammalian standards. The platypus has an average body temperature of about 32°C (90°F) rather than the 37°C (99°F) typical of placental mammals. Research suggests this has been a gradual adaptation to the harsh environmental conditions on the part of the small number of surviving monotreme species rather than a historical characteristic of monotremes [Dawson et al. 1979; Watson & Graves, 1988].

8.3. Taxonomy and Phylogeny

Monotremes are conventionally treated as comprising a single order **Monotremata**, though a recent classification proposes to divide them into the orders Platypoda (the platypus along with its fossil relatives) and Tachyglossa (the echidnas, or spiny anteaters). The entire grouping is also traditionally placed into a subclass Prototheria, which was extended to include several fossil orders, but these are no longer seen as constituting a natural group allied to monotreme ancestry. A controversial hypothesis now relates the monotremes to a different assemblage of fossil mammals in a clade termed Australosphenida [Luo et al. 2001, 2002].

The traditional "theria hypothesis" states that divergence of the monotreme lineage from the Metatheria (marsupial) and Eutheria (placental mammal) lineages happened prior to the divergence between marsupials and placental mammals, and this explains why monotremes retain a number of primitive traits presumed to have been present in the synapsid ancestors of later mammals, such as egg-laying. Most morphological evidence supports the theria hypothesis, but one possible exception is a similar pattern of tooth replacement seen in monotremes and marsupials, which originally provided the basis for the competing "marsupionata hypothesis" in which the divergence between monotremes and marsupials happened later than the divergence between these lineages and the placental mammals. An analysis by Van Rheede in 2006 concluded the genetic evidence favoring theria hypothesis, and this hypothesis continues to be the more widely accepted one.

The time when the monotreme line diverged from other mammalian lines is uncertain, but one survey of genetic studies gives an estimate of about 220 million years ago. Fossils of a jaw fragment 110 million years old were found at Lightning Ridge, New South Wales. These fragments, from the species *Steropodon galmani*, are the oldest known fossils of monotremes. Fossils from the genera *Kollikodon*, *Teinolophos*, and *Obdurodon* have also been discovered. In 1991, a fossil tooth of a 61-million-year-old platypus was found in southern Argentina (since named *Monotrematum*, though it is now considered to be an *Obdurodon* species). Molecular clock and fossil dating give a wide range of dates for the split between echidnas and platypuses, with one survey putting the split around 19–48 million years ago, but another putting it around 17–89 million years ago. All these dates are more recent than the oldest known platypus fossils, suggesting both the short-beaked and long-beaked echidna species are derived from a platypus-like ancestor [Benton 1997].

The precise relationships between extinct groups of mammals and modern groups such as monotremes are somewhat uncertain, but cladistic analyses usually put the last common ancestor (LCA) of placentals and monotremes close to the LCA of placentals and multituberculates, with a number of analyses giving a more recent LCA for placentals and monotremes, but some also suggesting the LCA of placentals and multituberculates was more recent [Philips et al. 2009].

8.4. Physiology

Monotremes may have less developed thermoregulation than other mammals, but recent research shows that they maintain a constant body temperature in a wide variety of circumstances without difficulty (for example, the platypus while living in an icy mountain stream). Early researchers were misled by two factors: firstly, monotremes maintain a lower average temperature than most mammals (around 32°C), compared to about 35°C (95°F) for marsupials and 37°C for most placentals); secondly, the short-beaked echidna (which is much easier to study than the reclusive platypus) maintains normal temperature only when it is active; during cold weather, it conserves energy by "switching off" its temperature regulation. Additional perspective came when reduced thermal regulation was observed in the hyraxes, which are placental mammals.

The echidna was originally thought to not enter rapid eye movement sleep, but a more recent study shows they do, amounting to about 15% of the total sleep time observed for subjects at an environmental temperature of 25°C (77°F). Surveying a range of environmental temperatures, the study observed very little REM at reduced temperatures of 15°C (59°F) and

20°C (68°F), and also a substantial reduction at the elevated temperature of 28°C (82°F) [Dawson et al. 1979; Watson and Graves, 1988].

8.5. Extreme Monotremes: Why Do Egg-Laying Mammals Still Exist?

Ancestors of the duck-billed platypus and the echidna may have survived their live-birthing competitors by taking to the water. Only two kinds of egg-laying mammals are left on the planet today—the duck-billed platypus and the echidna, or spiny anteater. These odd "monotremes" once dominated Australia, until their pouch-bearing cousins, the marsupials, invaded the land down under 71 million to 54 million years ago and swept them away. New research suggests these two kinds of creatures managed to survive because their ancestors took to the water.

Before making their way to Australia, the marsupials had migrated from Asia to the Americas to Antarctica. Forced to contend with all the animals along the way, marsupials may have been primed for competition, hence accounting for their overwhelming success in Australia.

Phillips and his colleagues suggest that platypuses and echidnas made it through the marsupial invasion because their ancestors sought refuge where marsupials could not follow—the water. When marsupials are born, they need to constantly suckle milk for weeks, and as such, newborns could drown if their mothers ever had to swim for long. The theory seems plausible for platypuses, which are amphibious creatures. Echidnas, however, dwell solely on land. Echidnas diverged from platypuses only 19 million to 48 million years ago, meaning that echidnas recently had semiaquatic ancestors and only later recolonized the land. A number of features of echidnas indicate that they may have once had an amphibious platypus like forerunner like streamlined bodies, rearward-projecting hind limbs that could serve as rudders, and the contours of a duck like bill during embryonic development [Philips et al. 2009].

A previous study of early monotreme fossils had suggested that platypus and the echidna diverged more than 110 million years ago, far longer than the genetic analysis indicates. But Phillips's team reanalyzed 439 features of those early fossils and found that echidnas and recently evolved platypuses were better grouped together instead of with more ancient fossils. The reconstructed lineages support the researchers' genetic findings, which they reported online September 23 in the Proceedings of the National Academy of Sciences USA2009. "The genes and the bones now appear to be telling the same story, which is encouraging," says mammalogist Robin Beck of the American Museum of Natural History in New York City, who did not take part in this research.

So far scientists have found no fossil evidence of an echidna transition from water; the fossil record of monotremes remains quite incomplete. But a number of fossil sites in Australia are 20 million to 25 million years old, about when the researchers think echidnas evolved. The presence of primitive traits in monotremes from distant ancestors, such as reptile like shoulders, is often offered as reasons for their apparent inferiority. These new findings help to recast these archaic features in a positive light—for instance, whereas the reptile like shoulders are poor for running fast, they provide strong bracing, allowing for huge shoulder and arm musculature to help echidnas dig into the dirt and platypuses maneuver in the water. “Most reptiles also possess these ‘primitive’ traits and yet, in terms of species numbers, are more successful than any mammal group,” Phillips adds [Philips et al. 2009].

8.6. Egg-Laying Mammal: Scientists Discover That for Australia the Long-Beaked Echidna May Not be a Thing of the Past

The western long-beaked echidna, one of the world's five egg-laying species of mammal, became extinct in Australia thousands of years ago…or did it? Smithsonian scientists and colleagues have found evidence suggesting that not only did these animals survive in Australia far longer than previously thought, but that they may very well still exist in parts of the country today.

The team's findings are published in the Dec. 28, 2012 issue of the journal *ZooKeys*.

With a small and declining population confined to the Indonesian portion of the island of New Guinea, the western long-beaked echidna (*Zaglossus bruijnii*) is listed as "Critically Endangered" on the International Union for Conservation of Nature's Red List of Threatened Species. It is also considered extinct in Australia, where fossil remains from the Pleistocene epoch demonstrate that it did occur there tens of thousands of years ago. Ancient Aboriginal rock art also supports the species' former presence in Australia. However, no modern record from Australia was known to exist until scientists took a closer look at one particular specimen stored in cabinets in the collections of the Natural History Museum in London. Previously overlooked, the specimen's information showed that it was collected from the wild in northwestern Australia in 1901—thousands of years after they were thought to have gone extinct there.

Long-beaked echidnas are known as monotremes—a small and primitive order of mammals that lay eggs rather than giving birth to live young. The platypus, the short-beaked echidna, and the three species of long-beaked echidna are the only monotremes that still exist. The platypus is found only in eastern Australia, the short-beaked echidna is found in Australia and New Guinea, and the long-beaked echidnas were previously known as living animals only from the island of New Guinea. Long-beaked echidnas, which grow to twice the size of the platypus or the short-beaked echidna, are beach-ball sized mammals covered in coarse blackish-brown hair and spines. They use their long, tubular snout to root for invertebrates in the forests and meadows of New Guinea. Among many peculiar attributes, reproduction is one of the most unique—females lay a single leathery egg directly into their pouch where it hatches in about 10 days.

The re-examined specimen in London reveals that the species was reproducing in Australia at least until the early 20th century. It was collected in the West Kimberley region of Western Australia by naturalist John T. Tunney in 1901, on a collecting expedition for the private museum of Lord L. Walter Rothschild in England. Despite collecting many species of butterflies, birds and mammals (some new to science at the time), no full report on his specimens has ever been published. The collection, including the long-beaked echidna specimen, was then transferred to the Natural History Museum in London in 1939 after Rothschild's death. It was another 70 years before Helgen visited the museum in London and came across the specimen with the original Tunney labels, which challenged previous thinking about the species' recent distribution and offered insight into where it may still occur.

The discovery of the western long-beaked echidna in Australia is astonishing. It highlights the importance of museum collections, and how much there is still to learn about Australia's fauna.

Learning whether the western long-beaked echidna still exists in Australia today will take time. "The next step will be an expedition to search for this animal," Helgen said. "We'll need to look carefully in the right habitats to determine where it held on, and for how long, and if any are still out there." To find it, Helgen hopes to draw on his experience with the species in New Guinea and to interview those who know the northern Australian bush best. "We believe there may be memories of this animal among Aboriginal communities, and we'd like to learn as much about that as we can," he said [Helgen 2003].

With the species in danger of extinction, finding Australian survivors or understanding why and when they vanished is an important scientific goal. "We hold out hope that somewhere in Australia, long-beaked echidnas still lay their eggs," said Helgen (2003).

8.7. Number of Species

Only five species of monotremes are extant, one of which is duck-billed platypus and remaining four belong to the echidna family. These echidnas are also called as spiny anteaters because they like to have ants and termites in their diet. Also, due to their pointing spines, they are called spiny anteaters. They are quite similar with other mammals in few things, but differ in most of the characteristics. Sometimes, they behave like reptiles, while at other times; they show the characteristics of birds.

8.8. Natural Habitat

Monotremes are, primarily, found in Australia, Tasmania and New Guinea. The duck-billed platypus is purely aquatic while the echidnas are land-dwelling animals. The echidnas can easily be found in regions having moderate temperature like forests, hilly areas, high mountains, desserts, green lands and so on. Just like platypuses, the echidnas do not have teeth, and they remain active both during the day and the night time.

8.9. Reproduction

Echidna

A female echidna possesses a pouch in the body where she lays an egg and it takes almost 10 days for the egg to hatch. After hatching, a very small sized baby echidna comes out of the egg. The baby is so small that it needs to spend three more months with mother in the pouch. Their mother feeds them with the breast milk. Since female echidna does not have nipples, it is a little problematic for the baby to suck milk. The echidnas like to live alone except during the breeding season.

Platypus

Coming towards the reproduction side of platypus, female lays a couple of eggs in a hole that is especially dug for this purpose. All the platypuses are found in and around water because they mainly feed on the marine insects, their larvae and other small organisms. Male platypus becomes more active at the time of breeding as compared to normal days. Feeling any kind of danger during breeding season, they release poison for killing attackers to save the female platypus from any sort of harm. It takes almost a month for the female platypus to get pregnant.

8.10. Fun Facts

Echidna

- Though many people may get scared from these spiny creatures, they themselves are, basically, very introvert and fearful.
- When feeling scared, they fold their legs into the body and assume a ball like shape.
- In the folded position, they cannot see anything, and think that nobody is watching them, which is one of the funny and interesting facts about echidna.
- If an echidna is on the ground, it tries to dig a hole and hide itself in it.

Platypus

- Introduced in 1798 in Europe, the animals have several features like those of reptiles and birds.
- It possesses a bill just like ducks have beaks, and they also have feet.
- They have a tail like that of beavers.
- The platypus uses its tail to swim in water and search for food, which includes various insects and fishes.
- Platypuses are pretty poisonous as well and they use poison to protect themselves against the attackers.
- According to an estimate, the amount of poison released by a platypus is more than enough to kill a dog.
- An interesting fact about platypus is that their fur makes them waterproof.
- They have smaller eyes and do not use them under water.
- Platypuses have no teeth and, that is why, their food is crushed by several plates in their mouth.

8.11. DIFFERENCES BETWEEN PLATYPUS AND ECHIDNA

Though the echidnas and platypuses share similarities as they both are monotremes, but there do exist certain differences between them. A female platypus lays eggs in the special holes or burrows while a female echidna, on the other hand, lays egg in her pouch. Female platypus can lay two eggs at a time while an echidna can lay only one egg at a time. There are many other differences in their body structure, feeding habits, diet and so on (Figure 3-4).

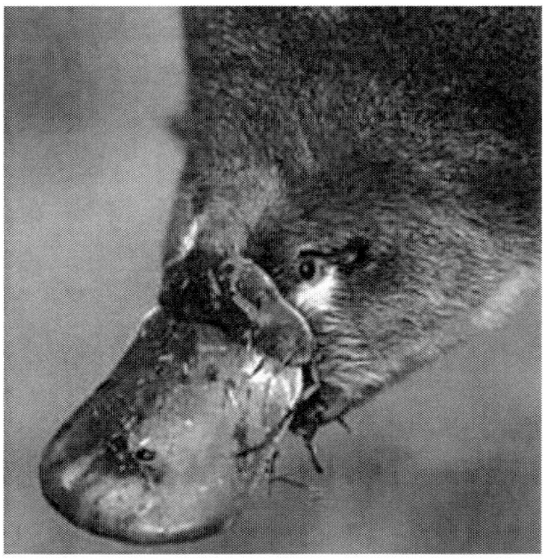

Figure 3. Duck-Billed Platypus and the other egg-laying mammal, the echidna, may owe their existence to their ancestors adapting to life in water, thereby avoiding marsupial competition.

Figure 4. The short-beaked echidna has fared better than its long-beaked relative. It is found throughout Australia, where it is the most widespread native mammal.

The duck-billed platypus and the spiny anteater are mammals that lay eggs rather than giving live birth. The platypus has a bill like a duck, fur like a mammal and webbed feet. It lays eggs that look like reptile eggs. Its young feed from the mother's mammary glands. The spiny anteater's covering looks like that of the porcupine. It has a long tongue to catch insects and a bill. It lays several eggs which it keeps in a pouch on its side.

Marsupials

Marsupials are an infraclass of mammals living primarily in the Southern Hemisphere. A distinctive characteristic, common to most species, is that the youngs are carried in a pouch. Well-known marsupials include kangaroos, wallabies, the koala, possums, opossums (Figure 5-6), wombats and the Tasmanian devil. Less well-known species of marsupials include the numbat, bandicoots, the bilby and quolls.

Marsupials are the group of mammals commonly thought of as pouched mammals. They give live birth, but they do not have long gestation times like placental mammals. Instead, they give birth very early and the young animal, essentially a helpless embryo, climbs from the mother's birth canal to the nipples. There it grabs on with its mouth and continues to develop, often for weeks or months depending on the species. The short gestation time is due to having a yolk-type placenta in the mother marsupial. Placental mammals nourish the developing young embryo using the mother's blood supply, allowing a longer gestation time.

Figure 5. Common brushtail possum_(*Trichosurus vulpecula*).

Figure 6. Virginia opossum (*Didelphis virginiana*), a North American marsupial.

Marsupials represent the clade originating with the last common ancestor of extant metatherians. Like other mammals in the Metatheria, they are characterized by giving birth to relatively undeveloped young, often residing in a pouch with the mother for a certain time after birth. Close to 70% of the 334 extant species occur in Australia, New Guinea, and nearby islands, with the remaining 100 found in the Americas, primarily in South America, but with 13 in Central America, and one in North America north of Mexico.

Like other mammals, the marsupials are covered with hair. Mothers nurse their young — a young kangaroo may nurse even when it has grown almost to the mother's size (Hunsakeer 1977).

The only naturally occurring marsupial in the United States is the opossum, *Didelphis marsupialis*. In the past, however, marsupials were quite common. During the Mesozoic marsupials were very common in North America; more common, in fact, than placental mammals. They persisted here until the mid- to late-Tertiary.

In South America and Australia, however, marsupials continued to be an important group of land mammals. Many South American forms are similar to the North American opossum. The marsupials of South America began to go extinct in the late Miocene and early Pliocene when a land connection with North America formed, allowing placental mammals to cross into South America. In Australia, though, marsupials continue to be very diverse, and are the dominant native mammals. They include koalas, wombats (Figure 7), kangaroos (Figure 8), tasmanian devils, (Figure 9), and other typical Australian mammals. Until recently, they also included the marsupial wolf, *Thylacinus* (Figure 10). Like the quagga, the marsupial wolf is now extinct. The last individual was seen in Tasmania in the 1950s (Paddle, 2002; Beck et al. 2008).

Though marsupials today do not have as many species as do the placental mammals, they are quite structurally diverse. They range from small four-footed forms like the marsupial mole, *Notoryctes*, to the large two-legged kangaroos.

Figure 7. Koala. Photo by Gerald and Buff Corsi, © 2002 California Academy of Sciences. Wombat. Photo © 2005 Sharon Chester.

Figure 8. Female eastern grey kangaroo with a joey in her pouch.

Figure 9. Thylacine (*Thylacinus cynocephalus*), an extinct carnivorous marsupial found in Tasmania until the 1930s.

There are several cases of *convergent evolution* between marsupials and placental mammals, in which the two animals have evolved to fill the same ecological niche in different parts of the world. There are burrowing forms, grazing forms, gliding forms, and even long-snouted ant-eating forms which have evolved independently in the two groups (Jessie, 2010; Hand et al., 2002).

Figure 10. *Thylacinus*, an extinct marsupial wolf.

8.12. EVOLUTION

The relationships between the three extant divisions of mammals (monotremes, marsupials, and placental mammals) was long a matter of debate among taxonomists. Most morphological evidence comparing traits such as number and arrangement of teeth and structure of the reproductive and waste elimination systems favors a closer evolutionary relationship between marsupials and placental mammals than either with the monotremes, as does most genetic and molecular evidence [Kemp, 2005; Luo et al., 2011].

The ancestors of marsupials split from those of placental mammals during the Jurassic period. In the absence of soft tissues, such as the pouch and reproductive system, fossil marsupials can be distinguished from placentals by the form of their teeth. Primitive marsupials possess four pairs of molar teeth in each jaw, whereas placental mammals never have more than three pairs. Using this criterion, the earliest known marsupial is *Sinodelphys szalayi*, which lived in China around 125 million years ago (mya). This makes it almost contemporary to some early eutherian fossils, which have been found in the same area [Leary et al., 2013].

The oldest metatherian fossils (Metatheria being a larger clade that groups marsupials with some of their extinct relatives) are found in present-day China. About 100 million years ago (mya), the supercontinent Pangaea was in the process of splitting into the northern continent Laurasia and the southern continent Gondwana, with what would become China and Australia already separated by the Tethys Ocean. Metatherians spread westward into modern North America (still attached to Eurasia) and then to South America, which was connected to North America until around 65 mya. Laurasian marsupials eventually died off, possibly due to competition from placental mammals for their ecological niches.

In South America, the opossums retained a strong presence, and the Tertiary saw the evolution of shrew opossums (Paucituberculata) and metatherian predators such as the borhyaenids and the saber-toothed *Thylacosmilus*. South American niches for mammalian carnivores were dominated by these marsupial and sparassodont metatherians. While placental predators were absent, the metatherians did have to contend with avian (terror bird) and terrestrial crocodilian competition. South America and Antarctica remained connected until 35 mya, as shown by the unique fossils found there. North and South America were disconnected until about three million years ago, when the Isthmus of Panama formed. This led to the Great American Interchange. Competition from placental mammals from the north drove sparassodonts to extinction, while didelphimorphs (opossums) invaded Central America, with the Virginia opossum reaching as far north as Canada.

Marsupials reached Australia via Antarctica about 50 mya, shortly after Australia had split off. This suggests a single dispersion event of just one species, most likely a relative to South America's monito del monte (a microbiothere, the only New World australidelphian). This progenitor may have rafted across the widening, but still narrow, gap between Australia and Antarctica. In Australia, they radiated into the wide variety seen today. Modern marsupials appear to have reached the islands of Borneo and Sulawesi relatively recently via Australia. A 2010 analysis of retrotransposon insertion sites in the nuclear DNA of a variety of marsupials has confirmed all living marsupials have South American ancestors. The branching sequence of marsupial orders indicated by the study puts Didelphimorphia in the most basal position, followed by Paucituberculata, then Microbiotheria, and ending with the radiation of Australian marsupials. This indicates that Australidelphia arose in South America, and reached Australia after Microbiotheria split off.

In Australia, terrestrial placental mammals disappeared early in the Cenozoic (their most recent known fossils being 55 million-year-old teeth resembling those of condylarths) for reasons that are not clear, allowing marsupials to dominate the Australian ecosystem. Extant native Australian terrestrial placental mammals (such as hopping mice) are relatively recent immigrants, arriving via island hopping from Southeast Asia.

Genetic analysis suggests a divergence date between the marsupials and the placentals at 160 million years ago. The ancestral number of chromosomes has been estimated to be 2n = 14 [Jessie, 2010; Hand et al. 2002].

8.13. Early Development

An early birth removes a developing marsupial from its mother's body much sooner than in placental mammals, thus marsupials have not developed a complex placenta to protect the embryo from its mother's immune system. Though early birth puts the tiny newborn marsupial at a greater environmental risk, it significantly reduces the dangers associated with long pregnancies, as there is no need to carry a large fetus to full-term in bad seasons.

Because newborn marsupials must climb up to their mother's nipples, their front limbs are much more developed than the rest of their bodies at the time of birth. This requirement possibly has resulted in the limited range of locomotor adaptations in marsupials compared to placentals. Marsupials must develop grasping forepaws during their early youth, making the transition from these limbs into hooves, wings, or flippers, as some groups of placental mammals have done, far more difficult.

An infant marsupial is known as a **joey**. Marsupials have a very short gestation period (about four to five weeks), and the joey is born in an essentially fetal state. The blind, furless, miniature newborn, the size of a jelly bean, crawls across its mother's fur to make its way into the pouch, where it latches onto a teat for food. It will not re-emerge for several months, during which time it develops fully. After this period, the joey begins to spend increasing lengths of time out of the pouch, feeding and learning survival skills. However, it returns to the pouch to sleep, and if danger threatens, it will seek refuge in its mother's pouch for safety.

Joeys stay in the pouch for up to a year in some species, or until the next joey is born. A marsupial joey is unable to regulate its own body temperature and relies upon an external heat source. Until the joey is well-furred and old enough to leave the pouch, a pouch temperature of 30–32 °C (86–90 °F) must be constantly maintained [Luckett, 1977; Smith, 2001; Freyer et al. 2009].

8.14. Reproductive System

Marsupials' reproductive systems differ markedly from those of placental mammals. The female develops a kind of yolk sac in her womb, which delivers nutrients to the embryo. Embryos of some marsupials additionally form placenta-like organs that connect them to the uterine wall, although it is not certain that they transfer nutrients from the mother to the embryo. Pregnancy is very short, typically 4 to 5 weeks, and the embryo is born at a very young stage of development [Renfree et al. 1987; Biggers 1996].

The evolution of reproduction in marsupials, and speculation about the ancestral state of mammalian reproduction, have engaged discussion since the end of the 19th century. Both sexes possess a cloaca, which is connected to a urogenital sac used to store waste before expulsion. The bladder of marsupials functions as a site to concentrate urine and empties into the common urogenital sinus in both females and males [Rodgers 1973].

Most male marsupials, except for the two largest species of kangaroos, have a bifurcated penis, separated into two columns, so that the penis has two ends corresponding to the females' two vaginas. The penis is used only for discharging semen into females, and is separate from the urinary tract. When not erect, it is retracted into the body in an S-shaped curve. Neither marsupials nor monotremes possess a baculum. The shape of the glans penis varies among marsupial species. A male koala's foreskin contains naturally occurring bacteria that play an important role in fertilization [Sadlier, 1965; Inns, 1982; Biggers 1996].

The male thylacine had a pouch that acted as a protective sheath, covering the male's external reproductive organs while he ran through thick brush [Sharman et al. 1964].

Female marsupials have two lateral *vaginas*, which lead to separate *uteri*, but both open externally through the same orifice. A third canal, the median vagina, is used for birth. This canal can be transitory or permanent. The definitive placenta in all marsupials is generated by the yolk sac. Among three fetal membranes in mammals, the yolk sac, allantois, and amnion, only the first two form a placenta. The evolution of placentation in vertebrates is linked to the evolution of viviparity, a reproductive system in which the females retain their eggs to give birth to their young. Marsupials give birth at a very early stage of development (about four to five weeks). After birth, newborn marsupials crawl up the bodies of their mothers and attach themselves to a nipple, which is located on the underside of the mother either inside a pouch called the *marsupium* or open to the environment. To crawl to the nipple and attach to it, the marsupial must have well-developed forelimbs and facial structures. This is accomplished by accelerating forelimb and facial development in marsupials compared to placental mammals, which results in decelerated development of such structures as the hindlimb and brain. There they remain for a number of weeks, attached to the nipple. The offspring are eventually able to leave the marsupium for short periods, returning to it for warmth, protection, and nourishment [Sadlier 1965].

EUTHERIA, THE PLACENTAL MAMMALS

18.15. GENERAL FEATURES, MORPHOLOGY, PHYLOGENY AND HISTORY

Placental mammals are a rather diverse group, with nearly 4000 described species, mostly rodents and bats. The placental mammals have a enormous range of body forms and complex social interactions. The 3782 species are divided among 18 orders. The placental mammals include such diverse forms as whales, elephants, shrews, and armadillos. They are also some of the most familiar organisms to us, including pets such as dogs and cats, as well as many farm and work animals, such as sheep, cattle, and horses. And humans, of course, are also placental mammals.

Placental mammals all bear live young, which are nourished before birth in the mother's uterus through a specialized embryonic organ attached to the uterus wall, the *placenta*. The placenta is derived from the same membranes that surround the embryos in the amniote eggs of reptiles, birds, and monotreme mammals. The term "placental mammals" is somewhat of a misnomer because marsupials also have *placentae*. The difference is that the placenta of marsupials is very short-lived and does not make as much of a contribution to fetal nourishment as it does in *eutherians*, as "placental mammals" are known scientifically (Kenneth, 2006).

Eutheria (Ancient Greek "true/good" + "beast" hence "true beasts") is one of two mammalian clades with extant members that diverged in the Late Jurassic. The other is the Metatheria, which includes kangaroos and the other mammals (most of the marsupials) that accommodate their neonates in pouches. Except for the opossum, which is a metatherian, all mammals indigenous to North America, Europe, Africa, and Asia are eutherians. Extant eutherians, their last common ancestor, and all extinct descendants of that ancestor are placentals, in the infraclass **Placentalia**.

Eutherians constitute any member of the mammalian group characterized by the presence of a placenta, which facilitates exchange of nutrients and wastes between the blood of the mother and that of the fetus. The placentals include all living mammals except marsupials and monotremes. Although some consider the marsupials (cohort Marsupialia) to be placental mammals, these animals have a less-developed, less-efficient type of placenta that limits the gestation period. The true placenta of the placentals allows for a longer developmental period within the protection of the womb, a factor considered to have contributed to the evolutionary success of the group. Fossil evidence show that the first placental mammals evolved during the early Paleogene Period (65.6 million to 23 million years ago), some 200,000 to 400,000 years after the extinction of the dinosaurs at the end of the Cretaceous Period (145.5 million to 65.5 million years ago).

Eutherians are distinguished from non-eutherians by various features of the feet, ankles, jaws and teeth. One of the major differences between placental and nonplacental eutherians is that placentals lack epipubic bones, which are present in all other fossil and living mammals (marsupials and monotremes).

The oldest known eutherian species is *Juramaia sinensis*, dated at 160 million years ago from the Jurassic in China (Figure 11) [Ji et al. 2002; Wible et al. 2007, 2009; Luo et al. 2011; Rook et al. 2013].

"Eutheria" was introduced by Thomas Henry Huxley in 1880, meant to be broader in definition than "Placentalia", the term previously in use.

Figure 11. Life restoration of Juramaia sinensis, the earliest known eutherian.

Difference from Metatherians

The features of Eutheria that distinguish them from *metatherians*, a group that includes modern marsupials, are:

- an enlarged malleolus ("little hammer") at the bottom of the tibia, the larger of the two shin bones.
- the joint between the first metatarsal bone and the entocuneiform bone in the foot is offset further back than the joint between the second metatarsal and middle cuneiform bones – in metatherians these joints are level with each other.
- various features of jaws and teeth [Ji et al. 2002].

Placental mammals are distinguished from other eutherians by:

- the presence of a malleolus at the bottom of the *fibula*, the smaller of the two shin bones.
- a complete mortise and tenon upper ankle joint, where the rearmost bones of the foot fit into a socket formed by the ends of the tibia and fibula [Ji et al. 2002].
- a wide opening at the bottom of the pelvis, which allows the birth of large, well-developed offspring. Marsupials have and nonplacental eutherians had a narrower opening that allows only small, immature offspring to pass through [Kenneth 2006].
- the absence of epipubic bones extending forward from the pelvis, which are not found in any placental, but are found in all other mammals – nonplacental eutherians, marsupials, monotremes, and earlier mammaliaforms as well as in other cynodonts that are closest to mammals. Their function is to stiffen the body during locomotion [Wible et al. 2007]. This stiffening would be harmful in pregnant placentals, whose abdomens need to expand [Wible et al. 2009].

Placental Mammal

The Placental or True mammals

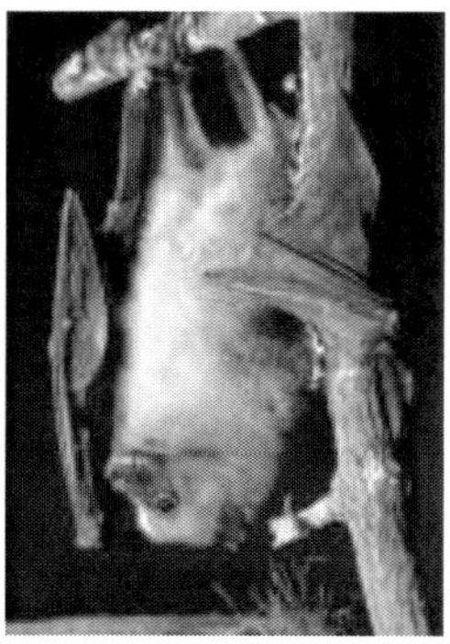

Figure 12. Flying fox.

8.16. Reproduction and Care of the Young

Placental mammals differ from the marsupials in that their young develop to a relatively mature stage within a uterus attached to the mother by an **allantoic placenta.** The placenta allows the foetus to plug into its mother's circulation and share her respiratory and excretory systems, and the nutrients carried in her blood. Mammals, like marsupials, feed their young with milk and care for them for prolonged periods until they can look after themselves [Freyer et al. 2009].

8.17. Skulls and Dentition

Dentition and the way the jaw moves to cope with different foods - insects, meat or vegetable matter - have been the dominant influences on the evolution of the jaw, its articulation and musculature and the overall form of the skull in the different eutherian orders. A mammal's life style is revealed more by its teeth than by any other feature.

Mammals have four types of teeth, each with special functions:

- incisors (I) snip and bite;
- canines (C) pierce and tear;
- premolars (PM) shear and slice;
- molars (M) crush and grind.

Figure 13. Addax with Calves.

Figure 14. Tapir with Young.

Only some anteaters and whales lack teeth, due to their diet.

A mammal's dental formula expresses the numbers of each type of tooth in one half of the upper jaw and one half of the lower jaw. The lower teeth are borne on the dentary bone, the upper on the premaxilla and the maxilla.

The most primitive formula is I3/3; C1/1; PM 4/4; M 3/3. Early eutherians were insectivorous and members of the Order Insectivora, some omnivores and more unspecialised carnivores are closest to this arrangment. Placental mammals typically have two sets of teeth - the milk teeth are replaced by deciduous teeth as the animals mature [Ji et al. 2002].

Early eutherians were insectivorous and modern insectivores retain the primitive dental formula from which more specialised dentitions evolved. The four types of teeth have special functions: incisors snip and bite; canines pierce and tear; premolars shear and slice; molars crush and grind. Only some whales and anteaters lack teeth, due to their diets.

8.18. Relationships of Modern Eutherian Mammals

Earlier accepted classification placed a number of small insectivorous mammals together in the order Insectivora and seven orders of hoofed animals plus the whales (order Cetacea) together as 'Ungulates'. Since then phylogenetic studies using molecular analysis suggest that eighteen orders of modern eutherians represent four clade or lineages, which diverged in the late Cetaceous period after the early eutherians had split from the early marsupials.

These findings indicate that members of the order Insectivora should be split among two orders, the order Afrosoricida and the Eulipotyphla, and that four orders of hoofed mammals belong to one clade, the Afrotheria, and three orders – the Perissodatyla, Artiodactyla and Cetacea (whales) - belong to another clade, the Laurasiatheria. As whales now appear to be related to the lineage that gave rise to the hippopotami, a new order has been created to accommodate both cetaceans and artiodactyls - the order Cetartiodactyla. The order Perissodactyla remains as before. The other mammals are split between two other clades - the Xenarthra and Euarchontoglires. The clades Xenarthra and Afrotheria appear to have evolved in the Southern Hemisphere in South America and Africa respectively; the clades Euarchontoglires and the Laurasiatheria in the Northern Hemisphere. The web pages entitled 'the Tree of Life' explain how molecular analyses led to the order Insectivora being split.

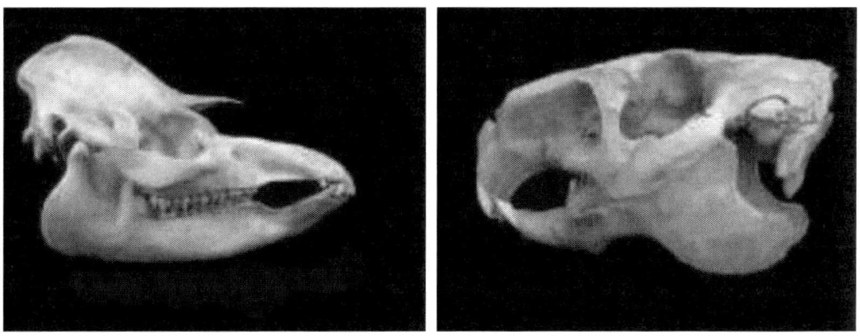

Figure 15. Tapir Skull and Capybara skull.

Aquatic Mammals

Aquatic and **semi-aquatic mammals** are a diverse group of mammals that dwell partly or entirely in bodies of water. They include the various marine mammals that dwell in oceans, as well as various freshwater species, such as the Platypus and the European Otter.

8.19. ADAPTATIONS

Mammals evolved on land, so all aquatic and semi-aquatic mammals have brought many terrestrial adaptations into the waters. They do not breathe underwater as fishes do, so their respiratory systems had to protect the body from the surrounding water; valvular nostrils and an intranarial larynx exclude water while breathing and swallowing.

To deal with water salinity, cetaceans and pinnipeds have evolved reniculate kidneys also found in bears. To navigate and detect prey in murky and turbid waters, aquatic mammals have developed a variety of sensory organs, for example, pinnipeds have elongated and highly sensitive whiskers that can pick up vibrations from fishes and reveal information about water currents, and toothed whales have evolved echolocation [Reidenberg et al. 2007].

Aquatic mammals also display a variety of locomotion styles. Cetaceans excel in streamlined body shape and the up-and-down movements of their flukes make them the fastest swimmers. The considerably slower sirenians can also propel themselves with their fluke but they can also walk on the bottom with their forelimbs. Seals sweep their hind flippers horizontally while fur seals and sea lions use their fore flippers. Walruses can grip the sea floor with their tusks and propel forward by a downward nod [Reidenberg et al. 2007].

Polar bears, otters, seals, sea lions, and beavers have fur, one of the defining mammalian features, that is oily and water proof in order to trap air to provide insulation. In contrast, other aquatic mammals, such as whales, dolphins, porpoises, manatees, dugongs, walruses, and hippopotamuses, have lost their fur in favour of a thick and dense epidermis and a thickened fat layer (the blubber) in response to hydrodynamic requirements [Reidenberg et al. 2007].

Wading and bottom-feeding animals (e.g., moose and manatee) need to be heavier than water in order to keep contact with the floor or to stay submerged, surface-living animals (e.g., sea otter) need the opposite, and free-swimming animals living in open waters (e.g., dolphins) need to be neutrally buoyant in order to be able to swim up and down. Typically, thick and dense bone is found in bottom feeders and low bone density is associated with mammals living in deep water [Reidenberg et al. 2007].

Some aquatic mammals have retained four weight-bearing limbs (e.g., polar bears, beaver, otter, muskrat) and can walk on land like fully terrestrial animals. The long and thin legs of a moose limit exposure to and friction from water in contrast to hippos that keep most of their body submerged and have short and thick legs. The semi-auatic pygmy hippopotamus can walk quickly on a muddy underwater surface thanks to robust muscles and because all toes are weight-bearing. They also retain several primitive muscles lost in other artiodactyls, suggesting that they diverged early from the other members of the group, perhaps together with the cetaceans. Some aquatic mammals with flippers (e.g., sirenians, pinnipeds) are amphibious and regularly leave the water, sometimes for extended periods, and they have developed a range of locomotor behaviours on land. Sea lions can raise their bodies and even run on land using both their hind- and forelimbs. Seals, in contrast, undulate their bodies to move on land, similar to the up-and-down body motion used underwater by many aquatic mammals (dolphins, manatees) (Figure 16) [Reidenberg et al. 2007].

Cetacean brains are among the largest of all animals, both in absolute mass and relative to body size. This, however, is neither an adaptation to an aquatic environment nor to accommodate echolocation since (1) these large brains evolved several million years after cetaceans became aquatic, and (2) larger brains did not evolve in other aquatic mammals (e.g., pinnipeds, sirenians), and (3) echolocation evolved independently in bats in the absence of both an aquatic environment and enlarged brains. The shape and function of the eyes in aquatic animals are dependent on water depth and light exposure: limited light exposure results in a retina similar to nocturnal terrestrial mammals. Additionally, cetaceans have two areas of high ganglion cell concentration ("best-vision areas"), where other aquatic mammals (e.g., pinnipeds, sirenians, sea otters) only have one.

Whales

Whales have streamlined bodies tapering toward the horizontal tail that distinguishes them from fish. Their front limbs have been modified into broad flippers. Whales lack back limbs and external ears. A thick layer of blubber beneath their skin helps to conserve heat. Whalebone whales do not have teeth, but plates of baleen or whalebone in their V-shaped upper jaw, which act as sieves or strainers for plankton. The coarse, frayed inner edges catch the tiny creatures on which the whale feeds. Whalebones are the largest whales -- the blue whale, the largest animal ever known to have lived, reaches up to 100 feet and 200 tons. Whalebones include the Californian gray whale, rorquals or fin whales, blue whales and the Greenland right whale.Toothed whales are much smaller than the whalebone whales. They have conical, pointed teeth in the lower or both jaws, or one tusk like tooth in the upper jaw. Toothed whales feed primarily on fish. Whales in this suborder include river dolphins, beaked whales, sperm whales, beluga or white whales, narwhal, killer whales, Atlantic bottlenose dolphins and the common or harbor porpoise.

Figure 16. Amazon River Dolphin (Inia geoffrensis) at Duisburg Zoo.

True or Earless Seals

Seals live in the oceans of the Northern and Southern hemispheres, primarily in polar, subpolar and temperate climates, with the exception of tropical monk seals. These semiaquatic carnivores have a streamlined shape, flippers and webbed feet. They have no visible external ears. Millions of years of adaptation to the sea have made seals expert swimmers and divers, but they are quite awkward on land. A thick layer of oily fat or blubber up to 3 inches thick insulates them by maintaining a body temperature around 100 degrees Fahrenheit. Among the 18 species are leopard seal, harp seal, common or harbor seal, gray seal, Weddell seal, spotted seal, ringed seal and Baikal seal.

Sea Lions

Sea lions have conspicuous external ears. Their fore limbs give them good mobility on land. The 13 species of sea lions are divided into two groups, sea lions and fur seals. The California sea lion, native to the southwest coast of North America, is the most abundant species of sea lion. The northern fur seal migrates in winter from the Bering Sea to California and Japan.

Walrus

The walrus is found in the Arctic Ocean and in subarctic seas. Walruses spend much of their time on land, but they are graceful in water. They dive up to 180 feet deep for clams and shellfish on the ocean floor, staying underwater for up to 30 minutes. The adult male walrus weighs more than 2,000 pounds. Its upper canines form tusks up to 27 inches long that it hooks into ice floes to pull itself out of the water. Walruses live up to 40 years in the wild.

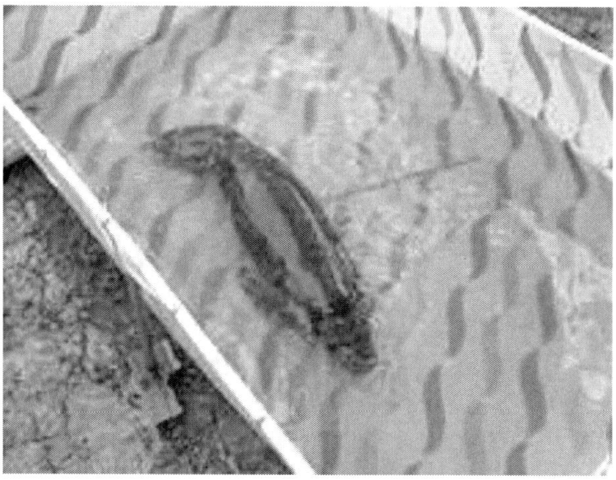

Figure 17. A baby Amazonian Manatee (Trichechus inunguis).

Manatees

Manatees are completely aquatic, but like all marine mammals they need to breathe (Figure 17). While resting they can stay submerged for up to 15 minutes; while swimming they need to breathe at the surface every three or four minutes. Found along the coasts and estuaries of Florida, Central America and the West Indies, the massive, slow-moving North American manatee is one of three species of manatee. It grows up to 13 feet long and weighs up to 1,300 pounds. Its diet is composed of water grasses, weeds and algae, and it eats as much as one-tenth of their body weight every day. The average life span in the wild is up to 40 years.

Dugong

Dugongs grow up to 10 feet long and weigh 510 to 1,100 pounds. They are closely related to manatees, and they never leave the water. Unlike the manatee they have a hairless body and a fluked horizontal tail like a whale's. They can stay underwater for up to six minutes. Frequenting the warm coastal waters of the western Pacific, Indian Ocean and Red Sea, dugongs are often observed alone or in pairs, and sometimes in herds of more than 100. Dugongs root for water grasses day and night. Docile dugongs are long-lived in the wild, with an average life span up to 70 years.

Hippopotamus

The hippopotamus loves water and stays submerged up to 16 hours of the day. With its eyes and nostrils set high up on its head, it is often tall enough to stay submerged and breathe at the same time. Hippopotami leave the water at dusk to feed on land vegetation. Found only in rivers in Africa, hippopotami are between 10 and 15 feet long including the tail. They weigh 5,000 to 8,000 pounds and live an average of 40 years in the wild (Figure 18).

Figure 18. A partially submerged hippopotamus (Hippopotamus amphibius) at the Memphis Zoo.

River and Sea Otters

Semiaquatic river otters, like other amphibious mammals, have dense, thick fur and streamlined bodies designed for life in the water. There are several species; the North American river otter lives along rivers, lakes and large creeks (Figure 19). River otters weigh up to 18 pounds and reach up to 4 feet long, including their tail. Sea otters are the largest member of weasel family, and the only one that lives almost entirely in the water -- hunting on the ocean floor, and coming to the surface to eat, groom, rest and socialize with other otters. The average adult grows up to 5 feet long and weighs up to 70 pounds -- females are slightly smaller. Populations throughout southeast Alaska, British Columbia and Washington are stable and increasing, while the numbers of sea otters in southwest Alaska are declining mainly because of predation by the killer whale. Sea otters can live up to 25 years in the wild (Figure 20).

Beavers, Nutria and Muskrat

Primarily aquatic, beavers are found in rivers and lakes in Europe, Asia and throughout North America, except northern Canada, the southern deserts of the United States and Mexico. They have a waterproof coat, closable ears, closable nostrils, paddle-shaped tail and webbed feet. These 30- to 50-pound rodents can greatly change whole environments with their dam-building activities.

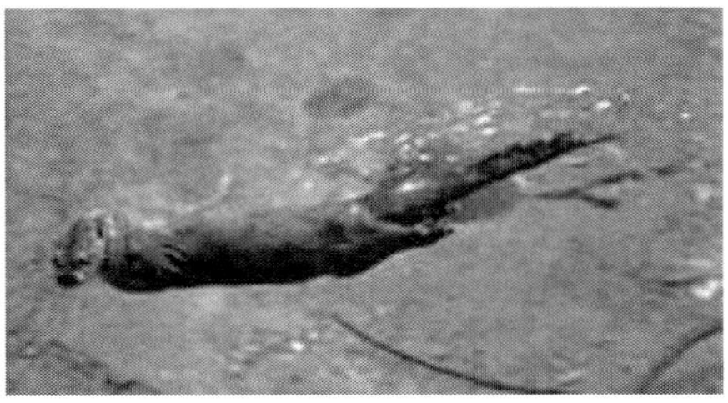

Figure 19. A North American River Otter (Lontra canadensis) at the Oregon Zoo.

Figure 20. Eurasian otter (Lutra lutra) inSouthwold, Suffolk, England.

Introduced intentionally and accidentally to North America, the nutria or coypus is native to Central and South America. It weighs up to 25 pounds. Well adapted to a semiaquatic life, nutria has small eyes and ears, and large, webbed hind feet for swimming (Figure 21). Muskrats are good swimmers and can stay underwater up to 17 minutes, preferring to live in 4 to 6 feet of water. Found in swamps, marshes, and wetlands from northern North America to the Gulf coast and the Mexican border, they are about 2 feet long and up to 4 pounds.

Duck-billed Platypus

The duck-billed platypus lives in the freshwater rivers and streams of eastern Australia, Tasmania and New Guinea. It lays eggs and has a long, leathery bill equipped with electroreceptors that help detect the magnetic fields of the aquatic invertebrates it feeds upon. Platypus are up to 2 feet long and weigh 4 pounds. Their average life span in the wild is up to 17 years (Figure 22).

Gliding / Flying Mammals

Gliding is defined as descent through the air at an angle of less than 45° to the horizontal, and this mode of locomotion has been achieved in a surprising diversity of animals through a number of impressive morphological and behavioral adaptations that generate the required aerodynamic forces (upward 'lift' exceeding air resistance, or 'drag'). Among the distinct types of gliding that are known, several provide excellent case studies in convergent evolution, as they have evolved repeatedly in very distantly related groups, including reptiles, mammals, **frogs**, **ants**, fish and even in some species of squid.

Figure 21. A North American Beaver (Castor canadensis).

Figure 22. A Platypus (Ornithorhynchus *anatinus*) at the Sydney Aquarium.

Figure 23. Gliding mammal.

Gliding mammals rely primarily on extensive skin membranes or 'patagia' that stretch between fore- and hind-limbs, creating a wing-like structure. Such large patagial membranes are perhaps most famously known from the so-called 'flying' squirrels, and they permit gliding as air pressure during descent from **trees** passively extends the membrane and generates lift to support the animal's body mass (Figure 23). Other mammalian adaptations to aerodynamic locomotion include a feather-like tuft or broad expansion at the end of the tail (e.g., in scaly-tailed possums), a dorso-ventrally flattened and light, slender body, skin membranes between fingers and toes ('interdigital webbing') and very accurate vision (e.g., in colugos). Each of these features is shared to a degree with adaptations of certain non-mammalian gliders, from several modern **lizards** and amphibians, to some unusual extinct forms of reptile.

18.20. Mammalian Gliding: The 'Wing' Patagium

Mammalian gliders are remarkably diverse, and typically have a patagium extending from the forelimb 'elbow' to the hindlimb 'ankle' or tail region, making an effective wing structure for maximising lift and minimising drag. Gliding has evolved independently at least seven times in the mammals (as described below), making them an excellent example of convergent evolution in their own right [Rayner 1981].

In the rodents both **'flying' squirrels** (Sciuridae: Pteromyini) and **anomalures, or scaly-tailed 'flying squirrels**' (Anomaluridae) have independently acquired a patagium that reaches from the wrist to the ankle, is controlled by cartilage spurs (these affect membrane tautness), and is supported as a gliding device by a flattened, broad tail (Figure 24). In the anomalures, two rows of scales are also present under the tail, possibly providing extra lift during gliding [Schufeldt 1911; Heissig 1979; Helgen 2003; Schunke et al. 2005].

Gliding squirrels are widespread in Eurasia and North America, and comprise 15 genera, including *Petaurista* (giant flying squirrels) in S.E. Asia, *Glaucomys* in N. America and *Petaurillus* (pygmy gliding squirrel) in Borneo-Malasia. They can glide over 200m, but usually glide from tree to tree, pursuing prey or escaping predators. Anomalures inhabit forests of Central Africa, and include the genera *Anomalurus, Idiurus* and *Zenkerella*. In spite of certain common names - e.g., Beecroft's flying squirrel (*Anomalurus beecrofti*) and the long-eared flying mouse (*Idiurus macrotis*) - they are not related to squirrels or mice, but represent convergence on a flying squirrel-like form [Helgen 2003; Brown et al. 2006].

The most stunning example of mammalian adaptation to gliding is found in the **colugos** (Dermoptera: Cynocephalidae), which are nocturnal tree-dwellers capable of gliding up to 70m between **trees** without any significant loss of height [Scmitz et al., 2002]. *Cynocephalus volans* and *Galeopterus variegatus* inhabit S.E. Asian rainforests, and two fossil colugo species (*Dermotherium major* and *D. chimaera*) are known from Eocene rocks <50 million years old. The colugo patagium covers as great a surface area as geometrically possible, its skeletal structure is adapted for maximising lift, and it possesses excellent binocular vision for precision landing. The patagium extends from the shoulder blade to the tip of the outer forelimb, then back to the tip of the toes and the tail; inter-digital webbing is present, and the limbs are long and thin, giving a light, aerodynamic build. Within the closely related **primates** we find limited gliding ability in **lemurs** (family Indriiidae) of the genus *Propithecus* (Figure 25). These **lemurs** are diurnal, arboreal **herbivores** from Madagascar, and are commonly known as **sifaka** [Key et al. 1990; Ducrocq et al. 1992].

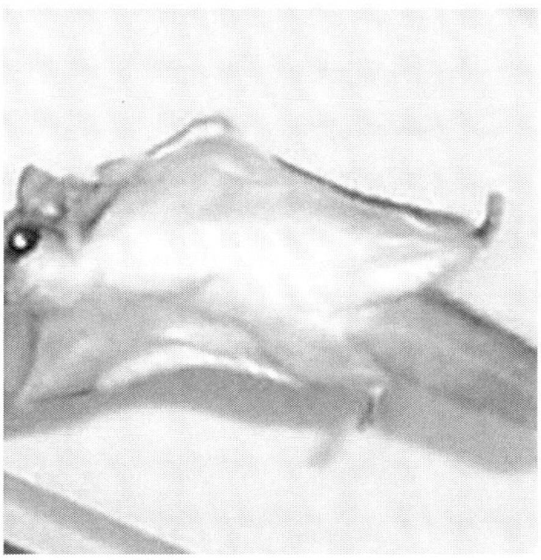

Figure 24. Gliding Squirrel.

Figure 25. Propithecus (lemur).

Propithecus diadema and its allies have thick hair on the forearm and a small membrane on inner forelimb, both of which provide limited aerofoil properties and degree of lift when jumping between the **trees**. The only other known gliding placental mammal is an extinct species, *Volaticotherium antiquus*, from the Lower Cretaceous (~125Ma) of Inner Mogolia [Rayner 1981].

Figure 26. Glaucomys volans.

It was a small insectivore (12-14cm long), similar in mass to the flying squirrel *Glaucomys volans* (Figure 26), and had a furry patagium stretched between fore- and hind-limbs, a smaller membrane ('uropatagium') between the hindlimbs and tail, and elongated limbs (as in colugo) for support. The tail was long, stiff and dorso-ventrally flattened, providing an excellent means of achieving control and stability during gliding. The relationship of *Volaticotherium* to other eutherian mammals is uncertain [Beard 1990, 1991].

Among the marsupial mammals gliding has evolved independently on three separate occasions; in the gliding possums, greater glider and feather-tailed possums. **Gliding possums or 'phalangers'** (Figure 27) comprise six species in the genus ***Petaurus*** (family Petauridae). They inhabit forests of Australia, New Guinea and Borneo, and are able to glide between **trees** for over 50m in search of sap, nectar or small insects to eat. In **petaurids** (e.g., *Petaurus biacensis* - the Biak glider, and *P. breviceps* - the sugar glider) the patagium stretches from the fifth finger to the first toe and is under a degree of muscular control, very similar to the eutherian group Pteromyini (flying squirrels). Stabilisation during gliding depends on a broad, flattened tail, as in most other flying mammals, and also in gliding **lizards** to a degree [Runestand & Ruff 1995; Crosby et al. 2004].

The **greater glider** ***Petauroides volans*** is the single species of the family Pseudochieridae, and resembles *Petaurus* in its habits and form, except that the patagium is more restricted, stretching from the fifth finger to the hindlimb 'knee'. **Feather-tailed possums** refer to two species of the family Acrobatidae, including the smallest known gliding mammal, the Australian 'feather-tailed glider' *Acrobates pygmaeus*, and the feather-tailed possum of New Guinea, *Distoechurus pennatus* (Figure 28). Both have a patagium between the fore- and hind-limbs, but are characterised by a tail that has long, stiffened hairs running down both sides, giving it a feather-like appearance and providing good steering ability when gliding [Russel 1980; Runestand & Ruff 1995; Crosby et al. 2004].

Figure 27. Gliding possums.

Figure 28. Feather tailed possums.

In addition to the convergent use of a large patagium in placental gliders (e.g., flying squirrels, anomalurids, colugos, sifaka, *Volaticotherium*) and marsupial gliders (e.g., phalangers, *Petauroides*, acrobatids), an extinct reptile called *Sharovipteryx* mirabilis also evolved a limb-associated 'wing' patagium, thus developing a similar adaptation to gliding in spite of millions of years of independent evolution from the last shared ancestor of both mammals and reptiles. ***Sharovipteryx*** is a small reptile from the Upper Triassic (225 million years ago), and was evidently tree-dwelling ('arboreal'). Presumably it was able to run up **trees** aided by claws on its two back legs, and then launch itself into a powerful, controlled glide [Baverstock 1984; Beard, 1990, 1991]. Its gliding mechanism is unique among the reptiles: a wide, triangular membrane (termed a 'delta wing') spreads between its outstretched hindlimbs at the posterior and joins the body just behind the shoulders to form an anterior point. In addition to the primary 'delta wing' ***Sharovipteryx*** is hypothesised to have had small membranes between the forelimbs and base of the neck, providing enhanced maneuverability. ***Sharovipteryx*** is an enigmatic reptile, and it's unusual 'delta wing gliding' mechanism is most similar to that seen in gliding mammals. Most other gliding reptiles using skin membranes spread between digits (e.g., geckkonid **lizards** ***Ptychozoon***) or as thoracic wings, supported on elongated ribs or dermal rods (e.g., agamid lizard ***Draco***) [Agrawal & Chakraborty 1970; Beard 1990, 1991; Runestand and Ruff 1995].

18.21. Mammalian Gliding: Inter-Digital Membranes

Well-developed patagial membranes between fingers and toes (interdigital webbing) are an important feature of the most extremely specialised of gliding mammals, the **colugos**. Colugos are tree-dwelling mammals that inhabit the South-East Asian rainforests, and the two species known, *Cynocephalus volans* and *Galeopterus varieagatus*, may be the closest relatives of **primates** alive today. Their interdigital webbing and expansive wing patagium permits them to glide between **trees** in search of food and mates [Schufelt 1911; Runestand & Ruff 1995].

Outside the colugos, webbed digits have evolved convergently in two types of lizard (geckos and lacertids) and also in two independent lineages of 'flying' **frogs**. Web-footed **lizards** (Figure 29) include gliding geckos, for example the six species of **Ptychozoon** from South-East Asia and Indo-Australia, and the distantly related West African lacertid, *Holaspis guentheri*. Flying **frogs** glide and maneuver in the air using extensively webbed digits, glide to reach mating sites on the rainforest floor and to escape predation. Such **frogs** live in tropical and neotropical forests, and include species of New World 'Hylidae' (e.g., *Hyla miliaria*) and Old World 'Rhacophoridae' (e.g., *Rhacophorus nigropalmatus*) [Runestand & Ruff 1995].

18.22. Mammalian Gliding: Tail Plumage

A laterally expanded tail plume is hypothesised to provide stability when gliding between **trees**, and as a morphological and functional adaptation, shows convergence in form both within the mammals and also between mammals and several gliding reptiles.

Figure 29. Web footed lizards.

Figure 30. Oviraptorosaurs.

Gliding in **scaly-tailed anomalures** (African rodents, e.g., *Anomalurus*, *Idiurus*) and **feather-tailed possums** (marsupials of the family Acrobatidae - e.g., *Acrobates*, *Distoechurus*) is dependent on projecting rows of scales or stiff, feather-like hairs respectively, projecting from either side of a flattened, broad tail. Anomalures and acrobatids are very distantly related - one being placental, the other marsupial, and yet some reptiles from within an extinct dinosaur group called the theropods provide an even more distantly related comparison and so a further example of convergence [Troughton 1941; Rayner 1981; Runestand & Ruff 1995].

Theropods include 'oviraptorosaurs', which are slightly less evolutionarily advanced than the 'Eumaniraptoria' (Figure 30) the latter group containing notable species such as ***Microraptor*** and *Sinovenator*, capable of gliding, and also ***Archaeopteryx*** and *Sapeornis*, primitive 'birds' capable of powered flight. The oviraptorosaur *Protarchaeopteryx robusta* had indisputably bird-like feathers in a tuft on the end of its tail, and Caudipteryx zoui had a well feathered tail and forelimbs; both species are Late Jurassic-Early Cretaceous (150-140 Ma) in age. ***Caudipteryx*** and *Protarchaeopteryx* evolved a 'tail-wing' independently from one another, as a gliding adaptation that appears to be only a small step behind the specialised 'four-wings plus tail-fan' of ***Microraptor*** and other eumaniraptors. As such, these broad-tailed reptiles share remarkable elements of convergence with dominant tail-fan bearing mammals such as *Anomalurus* and *Acrobates* [Troughton 1941; Rayner 1981; Runestand & Ruff 1995].

The only mammal capable of powered flight is the bat. All the Bats belong to the order Chiroptera. The word Chiroptera comes from the Greek words *Cheir* meaning "hand" and *pteron*, which means "wing" because the structure of the open wing is very similar to an outspread human hand (Figure 31). But the bones of the arm and four of the fingers are light, slender and lengthened as an adaptation for flight. They are used to support, spread and manipulate the membrane. The only digit not attached to the wing is the thumb. It remains free for clinging to various surfaces. The kness on the hind limbs bend backward, but the feet face forward [Schufelt 1911; Szalay and Lucas 1993].

Bats wing is an extension of the skin of the abdomen that runs to the tip of each digit, uniting the forelimb with the body (Figure 32). It is two tightly stretched layers of skin membranes connected together by connective tissue without any flesh between the layers. This is why bat's wings are translucent, which means you can see through them. This formation of skin membranes is called patagium. Bat wings usually run from the shoulder region to the ankle, or in some cases, to the toes themselves. The wing membrane joins the body along the sides, except in a few cases in which it arises near the middle of the back. When the wing is not extended the membrane folds up along countless creases more efficiently than an umbrella [Schuflt 1911; Szalay & Lucas 1993; Runestand & Ruff 1995; Simmon et al. 2008].

According to the bat biologists, the patagium has four distinct parts:

1. **Propatagium:** the leading edge of the patagium from the shoulder to the first digit. The thumb sits along the propatagium, but it is used for climbing, food handling and fighting rather than for flight.
2. **Dactylopatagium:** the portion found between the second and fifth digits.
3. **Plagiopatagium:** the portion found between the body and the fifth digit.
4. **Uropatagium:** the posterior portion of the body that runs between the two hindlimbs and encloses the tail of some species, such as vespertilionids.

Figure 31. Flying bat.

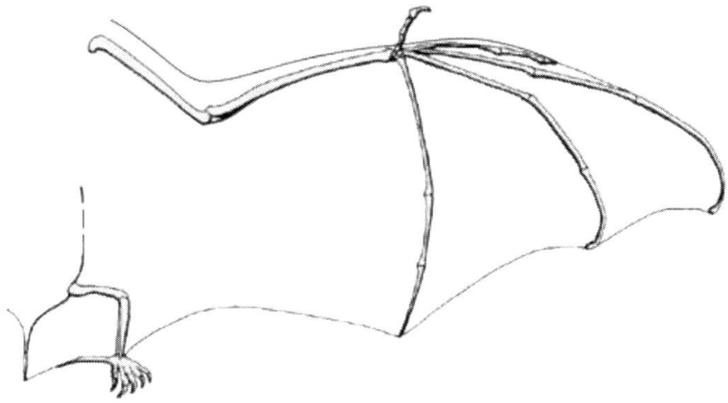

Figure 32. Bats patagium.

Wing and tail membranes appear naked on most bats. However on close look they appear covered with tiny hairs. Some species have distinct tufts and fringes of hairs. A few species have a thick layer of fur on their tail membranes and some have thick hairs on parts of their wings [Szalay & Lucas 1993; Simmon et al. 2008].

While flapping their wings, bats move upward and downward through the air by moving the membrane between the body and fifth finger (plagiopatagium). This is called 'lift'. Bats move forward (thrusts) by changing the shape of the membrane between second and fifth fingers (dactylopatagium). With their flexible wings, bats can change the degree and direction of lift and thrust very quickly by positioning their wings into different shapes. They can also fold one of their wings to steer and brake. This is how they are able to weave and dive in air like no other animal can. The rigid bird wing is more efficient at providing lift, but the flexible bat wing allows for greater maneuverability (being able to make 180 degree turns in less than half a wingspan length). Bats fly by using their hands and wings in a 'rowing motion' through air.

Bats fold their wings on the upstroke. They uses 35% less energy and reduces aerodynamic drag, which compensates for their heavier, more muscular wings. Both thrust and lift occur during the down stroke.

Bat's wings are different from bird's wings in that most of the flight muscles controlling the wing beat are attached to the shoulder blades. With birds, the muscles are braced against the rib cage, and they typically have a prominent breast keel where heavy muscles are attached. Bats have only one muscle attached to the breastbone, the ribs are flattened, and there is some fusing of the vertebral joints. These additional adaptations help make the bat incredibly agile in the air [Simmon et al. 2008].

Can Bats Take off from the Ground?

Bats with longer, narrower wings can not take off from the ground. They will take off by dropping from a hanging position. But some bats, like some species of horseshoe bat and the common vampire bat, can take off from the ground. Most bats land by slowing down until they stall and grab hold of a branch or the surface. Common vampire bats land on the ground, so they can sneak up on their prey from the ground.

Bats Can Also Move Around on the Ground

The first digit of the bat wing is small and clawed. Bats use it to climb or to walk on the ground. Several species of bat are known to use their strong forelimbs and weak hind legs to crawl along the ground. Some insectivorous species feed on the ground dwelling insects more than they feed on inscets captured in flight. Common vampire bats also get around quite well on the ground.

In an experiment a vampire bat broke into a bounding run when dropped on tread mill. Vampire bats can run up to about 2.5 miles per hour. Bats can also swim.

18.23. Flying Mammals Are a Big Waste of Energy

Bats are the only mammals that can technically fly, but this behavior makes no sense. Lots of small mammasl have perfected the art of gliding, using folds of skin to maintain lift over long distances, sometimes as much as hundred feet across the rainforest canopy. However it would be a much better use of energy for the creatures to simply run across the floor of the rainforest as any other creatures would. Gliding may actually be very taxing in terms of energy. An experiment with lemurs showed that it consistently took 1.5 times the amount of energy for the lemurs to glide over a given distance than it would for them to travel that far on foot. It is not just a question of why the lemurs would bother gliding when it is so energy intensive more to the point, why did it and other gliding species evolve that ability at all if it is such an inefficient adaptation? The researchers believe that simple survival may have trumped energy concerns for these mammals- being able to glide over dangerous terrain would have protected them from predators, and the ability to glide would give them a built-in safety net if they fell off a branch while trying to eat from leaves in the canopy.

Primate

A **primate** is a mammal of the order **Primates** (Latin: "prime, first rank"), which contains prosimians and simians [Goodman et al. 1990]. Primates arose from ancestors that lived in the trees of tropical forests; many primate characteristics represent adaptations to life in this challenging three-dimensional environment. Most primate species remain at least partly arboreal.

With the exception of humans, who inhabit every continent, most primates live in tropical or subtropical regions of the Americas, Africa and Asia. They range in size from Madame Berthe's mouse lemur, which weighs only 30 g (1 oz), to the eastern lowland gorilla, weighing over 200 kg (440 lb). According to fossil evidence, the primitive ancestors of primates may have existed in the late Cretaceous period around 65 million years ago. An early close primate relative known from abundant remains is the Late Paleocene *Plesiadapis*, *circa* 55–58 million years ago. Molecular clock studies suggest the primate branch may be even older, originating in the mid-Cretaceous period around 85 mya [Chatterjjee et al. 2009].

The order Primates has traditionally been divided into two main groupings: prosimians and anthropoids (simians). Prosimians have characteristics more like those of the earliest primates, and include the lemurs of Madagascar, lorisoids, and tarsiers. Simians include monkeys, apes and hominins. More recently, taxonomists have preferred to split primates into the suborder Strepsirrhini, or wet-nosed primates, consisting of nontarsier prosimians, and the suborder Haplorhini, or dry-nosed primates, consisting of tarsiers and the simians. Simians are divided into two groups: catarrhine (narrow-nosed) monkeys and apes of Africa and southeastern Asia and platyrrhine ("flat-nosed") or New World monkeys of South and Central America. Catarrhines consist of Old World monkeys (such as baboons and macaques), gibbons and great apes. New World monkeys include the capuchin, howler and squirrel monkeys. Humans are the only extant catarrhines to have spread successfully outside of Africa, South Asia, and East Asia, although fossil evidence shows many other species were formerly present in Europe. New primate species are still being discovered. More than 25 species were taxonomically described in the decade of the 2000s and eleven have been described since 2010.

Considered generalist mammals, primates exhibit a wide range of characteristics. Some primates (including some great apes and baboons) are primarily terrestrial rather than arboreal, but all species possess adaptations for climbing trees. Locomotion techniques used include leaping from tree to tree, walking on two or four limbs, knuckle-walking, and swinging between branches of trees (brachiation). Primates are characterized by large brains relative to other mammals, as well as an increased reliance on stereoscopic vision at the expense of smell, the dominant sensory system in most mammals. These features are more developed in monkeys and apes and noticeably less so in lorises and lemurs. Three-color vision has developed in some primates. Most also have opposable and some have prehensile tails. Many species are sexually dimorphic; differences include body mass, canine tooth size, and coloration. Primates have slower rates of development than other similarly sized mammals and reach maturity later, but have longer lifespans. Depending on the species, adults may live in solitude, in mated pairs, or in groups of up to hundreds of members.

18.24. History

The relationships among the different groups of primates were not clearly understood until relatively recently, so the commonly used terms are somewhat confused. For example, "ape" has been used either as an alternative for "monkey" or for any tailless, relatively humanlike primate.

Sir Wilfrid Le Gros Clark was one of the primatologists who developed the idea of trends in primate evolution, and the living members of the order could be arranged in an ascending series leading to humans [Dixon 1981]. Commonly used names for groups of primates, for example "prosimians", "monkeys", "lesser apes" and "great apes", reflect this arrangement. According to our current understanding of the evolutionary history of the primates, several of these groups are paraphyletic, meaning although all the species in the group descend from a common ancestor, the group does not include all the descendants of that ancestor [Stace & Clive 2010].

By contrast with approaches such as those of Le Gros Clark, modern classifications typically use groups that are monophyletic, since they include *all* the descendants of a common ancestor. The diagram below shows one possible classification of the living primates, with groups with commonly used names shown on the right [Mishler & Brent 2009; Cartmill & Smith 2011].

- All the groups that have scientific names are monophyletic (i.e., they are clades), so the scientific classification reflects evolutionary history. Some of the traditional groups which form an "ascending series", are paraphyletic:
- "prosimians" contain two monophyletic groups, the suborder Strepsirrhini (lemurs, lorises and allies), as well as the tarsiers, which are a sister group to the infraorder Anthropoidea (also called Simiiformes).
- "monkeys" consist of two monophyletic groups, New World monkeys and Old World monkeys, but exclude hominoids (superfamily Hominoidea).
- "apes" as a whole, and the "great apes" in particular, are not monophyletic because they exclude humans.

Thus, the two sets of groups, and hence names, do not match, which causes problems in relating scientific names to common names. Consider the superfamily Hominoidea. In terms of the common names, this group consists of apes and humans, and there is no single common name for all the members of the group. One possibility is to create a new common name, in this case "hominoids". Another possibility is to expand the use of one of the traditional terms. For example, in a 2005 book, the vertebrate palaeontologist Benton wrote, "The apes, Hominoidea, today include the gibbons and orang-utan ... the gorilla and chimpanzee ... and humans" (Figure 33), thereby using "apes" to mean "hominoids". The group traditionally called "apes" must then be called the "nonhuman apes".

As of July 2011, there is no consensus as to which approach to follow, whether to accept traditional paraphyletic common names or whether to use monophyletic names, either new ones or adaptations of old ones. Both approaches will be found in biological sources, often in the same work. Thus, although Benton defines "apes" to include humans, he also repeatedly uses "ape-like" to mean "like an ape rather than a human", and when discussing the reaction of others to a new fossil writes of "claims that *Orrorin* ... was an ape rather than a human" [Benton 2005].

18.25. Classification of Living Primates

A list of the families of the living primates is given below, together with one possible classification into ranks between order and family. Other classifications are also used. For example, an alternative classification of the living Strepsirrhini divides them into two infraorders, Lemuriformes and Lorisiformes [Hartwing 2011].

Figure 33. A 1927 drawing of chimpanzees, a gibbon (top right) and two orangutans (center and bottom center): The chimp in the upper left is brachiating; the orangutan at the bottom center is knuckle-walking.

Figure 34. Homo sapiens is the only living primate species that is fully bipedal.

Order Primates

Suborder Strepsirrhini: lemurs, galagos and lorisids
Infraorder Lemuriformes
Superfamily Lemuroidea
Family Cheirogaleidae: dwarf lemurs and mouse-lemurs (31 species)
Family Daubentoniidae: aye-aye (one species)
Family Lemuridae: ring-tailed lemur and allies (22 species)
Family Lepilemuridae: sportive lemurs (26 species)
Family Indriidae: woolly lemurs and allies (19 species)
Superfamily Lorisoidea

Family Lorisidae: lorisids (10 species)
Family Galagidae: galagos (20 species)

Suborder Haplorhini: tarsiers, monkeys and apes
Infraorder Tarsiiformes
Family Tarsiidae: tarsiers (10 species)
Infraorder Simiiformes (or Anthropoidea)
Parvorder Platyrrhini: New World monkeys
Family Callitrichidae: marmosets and tamarins (42 species)
Family Cebidae: capuchins and squirrel monkeys (17 species)
Family Aotidae: night or owl monkeys (douroucoulis) (10 species)
Family Pitheciidae: titis, sakis and uakaris (42 species)
Family Atelidae: howler, spider, woolly spider and woolly monkeys (28 species)
Parvorder Catarrhini
Superfamily Cercopithecoidea
Family Cercopithecidae: Old World monkeys (135 species)
Superfamily Hominoidea
Family Hylobatidae: gibbons or "lesser apes" (17 species)
Family Hominidae: great apes, including humans (seven species)

Order Primates was established by Carl Linnaeus in 1758, in the tenth edition of his book *Systema Naturae*, for the genera Homo (humans), *Simia* (other apes and monkeys), Lemur (prosimians) and *Vespertilio* (bats). In the first edition of the same book (1735), he had used the name Anthropomorpha for *Homo*, *Simia* and *Bradypus* (sloths). In 1839, Henri Marie Ducrotay de Blainville, following Linnaeus and imitating his nomenclature, established the orders Secundates (including the suborders Chiroptera, Insectivora and Carnivora),Tertiates (or Glires)and Quaternates (including Gravigrada, Pachydermata and Ruminantia), but these new taxa were not accepted [Linnaeus 1839, 1858; Blainville 1839].

Figure 35. The Philippine tarsier, once considered a prosimian, is now predominantly considered a haplorrhine.

Before Anderson and Jones introduced the classification of Strepsirrhini and Haplorhini in 1984, (followed by McKenna and Bell's 1997 work, *Classification of Mammals: Above the species level*), the Primates were divided into two superfamilies: Prosimii and Anthropoidea. Prosimii included all of the prosimians: Strepsirrhini plus the tarsiers (Figure 35). Anthropoidea contained all of the simians (Thorington & Anderson 1984; McKenna & Bell 1997; Strier 2007).

18.26. Evolutionary History

Order Primates is part of the clade Euarchontoglires, which is nested within the clade Eutheria of Class Mammalia. Recent molecular genetic research on primates, colugos, and tree shrews has shown the two species of colugos are more closely related to primates than tree shrews [Janecka et al. 2007] even though tree shrews were at one time considered primates [Kavanagh 1983]. These three orders make up the clade Euarchonta. The combination of this clade with the clade Glires (composed of Rodentia and Lagomorpha) forms the clade Euarchontoglires. Variously, both Euarchonta and Euarchontoglires are ranked as superorders. Some scientists consider Dermoptera a suborder of Primates and call the "true" primates the suborder Euprimates [McKenna & Bell 1997].

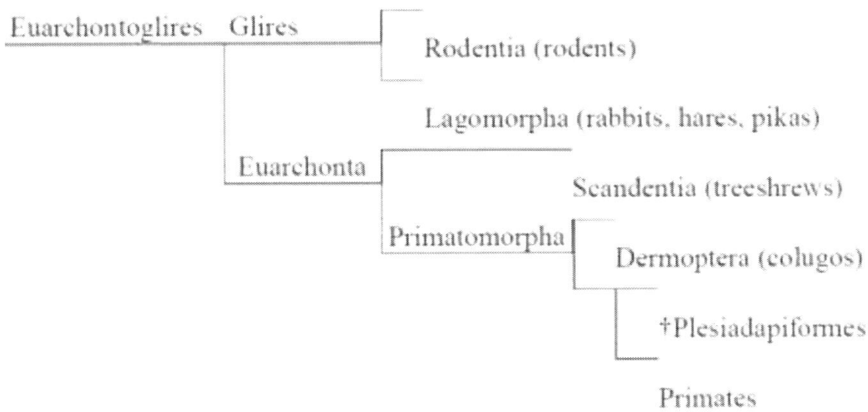

18.27. Evolution

The primate lineage is thought to go back at least 65 million years ago (mya), even though the oldest known primates from the fossil record date to the Late Paleocene of Africa (*Altiatlasius*) or the Paleocene-Eocene transition in the northern continents, *circa* 55 mya (*Cantius*, *Donrussellia*, *Altanius*, and *Teilhardina*). Other studies, including molecular clock studies, have estimated the origin of the primate branch to have been in the mid-Cretaceous period, around 85 mya [Lee 1999; Tavare et al. 2002; Williams et al. 2010].

By modern cladistic reckoning, order Primates is monophyletic. Suborder Strepsirrhini, the "wet-nosed" primates, is generally thought to have split off from the primitive primate line about 63 mya, although earlier dates are also supported. The seven strepsirrhine families are the five related lemur families and the two remaining families that include the lorisids and the galagos. Older classification schemes wrap Lepilemuridae into Lemuridae and Galagidae into Lorisidae, yielding a four-one family distribution instead of five-two as presented here. During the Eocene, most of the northern continents were dominated by two groups, the adapiforms and the omomyids. The former are considered members of Strepsirrhini, but did not have a toothcomb like modern lemurs. Recent analysis has demonstrated *Darwinius masillae* fits into this grouping. The latter was related closely to tarsiers, monkeys, and apes. How these two groups relate to extant primates is unclear. Omomyids perished about 30 mya, while adapiforms survived until about 10 mya [Seller 2000; Hartwig 2007].

According to genetic studies, the lemurs of Madagascar diverged from the lorisoids approximately 75 mya. These studies, as well as chromosomal and molecular evidence, also show that lemurs are more closely related to each other than to other strepsirrhine primates (Figure 36). However, Madagascar split from Africa 160 mya and from India 90 mya. To account for these facts, a founding lemur population of a few individuals is thought to have reached Madagascar from Africa via a single rafting event between 50 and 80 mya. Other colonization options have been examined, such as multiple colonizations from Africa and India, but none are supported by the genetic and molecular evidence [Seller 2000; Hartwig 2007].

Until recently, the aye-aye has been difficult to place within Strepsirrhini. Theories had been proposed that its family, Daubentoniidae, was either a lemuriform primate (meaning its ancestors split from the lemur line more recently than lemurs and lorises split) or a sister group to all the other strepsirrhines. In 2008, the aye-aye family was confirmed to be most closely related to the other Malagasy lemurs, likely having descended from the same ancestral population that colonized the island [Seller 2000; Hartwig 2007].

Figure 36. Ring-tailed lemur, a strepsirrhine primate.

Figure 37. Emperor tamarin, a New World monkey.

Suborder Haplorhini, the simple-nosed or "dry-nosed" primates, is composed of two sister clades. Prosimian tarsiers in the family Tarsiidae (monotypic in its own infraorder Tarsiiformes), represent the most basal division, originating about 58 mya. The earliest known haplorhine skeleton that of 55 MA old tarsier-like *Archicebus*, was found in central China, supporting an already suspected Asian origin for the group. The infraorder Simiiformes (simian primates, consisting of monkeys and apes) emerged about 40 mya, possibly also in Asia; if so, they dispersed across the Tethys Sea from Asia to Africa soon afterwards. There are two simian clades, both parvorders:Catarrhini, which developed in Africa, consisting of Old World monkeys, humans and the other apes, and Platyrrhini, which developed in South America, consisting of New World monkeys (Figure 37). A third clade, which included the eosimiids, developed in Asia, but went extinct millions of years ago [Schimdt et al. 2005; Shekelle 2005; Kay 2012].

As in the case of lemurs, the origin of New World monkeys is unclear. Molecular studies of concatenated nuclear sequences have yielded a widely varying estimated date of divergence between platyrrhines and catarrhines, ranging from 33 to 70 mya, while studies based on mitochondrial sequences produce a narrower range of 35 to 43 mya. The anthropoid primates possibly traversed the Atlantic Ocean from Africa to South America during the Eocene by island hopping, facilitated by Atlantic Ocean ridges and a lowered sea level. Alternatively, a single rafting event may explain this trans-oceanic colonization. Due to continental drift, the Atlantic Ocean was not nearly as wide at the time as it is today. Research suggests a small 1 kg (2.2 lb) primate could have survived 13 days on a raft of vegetation. Given estimated current and wind speeds, this would have provided enough time to make the voyage between the continents [Houle, 1999; Schrago and Russo, 2003].

Apes and monkeys spread from Africa into Europe and Asia starting in the Miocene. Soon after, the lorises and tarsiers made the same journey. The first hominid fossils were discovered in northern Africa and date back 5–8 mya. Old World monkeys disappeared from Europe about 1.8 mya. Molecular and fossil studies generally show modern humans originated in Africa 100,000–200,000 years ago.

Although primates are well studied in comparison to other animal groups, several new species have been recently discovered, and genetic tests have revealed previously unrecognised species in known populations. *Primate Taxonomy* listed about 350 species of primates in 2001; the author, Colin Groves, increased that number to 376 for his contribution to the third edition of *Mammal Species of the World* (MSW3). However, publications since the taxonomy in MSW3 was compiled in 2003 have pushed the number to 424 species, or 658 including subspecies [Pough et al., 2007].

18.28. HYBRIDS

Primate hybrids usually arise in captivity, but there have also been examples in the wild. Hybridization occurs where two species' range overlap to form hybrid zones; hybrids may be created by humans when animals are placed in zoos or due to environmental pressures such as predation. Intergeneric hybridizations, hybrids of different genera, have also been found in the wild. Although they belong to genera that have been distinct for several million years, interbreeding still occurs between the gelada and the Hamadryas baboon [Bernsteil, 1966; Sugawara, 1979; Tenaza, 1984].

18.29. DISTINGUISHING FEATURES

Primates have diversified in arboreal habitats (trees and bushes) and possess many characteristics that are adaptations to this environment. They are distinguished by:

- retention of the collar bone in the pectoral girdle
- shoulder joints which allow high degrees of movement in all directions
- five digits on the fore and hind limbs with opposable thumbs and big toes
- nails on the fingers and toes (in most species)
- a flat nail on the hallux (in all extant species)
- sensitive tactile pads on the ends of the digits
- orbits encircled in bone
- a trend towards a reduced snout and flattened face, attributed to a reliance on vision at the expense of olfaction (most notably in haplorrhines, and less so in strepsirrhines)
- a complex visual system with stereoscopic vision, high visual acuity and color vision
- a brain having a well-developed cerebellum with posterior lobe and a calcarine fissure
- a large brain in comparison to body size, especially in simians
- differentiation of an enlarged cerebral cortex
- reduced number of teeth compared to primitive mammals
- three kinds of teeth
- a well-developed caecum
- two pectoral mammary glands
- typically one young per pregnancy
- a pendulous penis and scrotal testes
- long gestational and developmental periods and
- a trend towards holding the torso upright leading to bipedalism.

Not all primates exhibit these anatomical traits, nor is every trait unique to primates. For example, other mammals have collar bones, three kinds of teeth and a pendulous penis, while spider monkeys have greatly reduced thumbs, ruffed lemurs have six mammary glands and strepsirrhines generally have longer snouts and a strong sense of smell.

In regard to behavior, primates are frequently highly social, with flexible dominance hierarchies. New World species form monogamous pair bonds, and show substantial paternal care of their young, unlike most Old World monkeys [Soligo, 1999; Pough et al., 2005; White and Kazlev, 2006].

18.30. Anatomy, Morphology and Physiology

Primates have forward-facing eyes on the front of the skull; binocular vision allows accurate distance perception, useful for the brachiating ancestors of all great apes. A bony ridge above the eye sockets reinforces weaker bones in the face which are put under strain during chewing. Strepsirrhines have a postorbital bar, a bone around the eye socket, to protect their eyes. In contrast, the higher primates, haplorrhines, have evolved fully enclosed sockets [Campbell and Loy, 2000].

The primate skull has a large, domed cranium, which is particularly prominent in anthropoids. The cranium protects the large brain, a distinguishing characteristic of this group (Figure 38). The endocranial volume (the volume within the skull) is three times greater in humans than in the greatest nonhuman primate, reflecting a larger brain size. The mean endocranial volume is 1201 cubic centimeters in humans, 469 cm^3 in gorillas, 400 cm^3 in chimpanzees and 397 cm^3 in orangutans. The primary evolutionary trend of primates has been the elaboration of the brain, in particular the neocortex (a part of the cerebral cortex), which is involved with sensory perception, generation of motor commands, spatial reasoning, conscious thought and, in humans, language. While other mammals rely heavily on their sense of smell, the arboreal life of primates has led to a tactile, visually dominant sensory system, a reduction in the olfactory region of the brain and increasingly complex social behavior [Aiello and Dean, 1990].

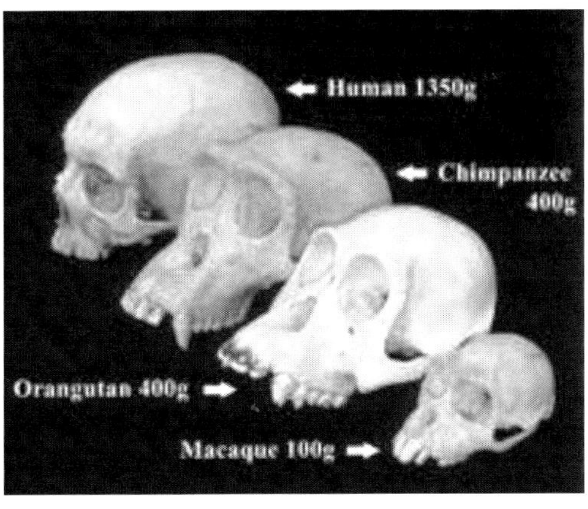

Figure 38. Primate crania with brain masses indicated.

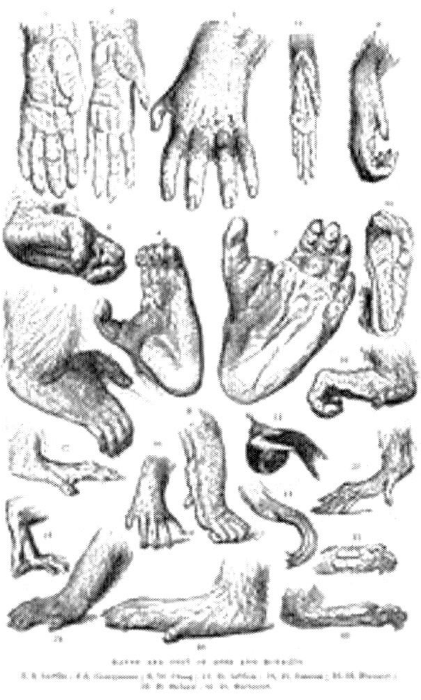

Figure 39. An 1893 drawing of the hands and feet of various primates.

Primates generally have five digits on each limb (pentadactyly), with keratin nails on the end of each finger. The bottom sides of the hands and feet have sensitive pads on the fingertips. Most have opposable thumbs, a characteristic primate feature, though not limited to this order (opossums, for example, also have them). Thumbs allow some species to use tools. In primates, the combination of opposing thumbs, short fingernails (rather than claws) and long, inward-closing fingers is a relict of the ancestral practice of gripping branches, and has, in part, allowed some species to develop brachiation (swinging by the arms from tree limb to tree limb) as a significant means of locomotion. Prosimians have claw like nails on the second toe of each foot, called toilet-claws, which they use for grooming (Figure 39-40) [Pough et al., 2005].

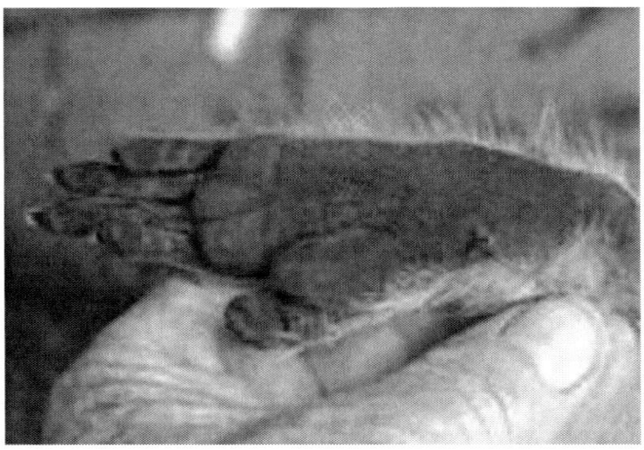

Figure 40. Vervet hindfoot showing fingerprint ridges on the sole.

The primate collar bone is retained as prominent element of the pectoral girdle; this allows the shoulder broad mobility. Apes have more mobile shoulder joints and arms due to the dorsal position of the scapula, broad ribcages that are flatter front-to-back, and a shorter, less mobile spine compared to Old World monkeys (with lower vertebrae greatly reduced, resulting in tail loss in some species). Old World monkeys are unlike apes in that most have tails. The only primate family with fully prehensile tails are the New World atelids, including the howler, spider, woolly spider and woolly monkeys (New World capuchins have partial prehensile tails) [White and Kazlev, 2006].

Primates show an evolutionary trend towards a reduced snout. Technically, Old World monkeys are distinguished from New World monkeys by the structure of the nose, and from apes by the arrangement of their teeth. In New World monkeys, the nostrils face sideways; in Old World monkeys, they face downwards. Dental pattern in primates vary considerably; although some have lost most of their incisors, all retain at least one lower incisor. In most strepsirrhines, the lower incisors and canines form a toothcomb, which is used in grooming and sometimes foraging, and the first lower premolar is shaped like a canine. Old World monkeys have eight premolars, compared with 12 in New World monkeys. The Old World species are divided into apes and monkeys depending on the number of cusps on their molars; apes have five, Old World monkeys have four, although humans may have four or five. The main hominid molar cusp (hypocone) evolved in early primate history, while the cusp of the corresponding primitive lower molar (paraconid) was lost. Prosimians are distinguished by their immobilized upper lips, the moist tip of their noses and forward-facing lower front teeth [Nelson and Wheeler, 2003].

The evolution of color vision in primates is unique among most eutherian mammals. While the remote vertebrate ancestors of the primates possessed three color vision(trichromaticism), the nocturnal, warm-blooded, mammalian ancestors lost one of three cones in the retina during the Mesozoic era. Fish, reptiles and birds are therefore trichromatic or tetrachromatic, while all mammals, with the exception of some primates and marsupials, are dichromats or monochromats (totally color blind). Nocturnal primates, such as the night monkeys and bush babies, are often monochromatic. Catarrhines are routinely trichromatic due to a gene duplication of the red-green opsin gene at the base of their lineage, 30 to 40 million years ago. Platyrrhines, on the other hand, are trichromatic in a few cases only. Specifically, individual females must be heterozygous for two alleles of the opsin gene (red and green) located on the same locus of the X chromosome. Males, therefore, can only be dichromatic, while females can be either dichromatic or trichromatic. Color vision in strepsirrhines is not as well understood; however, research indicates a range of color vision similar to that found in platyrrhines [Surridge and Osorio, 2003; Arrese et al., 2005].

Like catarrhines, howler monkeys (a family of platyrrhines) show routine trichromatism that has been traced to an evolutionarily recent gene duplication. Howler monkeys are one of the most specialized leaf-eaters of the New World monkeys; fruits are not a major part of their diets, and the type of leaves they prefer to consume (young, nutritive, and digestible) are detectable only by a red-green signal. Field work exploring the dietary preferences of howler monkeys suggests routine trichromaticism was selected by environment [Lucas and Dominy, 2003; Sussman, 2003].

Figure 41. Distinct sexual size dimorphism can be seen between the female and two male Hamadryas baboons.

Sexual Dimorphism

Sexual dimorphism is often exhibited in simians, though to a greater degree in Old World species (apes and some monkeys) than New World species. Recent studies involve comparing DNA to examine both the variation in the expression of the dimorphism among primates and the fundamental causes of sexual dimorphism. Primates usually have dimorphism in body mass and canine tooth size along with pelage and skin color. The dimorphism can be attributed to and affected by different factors, including mating system, size, habitat and diet (Figure 41) [Frisch, 1963; Crook, 1972; Kay, 1975; Lindsteadtand et al., 1985].

Comparative analyses have generated a more complete understanding of the relationship between sexual selection, natural selection, and mating systems in primates. Studies have shown that dimorphism is the product of changes in both male and female traits. Ontogenetic scaling, where relative extension of a common growth trajectory occurs, may give some insight into the relationship between sexual dimorphism and growth patterns. Some evidence from the fossil record suggests that there was convergence of dimorphism, and some extinct hominids probably had greater dimorphism than any living primate [Plavacan, 2001].

18.31. Locomotion

Primate species move by brachiation, bipedalism, leaping, arboreal and terrestrial quadrupedalism, climbing, knuckle-walking or by a combination of these methods. Several prosimians are primarily vertical clingers and leapers. These include many bushbabies, allindriids (i.e., sifakas, avahis and indris), sportive lemurs, and all tarsiers (Figure 42). Other prosimians are arboreal quadrupeds and climbers. Some are also terrestrial quadrupeds, while some are leapers. Most monkeys are both arboreal and terrestrial quadrupeds and climbers. Gibbons, muriquis and spider monkeys all brachiate extensively, with gibbons sometimes doing so in remarkably acrobatic fashion. Woolly monkeys also brachiate at times. Orangutans use a similar form of locomotion called quadramanous climbing, in which they use their arms and legs to carry their heavy bodies through the trees. Chimpanzees and gorillas knuckle walk, and can move bipedally for short distances. Although numerous species, such as australopithecines and early hominids, have exhibited fully bipedal locomotion, humans are the only extant species with this trait.

Figure 42. Diademed sifaka, a lemur that is a vertical clinger and leaper.

18.32. BEHAVIOUR

Social Systems

Primates are among the most social of animals, forming pairs or family groups, uni-male harems, and multi-male/multi-female groups. Richard Wrangham stated that social systems of non-human primates are best classified by the amount of movement by females occurring between groups [Wrangham, 1982]. He proposed four categories:

- **Female transfer systems** – females move away from the group in which they were born. Females of a group will not be closely related whereas males will have remained with their natal groups, and this close association may be influential in social behavior. The groups formed are generally quite small. This organization can be seen in chimpanzees, where the males, who are typically related, will cooperate in defense of the group's territory. Among New World Monkeys, spider monkeys and muriquis use this system [Fiore and Campbell, 2007].

Figure 43. Japanese macaques bathing together in Jigokudani Hot Springs.

- **Male transfer systems** – while the females remain in their natal groups, the males will emigrate as adolescents. Polygynous and multi-male societies are classed in this category. Group sizes are usually larger. This system is common among the ring-tailed lemur, capuchin monkeys and cercopithecine monkeys.
- **Monogamous species** – a male–female bond, sometimes accompanied by a juvenile offspring. There is shared responsibility of parental care and territorial defense. The offspring leaves the parents' territory during adolescence. Gibbons essentially use this system, although "monogamy" in this context does not necessarily mean absolute sexual fidelity.
- **Solitary species** – often males who defend territories that include the home ranges of several females. This type of organization is found in the prosimians such as the slow loris. Orangutans do not defend their territory but effectively have this organization [Bartlett, 2007].

Other systems are known to occur as well. For example, with howler monkeys both the males and females typically transfer from their natal group on reaching sexual maturity, resulting in groups in which neither the males nor females are typically related. Some prosimians, colobine monkeys and callitrichid monkeys use this system.

Primatologist Jane Goodall, who studied in the Gombe Stream National Park, noted fission-fusion societies in chimpanzees. There is *fission* where the main group splits up to forage during the day, then *fusion* when the group returns at night to sleep as a group. This social structure can also be observed in the Hamadryas baboon, spider monkeys and the bonobo. The gelada has a similar social structure in which many smaller groups come together to form temporary herds of up to 600 monkeys [Constable et al., 2001].

These social systems are affected by three main ecological factors: distribution of resources, group size and predation. Within a social group there is a balance between cooperation and competition. Cooperative behaviors include social grooming (removing skin parasites and cleaning wounds), food sharing, and collective defense against predators or of a territory. Aggressive behaviors often signal competition for availability of food, sleeping sites or mates. Aggression is also used in establishing dominance hierarchies (Figure 43-44) [Smuts et al., 1987].

Figure 44. Chimpanzees are social animals.

18.33. Interspecific Associations

Several species of primates are known to associate in the wild. Some of these associations have been extensively studied. In the Tai Forest of Africa several species coordinate anti-predator behavior. These include the Diana monkey, Campbell's mona monkey, lesser spot-nosed monkey, western red colobus, king colobus and sooty mangabey, which coordinate anti-predator alarm calls. Among the predators of these monkeys is the common chimpanzee [Bshary, 2007].

The red-tailed monkey associates with several species, including the Western red colobus, blue monkey, Wolf's mona monkey, mantled guereza, black crested mangabey and Allen's swamp monkey. Several of these species are preyed upon by the common chimpanzee [Stanford, 1998].

In South America, squirrel monkeys associate with capuchin monkeys. This may have more to do with foraging benefits to the squirrel monkeys than anti-predation benefits [Boinski, 2000].

18.34. Cognition and Communication

Primates have advanced cognitive abilities: some make tools and use them to acquire food and for social displays; [Westergaard, 1998] some have sophisticated hunting strategies requiring cooperation, influence and rank; they are status conscious, manipulative and capable of deception; [Paar et al., 2000] they can recognise kin and conspecifics [Fujita et al., 1997] and they can learn to use symbols and understand aspects of human language including some relational syntax and concepts of number and numerical sequence [Itakura and Tanaka, 1998; Goutex et al., 2001; Call, 2001]. Research in primate cognition explores problem solving, memory, social interaction, a theory of mind, and numerical, spatial, and abstract concepts. Comparative studies show a trend towards higher intelligence going from prosimians to New World monkeys to Old World monkeys, and significantly higher average cognitive abilities in the great apes [Tomasello and Call, 1997; Reader et al. 2011]. However, there is a great deal of variation in each group (e.g., among New World monkeys, both spider and capuchin monkeys have scored highly by some measures), as well as in the results of different studies [Tomasello and Call, 1997; Reader et al. 2011].

Lemurs, lorises, tarsiers, and New World monkeys rely on olfactory signals for many aspects of social and reproductive behavior. Specialized glands are used to mark territories with pheromones, which are detected by the vomero-nasal organ; this process forms a large part of the communication behavior of these primates. In Old World monkeys and apes this ability is mostly vestigial, having regressed as trichromatic eyes evolved to become the main sensory organ [Liman and Innan, 2003]. Primates also use vocalizations, gestures, and facial expressions to convey psychological state [Egnor et al., 2004]. The Philippine tarsier, has a high-frequency limit of auditory sensitivity of approximately 91 kHz with a dominant frequency of 70 kHz. Such values are among the highest recorded for any terrestrial mammal, and a relatively extreme example of ultrasonic communication. For Philippine tarsiers, ultrasonic vocalizations might represent a private channel of communication that subverts detection by predators, prey and competitors, enhances energetic efficiency, or improves detection against low-frequency background noise.

18.35. Life History

Primates have slower rates of development than other mammals. All primate infants are breastfed by their mothers (with the exception of some human cultures and various zoo raised primates which are fed formula) and rely on them for grooming and transportation (Figure 45). In some species, infants are protected and transported by males in the group, particularly males who may be their fathers. Other relatives of the infant, such as siblings and aunts, may participate in its care as well. Most primate mothers cease ovulation while breastfeeding an infant; once the infant is weaned the mother can reproduce again. This often leads to weaning conflict with infants who attempt to continue breastfeeding. Infanticide is common in polygynous species such as gray langurs and gorillas. Adult males may kill dependent offspring that are not theirs so the female will return to estrus and thus they can sire offspring of their own. Social monogamy in some species may have evolved to combat this behavior [Opie et al., 2013].

Primates have a longer juvenile period between weaning and sexual maturity than other mammals of similar size. Some primates such as galagos and new world monkeys use tree-holes for nesting, and park juveniles in leafy patches while foraging. Other primates follow a strategy of "riding", i,e. carrying individuals on the body while feeding. Adults may construct and/or use nesting sites, sometimes accompanied by juveniles, for the purpose of resting, a behavior which has developed secondarily in the great apes [Kappeler, 1998]. During the juvenile period, primates are more susceptible than adults to predation and starvation; they gain experience in feeding and avoiding predators during this time. They learn social and fighting skills, often through playing. Primates, especially females, have longer lifespans than other similarly sized mammals. Late in life, female catarrhine primates appear to undergo a cessation of reproductive function known as menopause; other groups are less studied [Walker and Herndon, 2008].

Figure 45. A Crab-eating macaque breastfeeding her baby.

18.36. Diet, Feeding and Hunting

Primates exploit a variety of food sources. It has been said that many characteristics of modern primates, including humans, derive from an early ancestor's practice of taking most of its food from the tropical canopy [Milton, 1993]. Most primates include fruit in their diets to obtain easily digested carbohydrates and lipids for energy. However, they require other foods, such as leaves or insects, foramino acids, vitamins and minerals. Primates in the suborder Strepsirrhini (non-tarsier prosimians) are able to synthesize vitamin C, like most other mammals, while primates of the suborder Haplorrhini (tarsiers, monkeys and apes) have lost this ability, and require the vitamin in their diet (Figure 46) [Pollock and Mullin, 1986].

Many primates have anatomical specializations that enable them to exploit particular foods, such as fruit, leaves, gum or insects. For example, leaf eaters such as howler monkeys, black-and-white colobuses and sportive lemurs have extended digestive tracts which enable them to absorb nutrients from leaves that can be difficult to digest. Marmosets, which are gum eaters, have strong incisor teeth, enabling them to open tree bark to get to the gum, and claws rather than nails, enabling them to cling to trees while feeding. The aye-aye combines rodent-like teeth with a long, thin middle finger to fill the same ecological niche as a woodpecker. It taps on trees to find insect larvae, then gnaws holes in the wood and inserts its elongated middle finger to pull the larvae out [Milliken et al., 1991]. Some species have additional specializations. For example, the grey-cheeked mangabey has thick enamel on its teeth, enabling it to open hard fruits and seeds that other monkeys cannot. The gelada is the only primate species that feeds primarily on grass (Figure 47).

18.37. Hunting

Tarsiers are the only extant obligate carnivorous primates, exclusively eating insects, crustaceans, small vertebrates and snakes (including venomous species) [Wright et al., 2003]. Capuchin monkeys can exploit many different types of plant matter, including fruit, leaves, flowers, buds, nectar and seeds, but also eat insects and other invertebrates, bird eggs, and small vertebrates such as birds, lizards, squirrels and bats. Crab-eating macaques typically do not consume crabs; although fruits and seeds make up 60 - 90% of their diet, they sometimes prey on vertebrates including bird chicks, nesting female birds, lizards, frogs and fish.

Figure 46. Leaf eating mantled guereza, a species of black-and-white colobus.

Figure 47. Foraging crab-eating macaques temporarily store food in their cheek pouches.

The common chimpanzee has a varied diet that includes predation on other primate species, such as the western red colobus monkey. This sometimes involves tool use. Common chimpanzees sharpen sticks to use as weapons when hunting mammals. This is considered the first evidence of systematic use of weapons in a species other than humans. Researchers documented 22 occasions where wild chimpanzees fashioned sticks into "spears" to hunt lesser bush babies (*Galago senegalensis*). In each case, chimpanzee modified a branch by breaking off one or two ends and, frequently using its teeth, sharpened the stick. The tools, on average, were about 60 cm (24 in) long and 1.1 cm (0.4 in) in circumference. The chimpanzee then jabbed the spear into hollows in tree trunks where bush babies sleep. There was a single case in which a chimpanzee successfully extracted a bush baby with the tool. The bonobo is an omnivorous frugivore. The majority of its diet is fruit, but supplements its diet with leaves, meat from small vertebrates such as anomalures, flying squirrels and duikers [Ihobe , 1992] and invertebrates [Rafert and Vineberg, 1997]. In some instances, bonobos have been shown to consume lower-order primates [Surbeck and Hohmann, 2008; Surbeck et al., 2009].

18.38. As Prey

Predators of primates include various species of carnivores, birds of prey, reptiles and other primates. Even gorillas have been recorded as prey. Predators of primates have diverse hunting strategies and as such, primates have evolved several different anti predator adaptations including crypsis, alarm calls and mobbing. Several species have separate alarm calls for different predators such as air-borne or ground-dwelling predators. Predation may have shaped group size in primates as species exposed to higher predation pressures appear to live in larger groups [Fichtel, 2012]. With their technology and increased intelligence, modern humans are nearly free of threats from predators and are themselves apex predators.

18.39. Tool Use and Manufacture

Tool Use

There are many reports of primates using tools, both in the wild or when captive. The use of tools by primates is varied and includes hunting (mammals, invertebrates, fish), collecting honey, processing food (nuts, fruits, vegetables and seeds), collecting water, weapons and shelter.

In 1960, Jane Goodall observed a chimpanzee poking pieces of grass into a termite mound and then raising the grass to his mouth. After he left, Goodall approached the mound and repeated the behaviour because she was unsure what the chimpanzee was doing. She found that the termites bit onto the grass with their jaws. The chimpanzee had been using the grass as a tool to "fish" or "dip" for termites. There are more limited reports of the closely related bonobo using tools in the wild; it has been claimed they rarely use tools in the wild although they use tools as readily as chimpanzees when in captivity. It has been reported that both female chimpanzees and bonobos use tools more avidly than males [Gruber et al., 2010]. Orangutans in Borneo scoop catfish out of small ponds. Anthropologist Anne Russon saw several animals on these forested islands learn on their own to jab at catfish with sticks, so that the panicked prey would flop out of ponds and into the orangutan's waiting hands. There are few reports of gorillas using tools in the wild. An adult female Western lowland gorilla used a branch as a walking stick apparently to test water depth and to aid her crossing a pool of water. Another adult female used a detached trunk from a small shrub as a stabilizer during food gathering, and another used a log as a bridge (Figure 48) [Breuer et al., 2005].

The black-striped capuchin was the first non-ape primate for which tool use was documented in the wild. Individuals were observed cracking nuts by placing them on a stone anvil and hitting them with another large stone [Fragaszy et al., 2004]. In Thailand and Myanmar, crab-eating macaques use stone tools to open nuts, oysters and other bivalves, and various types of sea snails [Gumert et al., 2009]. Chacma baboons use stones as weapons; stoning by these baboons is done from the rocky walls of the canyon where they sleep and retreat when they are threatened. Stones are lifted with one hand and dropped over the side whereupon they tumble down the side of the cliff or fall directly to the canyon floor [Hamilton et al., 1975].

Figure 48. A gorilla using a stick possibly to gauge the depth of water.

Although they have not been observed to use tools in the wild, lemurs in controlled settings have been shown to be capable of understanding the functional properties of the objects they had been trained to use as tools, performing as well as tool-using haplorrhines [Fichtel and Kappeler, 2010].

18.40. Tool Manufacture

Tool manufacture is much rarer than simple tool use and probably represents higher cognitive functioning. Soon after her initial discovery of tool use, Goodall observed other chimpanzees picking up leafy twigs, stripping off the leaves and using the stems to fish for insects. This change of a leafy twig into a tool was a major discovery. Prior to this, scientists thought that only humans manufactured and used tools, and that this ability was what separated humans from other animals [van Schaik, 1996]. Both bonobos and chimpanzees have also been observed making "sponges" out of leaves and moss that suck up water and are used as grooming tools. Sumatran orangutans have been observed making and using tools. They will break off a tree branch that is about 30 cm long, snap off the twigs, fray one end and then use the stick to dig in tree holes for termites. In the wild, mandrills have been observed to clean their ears with modified tools. Scientists filmed a large male mandrill at Chester Zoo (UK) stripping down a twig, apparently to make it narrower, and then using the modified stick to scrape dirt from underneath its toenails. Captive gorillas can made a variety of tools [van Schaik, 1996].

Habitat and Distribution

Primates evolved from arboreal animals, and many species live most of their lives in trees. Most primate species live in tropical rain forests. The number of primate species within tropical areas has been shown to be positively correlated to the amount of rainfall and the amount of rain forest area [Reed and Fleagle, 1995]. Accounting for 25% to 40% of the fruit-eating animals (by weight) within tropical rainforests, primates play an important ecological role by dispersing seeds of many tree species.

Some species are partially terrestrial, such as baboons and patas monkeys, and a few species are fully terrestrial, such as geladas and humans. Non-human primates live in a diverse number of forested habitats in the tropical latitudes of Africa, India, Southeast Asia, and South America, including rainforests, mangrove forests, and montane forests. There are some examples of non-human primates that live outside of the tropics. The mountain-dwelling Japanese macaque lives in the north of Honshū where there is snow-cover eight months of the year; the Barbary macaque lives in the Atlas Mountains of Algeria and Morocco. Primate habitats span a range of altitudes: the black snub-nosed monkey has been found living in the Hengduan Mountains at altitudes of 4,700 meters (15,400 ft) [Long et al., 1994], the mountain gorilla can be found at 4,200 meters (13,200 ft) crossing the Virunga Mountains [Schaller, 1963] and the gelada has been found at elevations of up to 5,000 m (16,000 ft) in the Ethiopian Highlands. Although most species are generally shy of water, a few are good swimmers and are comfortable in swamps and watery areas, including the proboscis monkey, De Brazza's monkey and Allen's swamp monkey, which has developed small webbing between its fingers. Some primates, such as the rhesus macaque and gray langurs, can exploit human-modified environments and even live in cities (Figure 49) [Wolfe and Fuentes, 2007].

18.41. INTERACTIONS BETWEEN HUMAN AND PRIMATES

Close interactions between humans and non-human primates (NHPs) can create pathways for the transmission of zoonotic diseases. Viruses such as *Herpesviridae* (most notably Herpes B Virus), *Poxviridae*, measles, ebola, rabies, the Marburg virus and viral hepatitis can be transmitted to humans. In some cases the viruses produce potentially fatal diseases in both humans and non-human primates [Renquits and Whiteny, 1987].

Figure 49. Rhesus macaque at Agra Fort, India.

Figure 50. Capuchin monkeys' manual dexterity is one reason they can assist quadriplegic humans.

18.42. Legal and Social Status

Only humans are recognized as persons and protected in law by the United Nations Universal Declaration of Human Rights. The legal status of NHPs, on the other hand, is the subject of much debate, with organizations such as the Great Ape Project (GAP) campaigning to award at least some of them legal rights. In June 2008, Spain became the first country in the world to recognize the rights of some NHPs when its parliament's cross-party environmental committee urged the country to comply with GAP's recommendations, which are that chimpanzees, bonobos, orangutans, and gorillas not be used for animal experiments.

Many species of NHP are kept as pets by humans, the Allied Effort to Save Other Primates (AESOP) estimates that around 15,000 NHPs live as exotic pets in the United States. The expanding Chinese middle class has increased demand for NHPs as exotic pets in recent years [Workmam, 2004]. Although NHP import for the pet trade was banned in the U.S. in 1975, smuggling still occurs along the United States – Mexico border, with prices ranging from US$3000 for monkeys to $30,000 for apes.

Primates are used as model organisms in laboratories and have been used in space missions [Bushnell, 1958]. They serve as service animals for disabled humans. Capuchin monkeys can be trained to assist quadriplegic humans. Their intelligence, memory, and manual dexterity make them ideal helpers (Figure 50).

NHPs are kept in zoos around the globe. Historically, zoos were primarily a form of entertainment, but more recently have shifted their focus to conservation, education and research. Many zoos now feature naturalistic exhibits and educational material for the public. In the United States many participate in the Species Survival Plan (SSP), developed by the Association of Zoos and Aquariums (AZA), to maximize genetic diversity through captive breeding. Zoos and other animal welfare supporters generally oppose animal rights initiatives and the GAP's insistence that all NHPs be released from captivity for two primary reasons. First, captive-born primates lack the knowledge and experience to survive in the wild if released. Second, zoos provide living space for primates and other animals threatened with extinction in the wild.

18.43. Role in Scientific Research

Thousands of non-human primates are used around the world in research because of their psychological and physiological similarity to humans [Chen and Li, 2001]. In particular, the brains and eyes of NHPs more closely parallel human anatomy than those of any other animals. NHPs are commonly used in preclinical trials, neuroscience, ophthalmology studies, and toxicity studies [Conlee et al., 2004]. In 2005, GAP reported that 1,280 of the 3,100 NHPs living in captivity in the United States were used for experiments. In 2004, the European Union used around 10,000 NHPs in such experiments; in 2005 in Great Britain, 4,652 experiments were conducted on 3,115 NHPs. Governments of many nations have strict care requirements of NHPs kept in captivity. In the US, federal guidelines extensively regulate aspects of NHP housing, feeding, enrichment, and breeding. European groups such as the European Coalition to End Animal Experiments are seeking a ban on all NHP use in experiments as part of the European Union's review of animal testing legislation.

18.44. CONSERVATION

The International Union for Conservation of Nature (IUCN) lists more than a third of primates as critically endangered or vulnerable. Trade is regulated, as all species are listed by CITES in Appendix II, except 50 species and subspecies listed in Appendix I, which gain full protection from trade. Common threats to primate species include deforestation, forest fragmentation, monkey drives (resulting from primate crop raiding) [Grubb, 1998] and primate hunting for use in medicines, as pets, and for food. Large-scale tropical forest clearing is widely regarded as the process that most threatens primates [Chapman and Peres, 2001; Southwick and Siddqi, 2001]. More than 90% of primate species occur in tropical forests [Cowlisha and Dunbar, 2000]. The main cause of forest loss is clearing for agriculture, although commercial logging, subsistence harvesting of timber, mining, and dam construction also contribute to tropical forest destruction. In Indonesia large areas of lowland forest have been cleared to increase palm oil production, and one analysis of satellite imagery concluded that during 1998 and 1999 there was a loss of 1,000 Sumatran orangutans per year in the Leuser Ecosystem alone (Figure 51) [van Schaik et al., 2001].

Primates with a large body size (over 5 kg) are at increased extinction risk due to their greater profitability to poachers compared to smaller primates. They reach sexual maturity later and have a longer period between births. Populations therefore recover more slowly after being depleted by poaching or the pet trade [Purvis et al., 2000]. Data for some African cities show that half of all protein consumed in urban areas comes from the bushmeat trade. Endangered primates such as guenons and the drill are hunted at levels that far exceed sustainable levels. This is due to their large body size, ease of transport and profitability per animal [Fa et al., 1995]. As farming encroaches on forest habitats, primates feed on the crops, causing the farmers large economic losses. Primate crop raiding gives locals a negative impression of primates, hindering conservation efforts [Hill 1997, 2002].

Figure 51. The critically endangered Sumatran orangutan.

Madagascar, home to five endemic primate families, has experienced the greatest extinction of the recent past. Since human settlement 1,500 years ago, at least eight classes and fifteen of the larger species have become extinct due to hunting and habitat destruction.

Among the primates wiped out were, *Archaeoindris* (a lemur larger than a silverback gorilla) and the families Palaeopropithecidae and Archaeolemuridae.

In Asia, Hinduism, Buddhism, and Islam prohibit eating primate meat; however, primates are still hunted for food. Some smaller traditional religions allow the consumption of primate meat [Choudhury, 2001; Kumar and Singh, 2004]. The pet trade and traditional medicine also increase demand for illegal hunting [Nijman, 2004; O'Brien et al., 2004]. The rhesus macaque, a model organism, was protected after excessive trapping threatened its numbers in the 1960s. The program was so effective that they are now viewed as a pest throughout their range.

In Central and South America, forest fragmentation and hunting are the two main problems for primates. Large tracts of forest are now rare in Central America [Estrada et al., 1994]. This increases the amount of forest vulnerable to edge effects such as farmland encroachment, lower levels of humidity and a change in plant life [Turner, 1996; Marsh, 2003]. Movement restriction results in a greater amount of inbreeding, which can cause deleterious effects leading to a population bottleneck, whereby a significant percentage of the population is lost [Pope, 1996; Chiarello, 2003].

There are 21 critically endangered primates, 7 of which have remained on the IUCN's "The World's 25 Most Endangered Primates" list since the year 2000: the silky sifaka, Delacour's langur, the white-headed langur, the gray-shanked douc, the Tonkin snub-nosed monkey, the Cross River gorilla and the Sumatran orangutan (Figure 52) [Mittermier et al., 2009]. Miss Waldron's red colobus was recently declared extinct when no trace of the subspecies could be found from 1993 to 1999 [Oates et al., 2000]. A few hunters have found and killed individuals since then, but the subspecies' prospects remain bleak [McGraw, 2005].

Figure 52. The critically endangered Cross River gorilla.

REFERENCES

Agrawal, V. C. & Chakraborty, S. (1970). Occurrence of the Woolly Flying Squirrel, Eupetaurus cinereus Thomas (Mammalia: Rodentia: Sciuridae) in North Sikkim. *J Bom Nat Hist Soc*, *66*, 615–616.

Aiello, L. C. & Dean, M. C. (1990). An Introduction to Human Evolutionary Anatomy. Academic Press, London.

Arrese, C. A. et al.. (2005). Cone topography and spectral sensitivity in two potentially trichromatic marsupials, the quokka (*Setonix brachyurus*) and quenda (*Isoodon obesulus*). *Proc Biol Sci*, *272*, 791–796.

Ash, M. M., Nelson, S. J. & Wheeler, R. C. (2003). Wheeler's Dental Anatomy, Physiology, and Occlusion. W.B. Saunders. p. 12.

Bartlett, T. Q. (2007). The Hylobatidae. In Campbell, C.J., Fuentes, A., MacKinnon, K.C., Panger, M. & Bearder, S.K. *Primates in Perspective*. Oxford University Press. pp. 283.

Baverstock, P. R. (1984). The Molecular Relationships of Australasian Possums and Gliders. In *Possums and Gliders*, ed. Smith, A. & Hume, I. 1–8. Sydney: Surrey Beatty.

Beard, K. C. (1990). Gliding Behaviour and Palaeoecology of the Alleged Primate Family Paromomyidae (Mammalia, Dermoptera). *Nature*, *345*, 340–341.

Beard, K. C. (1991). Vertical Postures and Climbing in the Morphotype of Primatomorpha: Implications for Locomotor Evolution in Primate History. In *Origine(s) de la Bipedie chez*

Beck, R. M. D., Godthelp, H., Weisbecker, V., Archer, M. & Hand, S. J. (2008). Australia's oldest marsupial fossils and their biogeographical implications. In Hawks, *J. PLoS ONE*, *3*, e1858.

Benton, M. J. (1997). Vertebrate Palaeontology. London: Chapman & Hall. 303–304.

Benton, M. J. (2004). *Vertebrate Palaeontology,* pp. 300.

Bernsteil, I. S. (1966). Naturally occurring primate hybrid. *Science*, *154*, 1559–1560.

Biggers, J. D. (1966). Reproduction in male marsupials. Comparative biology of reproduction in mammals (Rowlands, I.W. ed.). Academic Press, New York, 251-280.

Blainville, H. (1839). "Nouvelle classification des Mammifères". *Annales Françaises et Etrangères d'Anatomie et de Physiologie Appliquées à la Médicine et à l'Histoire Naturelle, 3*. 268–269.

Boinski, S. (2000). Social Manipulation Within and Between Troops Mediates Primate Group Movement. In: Boinski, S. & Garber, P. *On the Move : how and why animals travel in groups*. University of Chicago Press. 447–448.

Breuer, T., Ndoundou-Hockemba, M. & Fishlock, V. (2005). First observation of tool use in wild gorillas. *PLoS Biol*, *3*, e380.

Brown, M., Cooksley, H., Carthew, S. M. & Coopers, S. J. B. (2006). Conservation Units and Phylogenetic Structure of an Arboreal Marsupial, the Yellow-Bellied Glider (*Petaurus australis*). *Aust J Zool*, *54*, 305–317.

Bshary, R. (2007). Interactions between Red Colobus Monkeys and Chimpanzees. In McGraw, W., Zuberbuhler, K. & Noe, R. *Monkeys of Tai Forest, An African Primate Community*. Cambridge University Press. 155–170.

Bushnell, D. (1958). The Beginnings of Researcin Space Biology at the Air Force Missile Development Center, 1946–1952. *Hist Res Sp Biol Biodynam* NASA.

Butler, A. B. & William, H. (2005). Comparative Vertebrate Neuroanatomy: Evolution and Adaptation, p. 361

Call, J. (2001). Object permanence in orangutans (*Pongo pygmaeus*), chimpanzees (*Pan troglodytes*), and children (*Homo sapiens*). *J Comp Psychol*, *115*, 159–171.

Campbell, B. G. & Loy, J. D. (2000). Humankind Emerging (8th edition). Allyn & Bacon. p. 85.

Cartmill, M. & Smith, F. H. (2011). *The Human Lineage*. John Wiley & Sons.

Chapman, C. A. & Peres, C. A. (2001). Primate conservation in the new millennium: the role of scientists. *Evol Anthropol*, *10*, 16–33.

Chatterjee, H. J., Simon Y. W. H, Barnes, I. & Groves, C. (2009). Estimating the phylogeny and divergence times of primates using a supermatrix approach. *BMC Evol Biol*, *9*, 259.

Chen, F. C. & Li, W. H. (2001). Genomic divergences between humans and other hominoids and the effective population size of the common ancestor of humans and chimpanzees. *Am J Hum Gen*, *68*, 444–456.

Chiarello, A. G. (2003). Primates of the Brazilian Atlantic forest: the influence of forest fragmentation on survival. In Marsh, L.K. *Primates in Fragments: Ecology and Conservation*. New York: Kluwer Academic/Plenum Publishers. 99–121.

Choudhury, A. (2001). Primates in Northeast India: an overview of their distribution and conservation status. *En Bull: Wild Protect Ar*, *1*, 92–101.

Conley, A. J., Corbin, C. J., Mapes, S. M. & Heffelfinger, J. R. (2004). Pigs, peccaries and the evolution of aromatase. *Biol Reprod*, *102*.

Constable, J. L. et al. (2001). Noninvasive paternity assignment in Gombe chimpanzees. *Mol Ecol*, *10*, 1279–1300.

Cowlishaw, G. & Dunbar, R. (2000). Primate Conservation Biology. Chicago: University of Chicago Press.

Crook, J. H. (1972). Sexual selection, dimorphism, and social organization in the primates. In Campbell, B.G. *Sexual selection and the descent of man*. Aldine Transaction. p. 246.

Crosby, K., Bassarova, M., Archer, M. & Carbery, K. (2004). Fossil Possums in Australasia: Discovery, Diversity and Evolution. In: Biology of Possums and Gliders, ed. Goldingay, R.L. & Jackson, S.M. pp. 161–176. Sydney: Surrey Beatty.

Dawson, T. J., Grant, T. R. & Fanning, D. (1979). Standard Metabolism of Monotremes and the Evolution of Homeothermy. *Aust J Zool*, *27*, 511–515.

Di Fiore, A. & Campbell, C. J. (2007). The Atelines: variation in ecology, behavior, and social organization. In: Primates in Perspective, Campbell, C.J., Fuentes, A., MacKinnon, K.C., Panger, M. & Bearder, S.K. (eds.), Oxford University Press, New York, 155–185..

Dixson, A. F. (1981). The Natural History of the Gorilla, London: Weidenfeld & Nicholson.

Ducrocq, S., Buffetaut, E., Buffetaut-Tong, H., Jaeger, J. J., Jortgkanjanasoontorn, Y. & Suteethorn, V. (1992). First Fossil Flying Lemur: A Dermopteran from the Late Eocene of Thailand. *Palaeontol*, *35*, 373–380.

Egnor, R., Miller, C. & Hauser, M. D. (2004). Nonhuman Primate Communication. *Encycl Lang Ling* (2nd ed.). Elsevier.

Estrada, A., Coates-Estrada, R. & Meritt, D. (1994). Non-flying mammals and landscape changes in the tropical forest region of Los Tuxtlas, Mexico. *Ecography*, *17*, 229–241.

Fa, J. E., Juste, J., Perez de Val, J. & Castroviejo, J. (1995). Impact of market hunting on mammal species in Equatorial Guinea. *Conserv Biol*, *9*, 1107–1115.

Fichtel, C. & Kappeler, P. M. (2010). Chapter 19: Human universals and primate symplesiomorphies: Establishing the lemur baseline. In: Kappeler, P.M. & Silk, J.B. *Mind the Gap: Tracing the Origins of Human Universals*. Springer.

Fichtel, C. (2012). Predation. In Mitani, J.C., Call, J., Kappeler, P.M., Palombit, R.A. & Silk, J.B. *The Evolution of Primate Societies*. University of Chicago Press. 169–84.

Fragaszy, D., Izar, P., Visalberghi, E., Ottoni, E.B. & de Oliveira, M.G. (2004). Wild capuchin monkeys (Cebus libidinosus) use anvils and stone pounding tools. *Am J Primatol*, *64*, 359–366.

Freyer, C. & Renfree, M. B. (2009). The mammalian yolk sac placenta. *J Exp Zool Part B: Mol Dev Evol*, *312B*, 545–554.

Frisch, J. E. (1963). Sex-differences in thecanines of the gibbon (*Hylobates lar*). *Primates*, *4*, 1–10.

Fujita, K., Watanabe, K., Widarto, T. H. & Suryobroto, B. (1997). Discrimination of macaques by macaques: The case of sulawesi species. *Primates*, *38*, 233–245.

Goodman, M., Tagle, D. A., Fitch, D. H., Bailey, W., Czelusniak, J., Koop, B. F., Benson, P. & Slightom, J. L. (1990). Primate evolution at the DNA level and a classification of hominoids. *J Mol Evol*, *30*, 260–266.

Gouteux, S., Thinus-Blanc, C. & Vauclair, J. (2001). Rhesus monkeys use geometric and nongeometric information during a reorientation task. *J Exp Psychol: Gen*, *130*, 505–519.

Groves, C. P. (2005). Wilson, D.E. & Reeder, D.M. eds. *Mammal Species of the World* (3rd ed.). Baltimore: Johns Hopkins University Press. 1–2.

Grubb, P. (1998). The Sierra Leone monkey drives. Mammals of Ghana, Sierra Leone, and the Gambia. St. Ives: Trendrine. 214–219.

Gruber, T., Clay, Z. & Zuberbühler, K. (2010). A comparison of bonobo and chimpanzee tool use: evidence for a female bias in the *Pan* lineage. *Anim Behav*, *80*, 1023-1033.

Gumert, M. D., Kluck, M. & Malaivijitnond, S. (2009). The physical characteristics and usage patterns of stone axe and pounding hammers used by long-tailed macaques in the Andaman Sea region of Thailand. *Am J Primatol*, *71*, 594–608.

Hamilton, W. J., Buskirk, R. E. & Buskirk, W. H. (1975). Defensive stoning by baboons. *Nature*, *256*, 488-489.

Hand, S. J., Long, J., Archer, M. & Flannery, T. (2002). Prehistoric mammals of Australia and New Guinea: one hundred million years of evolution. Baltimore: Johns Hopkins University Press.

Hartwig, W. (2007). Primate Evolution. In Campbell, C., Fuentes, A., MacKinnon, K., Panger, M. & Bearder, S. *Primates in Perspective*. Oxford University Press. 13–17.

Hartwig, W. (2011). Primate evolution". In Campbell, C.J., Fuentes, A., MacKinnon, K.C., Bearder, S.K. & Stumpf, R.M. *Primates in Perspective* (2nd ed.). Oxford University Press. 19–31.

Heissig, K. (1979). The Earliest Flying Squirrels and Primitive Ailuravinae (Rodentia, Mammalia) from the South German Oligocene. *Mitteilungen der Bayerischen Staatssammlung für Paläontologie und Historische Geologie*, *19*, 139–170.

Helgen, K. M. (2003). The Feather-Tailed Glider (*Acrobates pygmeus* [sic]) in New Guinea. *Treubia*, *33*, 107–111.

Hill, C. M. (1997). Crop-raiding by wild vertebrates: The farmer's perspective in an agricultural community in western Uganda. *Int J Pest Manag*, *43*, 77–84.

Hill, C. M. (2002). Primate conservation and local communities: Ethical issues and debates. *Am Anthr, 104*, 1184–1194.

Houle, A. (1999). The origin of platyrrhines: An evaluation of the Antarctic scenario and the floating island model. *Am J Phy Anthr, 109*, 541–559.

Hugall, A. F. et al. (2007). Calibration choice, rate smoothing, and the pattern of tetrapod diversification according to the long nuclear gene RAG-1. *Syst Biol, 56*, 543-63.

Ihobe, H. (1992). Observations on the meat-eatinbehavior of wild bonobos (*Pan paniscus*) at Wamba, Republic of Zaire. *Primates, 33*, 247–250.

Inns, R. W. (1982). Seasonal changes in the accessory reproductive system and plasma testosterone levels of the male tammar wallaby, *Macropus eugenii*, in the wild. *J Repr Fert, 66*, 675–80.

Itakura, S. & Tanaka, M. (1998). Use of experimenter-given cues during object-choice tasks by chimpanzees (*Pan troglodytes*), an orangutan (*Pongo pygmaeus*), and human infants (*Homo sapiens*). *J Comp Psychol, 112*, 119–126.

Janečka, J. E., Miller, W., Pringle, T. H., Wiens, F., Zitzmann, A., Helgen, K. M., Springer, M. S. & Murphy, W. J. (2007). Molecular and Genomic Data Identify the Closest Living Relative of Primates. *Science, 318*, 792–794.

Ji, Q., Luo, Z. X., Yuan, C. X., Wible, J. R., Zhang, J. P. & Georgi, J. A. (2002). The earliest known eutherian mammal. *Nature, 416*, 816–822.

Kappeler, P. M. (1998). Nests, Tree Holes, and the Evolution of Primate Life Histories. *Am J Primatol* (Wiley-Liss, Inc.), *46*, 7–33.

Kavanagh, M. (1983). A Complete Guide to Monkeys, Apes and Other Primates. New York: Viking Press. p. 18.

Kay, R. F. (1975). The functional adaptations ofprimate molar teeth. *Am J Phy Anthr, 43*, 195–215.

Kay, R. F. (2012). Evidence for an Asian origin of stem anthropoids. *Proc Nat Acad Sci USA, 109*, 10132–10133.

Kay, R. F., Thorington, R. W. Jr. & Houde, P. (1990). Eocene Plesiadapiform Shows Affinities with Flying Lemurs Not Primates. *Nature, 345*, 342–343.

Kemp, T. S. (2005). The origin and evolution of mammals. Oxford : Oxford University Press.

Kumara, H. N. & Singh, M. (2004). Distribution and abundance of primates in rainforests of the Western Ghats, Karnataka, India and the conservation of *Macaca silenus*. *Int J Primatol, 25*, 1001–1018.

Lee, M. (1999). Molecular Clock Calibrations and Metazoan Divergence Dates. *J Mol Evol, 49*, 385–391.

les Hominides, ed. Coppens, Y. & Senut, B. pp. 79–87. Cahiers de Paleoanthropologie. Paris: Editions du CNRS.

Liman, E. R. & Innan, H. (2003). Relaxed selective pressure on an essential component of pheromone transduction in primate evolution. *Proc Nat Acad Sci USA, 100*, 3328–3332.

Lindstedtand, S. L. & Boyce, M. S. (1985). Seasonality, Fasting Endurance, and Body Size in Mammals. *The American Naturalist, 125* (6), 873-878.

Linnaeus, C. (1735). *Sistema naturae sive regna tria Naturae systematice proposita per classes, ordines, genera, & species*. apud Theodorum Haak, Lugduni Batavorum. pp. s.p.

Linnaeus, C. (1758). *Sistema naturae per regna tria Naturae, secundum classes, ordines, genera, species, cum characteribus differentiis, synonimis locis. Tomus I.* Impensis direct. Laurentii Salvii, Holmia. 20–32.

Long, Y. C., Kirkpatrick, R. C., Zhong, T. & Xiao, L. (1994). Report on the distribution, population, and ecology of the Yunnan snub-nosed monkey (*Rhinopithecus bieti*). *Primates*, *35*, 241–250.

Lucas, P. W. & Dominy, N. J. (2003). Evolutioand function of routine trichromatic vision in primates. *Evolution*, *57*, 2636–43.

Luckett, W. P. (1977). Ontogeny of amniote fetal membranes and their application to phylogeny. Major patterns in Vertebrate Evolution. New York, London: Plenum Publishing Corporation. 439-516

Luo, Z. X., Cifelli, R. L. & Kielan-Jaworowska, Z. (2002). In quest for a phylogeny of Mesozoic mammals. *Acta Palaeontol Pol*, *47*, 1–78.

Luo, Z. X., Cifelli, R. L. & Kielan-Jaworowska, Z. (2001). Dual origin of tribosphenic mammals. *Nature*, *409*, 53–57.

Luo, Z. X., Yuan, C. X., Meng, Q. J. & Ji, Q. (2011). A Jurassic eutherian mammal and divergence of marsupials and placentals. *Nature*, *476*, 442–445.

Luo, Z., Yuan, C., Meng, Q. & Ji, Q. (2011). A Jurassic eutherian mammal and divergence of marsupials and placentals. *Nature*, *476*, 42–45.

Marsh, L.K. (2003). The nature of fragmentation. In: Marsh, L.K. *Primates in Fragments: Ecology and Conservation*. New York: Kluwer Academic/Plenum Publishers. 1–10.

McGraw, W. S. (2005). Update on the search for Miss Waldron's red colobus monkey. *Int J Primatol*, *26*, 605–619.

McKenna, M. C. & Bell, S. K. (1997). Classification of Mammals Above the Species Level. New York: Columbia University Press. 631 pp.

McKenna, M. C. & Bell, S. K. (1997). Classification of Mammals: Above the species level. New York: Columbia University Press. p. 631.

Milliken, G. W., Ward, J. P. & Erickson, C. J. (1991). Independent digit control in foraging by the aye-aye (*Daubentonia madagascariensis*). *Folia Primatol*, *56*, 219–224.

Milton, K. (1993). Diet and Primate Evolution. *Sci Am*, *269*, 86–93.

Mishler, B. D. (2009). Species are not Uniquely Real Biological Entities, in Ayala, F.J. & Arp, R. *Contemp Deb Philos Biol*, 110–122.

Mittermeier, R. A., Wallis, J., Rylands, A. B., Ganzhorn, J. U., Oates, J. F., Williamson, E. A., Palacios, E., Heymann, E. W., Kierulff, M. C. M. & Supriatna, L. eds. (2009). *Primates in Peril: The World's 25 Most Endangered Primates 2008–2010*. Illustrated by Nash, S.D. Arlington, VA.: IUCN/SSC Primate Specialist Group (PSG), International Primatological Society (IPS), and Conservation International (CI). 23–26.

Nijman, V. (2004). Conservation of the Javan gibbon *Hylobates moloch*: population estimates, local extinction, and conservation priorities. *The Raff Bull Zool*, *52*, 271–280.

Oates, J. F., Abedi-Lartey, M., McGraw, W. S., Struhsaker, T. T. & Whitesides, G. H. (2000). Extinction of a West African Red Colobus Monkey. *The J Soc Conserv Biol*, *14*, 1526–1532.

O'Brien, T. G., Kinnaird, M. F., Nurcahyo, A., Iqbal, M. & Rusmanto, M. (2004). Abundance and distribution of sympatric gibbons in a threatened Sumatran rain forest. *Int J Primatol*, *25*, 267–284.

O'Leary, M. A., Bloch, J. I., Flynn, J. J., Gaudin, T. J., Giallombardo, A. G., Norberto, P., Goldberg, S. L., Kraatz, B. P., Luo, Z. X., Meng, J., Ni, M. J., Novacek, F. A., Perini, Z. S., Randall, G., Rougier, E. J., Sargis, M. T., Silcox, N., Simmons, M., Spaulding, P. M.,

Velazco, M., Weksler, J., Wible, A. L. & Cirranello, A. L. (2013). The Placental Mammal Ancestor and the Post–K-Pg Radiation of Placentals. *Science*, *339*, 662–667.
Opie, C., Atkinson, Q. D., Dunbar, R. I. M. & Shultz, S. (2013). Male infanticide leads to social monogamy in primates. *Proc Nat Acad Sci USA*, *110*, 13328-13332.
Paar, L. A., Winslow, J. T., Hopkins, W. D. & de Waal, F. B. M. (2000). Recognizing facial cues: Individual discrimination by chimpanzees (*Pan troglodytes*) and rhesus monkeys (*Macaca mulatta*). *J Comp Psychol*, *114*, 47–60.
Paddle, R. (2002). The last Tasmanian tiger : the history and extinction of the thylacine (Paperback ed.). Port Melbourne, Vic.: Cambridge University Press
Phillips, M. J., Bennett, T. H. & Lee, M. S. (2009). Molecules, morphology, and ecology indicate a recent, amphibious ancestry for echidnas. *Proc Nat Acad Sci USA*, *106*, 17089–17094.
Plavcan, J. M. (2001). Sexual dimorphism in primate evolution". *Am J Phy Anthr*, *33*, 25–53.
Pollock, J. I. & Mullin, R. J. (1986). Vitamin C biosynthesis in prosimians: Evidence for the anthropoid affinity of Tarsius. *Am J Phy Anth*, *73*, 65–70.
Pope, T. R. (1996). Socioecology, population fragmentation, and patterns of genetic loss in endangered primates. In Avise, J. & Hamrick, J. *Conservation Genetics: Case Histories from Nature*. Norwell: Kluwer Academic Publishers. 119–159.
Pough, F. W., Janis, C. M. & Heiser, J. B. (2005). Characteristics of Primates. *Vertebrate Life* (7th ed.). Pearson. p. 630.
Pough, F. W., Janis, C. M. & Heiser, J. B. (2005). Primate Evolution and the Emergence of Humans. *Vertebrate Life* (7th ed.). Pearson, pp. 650.
Purvis, A., Gittleman, J. L., Cowlishaw, G. & Mace, G. M. (2000). Predicting extinction risk in declining species. *Proc R Soc Lon*, *267*, 1947–1952.
Rafert, J. & Vineberg, E. O. (1997). Bonob Nutrition – relation of captive diet to wild diet, *Bonobo Husbandry Manual*, American Association of Zoos and Aquariums.
Rayner, J.M.V. (1981). Flight Adaptations in Vertebrates. *Symp Zool Soc Lon*, *48*, 137–172.
Reader, S. M., Hager, Y. & Laland, K .N. (2011). The evolution of primate general and cultural intelligence. *Philos Trans R Soc B: Biol Sci*, *366*, 1017–1027.
Reed, K. & Fleagle, J. (1995). Geographic and climatic control of primate diversity. *Proc Nat Acad Sci USA*, *92*, 7874–7876.
Reidenberg, J. S. (2007). Anatomical Adaptations of Aquatic Mammals. *Anat Rec (Hoboken)*, 290, 507–13.
Renfree, M. & Tyndale-Biscoe, H. (1987). Reproductive Physiology of Marsupials. Cambridge University Press.
Renquist, D. M. & Whitney, R. A. (1987). Zoonoses Acquired from Pet Primates. *Vet Cl N Am Sm An Prac*, *17*, 219–240.
Rich, T. H., Hopson, J. A., Musser, A. M., Flannery, T. F. & Vickers-Rich, P. (2005). Independent origins of middle ear bones in monotremes and therians. *Science*, *307*, 910 914.
Rodger, J. C. & Hughes, R. L. (1973). Studies of the accessory glands of male marsupials. *Aust J Zool*, *21*, 303.
Rook, D. L. & Hunter, J. P. (2013). Rooting Around the Eutherian Family Tree: the Origin and Relations of the Taeniodonta. *J Mamm Evol*, 1–17.
Rose, K. D. (2006). The beginning of the age of mammals. Baltimore: Johns Hopkins University Press.

Runestad, J. A. & Ruff, C. B. (1995). Structural Adaptations for Gliding in Mammals with Implications for Locomotor Behaviour in Paromomyids. *Am J Phy Anth*, *98*, 101–119.

Russell, R. (1980). Spotlight on Possums, published by University of Queensland Press, St. Lucia, Queensland.

Sadleir, R. M. F. S. (1965). Reproduction in two species of kangaroo (*Macropus robustus* and *Megaleia* rufa in the arid Pilbara region of Western Australia. *Proc Zool Soc Lon*, *145*, 239–261.

Schaller, G. B. (1963). The Mountain Gorilla: Ecology and Behavior. Chicago: University Chicago Press.

Schiewe, J. (2010). Australia's marsupials originated in what is now South America, study says. *LATimes.Com*. Los Angeles Times.

Schmidt, T. et al. (2005). Rapid electrostatievolution at the binding site for cytochrome c on cytochrome c oxidase in anthropoid primates. *Proc Nat Acad Sci USA*, *102*, 6379–6384.

Schmitz, J., Ohme, M., Suryobroto, B. & Zischler, H. (2002). The Colugo (*Cynocephalus variegatus*, Dermoptera): The Primates' Gliding Sister? *Mol Ecol Evol*, *19*, 2308–2312.

Schrago, C. G. & Russo, C. A. M. (2003). Timing the Origin of New World Monkeys. *Mol Biol Evol*, *20*, 1620–1625.

Schunke, A. C. (2005). Systematics and Biogeography of the African Scaly-Tailed Squirrels (Mammalia: Rodentia: Anomaluridae). Ph.D. thesis, Rheinischen Friedrich-Wilhelms-Universitat Bonn, Bonn.

Sellers, B. (2000). Primate Evolution. University of Edinburgh. pp. 13–17.

Sharman, G. B. & Pilton, P. E. (1964). The life history and reproduction of the red kangaroo (*Megaleia rufa*). *Proc Zool Soc Lon*, *142*, 29–48.

Shufeldt, R. W. (1911). The Skeleton in the Flying Lemurs, Galeopteridae. *The Philip J Sci*, *6*, 185–211.

Simmons, N. B., Seymour, K. L., Habersetzer, J. & Gunnell, G. F. (2008). Primitive Early Eocene bat from Wyoming and the evolution of flight and echolocation. *Nature*, *451*, 818–822.

Smith, K. K. (2001). Early development of the neural plate, neural crest and facial region of marsupials. *J Anat*, *199*, 121–131.

Smuts, B. B., Cheney, D. L. Seyfarth, R. M., Wrangham, R. W. & Struhsaker, T. T. (Eds.) (1987). *Primate Societies*. Chicago: University of Chicago Press for articles on the structure and function of various primate societies.

Soligo, C., Müller, A. E. (1999). Nails and claws in primate evolution. *J Hum Evol*, *36*, 97–114.

Southwick, C. H. & Siddiqi, M. F. (2001). Status, conservation and management of primates in India. *En Bull: Wild Prot Ar*, *1*, 81–91.

Stace, C. A. (2010). Classification by molecules: What's in it for field botanists?, *Watsonia*, *28*, 103–122.

Strier, K. (2007). Primate Behavioral Ecology (Third ed.), pp. 50–53.

Sugawara, K. (1979). Sociological study of a wild group of hybrid baboons between *Papio anubis* and *P. hamadryas* in the Awash Valley, Ethiopia. *Primates*, *20*, 21–56.

Surbeck, M. & Hohmann, G. (2008). Primate hunting by bonobos at LuiKotale, Salonga National Park. *Curr Biol*, *18*, 906–7.

Surbeck, M., Fowler, A., Deimel, C. & Hohmann, G. (2008). Evidence for the consumption of arboreal, diurnal primates by bonobos (*Pan paniscus*). *Am J Primatol*, *71*, 171–4.

Surridge, A. K. & Osorio, D. (2003). Evolution and selection of trichromatic vision in primates. *Tr Ecol Evol*, *18*, 198–205.

Sussman, R. W. (1999). Primate Ecology and Social Structure Volume 1: Lorises, Lemurs and Tarsiers. Needham Heights, MA: Pearson Custom Publishing & Prentice Hall. pp. 78, 89–90, 108, 121–123, 233.

Szalay, F. S. & Lucas, S. G. (1993). Cranioskeletal Morphology of Archontans and Diagnoses of Chiroptera, Volitantia and Archonta. In: *Primates and Their Relatives in Phylogenetic Perspective*, ed. MacPhee, R.D.E. 187–226. New York: Plenum Press.

Tavaré, S., Marshall, C. R., Will, O., Soligo, C. & Martin R. D. (2002). nUsing the fossil record to estimate the age of the last common ancestor of extant primates. *Nature*, *416*, 726–729.

Tenaza, R. (1984). Songs of hybrid gibbons (*Hylobates lar* × *H. muelleri*). *Am J Primatol* 8, 249–253.

Thorington, R. W. Jr. & Anderson, S. (1984). Primates. In Anderson, S. & Jones, J.K. Jr. *Orders and Families of Recent Mammals of the World*. New York: John Wiley and Sons. 187–217.

Tomasello, M. & Call, J. (1997). *Primate Cognition*. Oxford University Press US.

Troughton, E. (1941). Furred Animals of Australia, published by Angus and Robertson (Publishers) Pty. Ltd, Sydney (revised edition 1973).

Turner, I. M. (1996). Species loss in fragments of tropical rain forest: a review of the evidence. *J App Ecol*, *33*, 200–209.

van Rheede, T., Bastiaans, T., Boone, D. N., Hedges, S. B., de Jong, W. W. et al. (2006). The platypus is in its place: nuclear genes and indels confirm the sister group relation of monotremes and therians. *Mol Biol Evol*, *23*, 587-597.

Van Schaik, C. P., Monk, K. A. & Robertson, J. M. Y. (2001). Dramatic decline in orangutan numbers in the Leuser Ecosystem, northern Sumatra. *Oryx*, *35*, 14–25.

van Schaik, C., Fox, E. & Sitompul, A. (1996). Manufacture and use of tools in wild Sumatran orangutans. *Naturwissenschaften*, *83*, 186–188.

Walker, M. L. & Herndon, J. G. (2008). Menopause in nonhuman primates? *Biol Repr*, *79*, 398–406.

Watson, J. M. & Graves, J. A. M. (1988). Monotreme Cell-Cycles and the Evolution of Homeothermy. *Aust J Zool*, *36*, 573–584.

Westergaard, G.C. et al.. (1998). Why some capuchin monkeys (*Cebus apella*) use probing tools (and others do not). *J Comp Psychol*, *112*, 207–211.

Wible, J. R., Rougier, G. W., Novacek, M. J. & Asher, R. J. (2007). Cretaceous eutherians and Laurasian origin for placental mammals near the K/T boundary. *Nature*, *447*, 1003–1006.

Wible, J. R., Rougier, G. W., Novacek, M. J. & Asher, R. J. (2009). The Eutherian Mammal *Maelestes gobiensis* from the Late Cretaceous of Mongolia and the phylogeny of cretaceous eutheria. *Bull Am Mus Nat Hist*, *327*, 1–123.

Williams, B. A., Kay, R. F. & Kirk, E. C. (2010). New perspectives on anthropoid origins. *Proc Nat Acad Sci USA*, *107*, 4797–4804.

Wolfe, L. D. & Fuentes, A. (2007). Ethnoprimatology. In Campbell, C.J., Fuentes, A., MacKinnon, K. C., Panger, M. & Bearder, S. K. *Primates in Perspective*. Oxford University Press. p. 692.

Workman, C. (2004). Primate conservation in Vietnam: toward a holistic environmental narrative. *Am Anthr*, *106*, 346–352.

Wrangham, R. W. (1982). Mutualism, kinship and social evolution. *Curr Prob Sociobiol* Cambridge University Press. 269–289.

Wright, P., Simmons, E. & Gursky, S. (2003). Introduction. In Wright, P., Simmons, E. & Gursky, S. *Tarsiers Past, Present and Future*. Rutgers University Press. p. 1.

INDEX

A

B

C

D

E

F

G

H

I

J

K

L

M

N

O

P

Q

R

W

X

Y

Z